Essentials of
Anatomy and
Physiology
Second Edition

Valerie C. Scanlon, PhD
College of Mount Saint Vincent
Riverdale, New York

Tina Sanders
Medical Illustrator
Castle Creek, New York
Formerly
Head Graphic Artist
Tompkins Courtland Community College
Dryden, New York

F. A. DAVIS COMPANY • Philadelphia

F. A. Davis Company
1915 Arch Street
Philadelphia, PA 19103

Printed in the United States of America

Last digit indicates print number: 10 9 8 7 6 5

Publisher, Nursing: Robert G. Martone
Nursing Editor: Alan Sorkowitz
Production Editor: Crystal S. McNichol
Cover Designer: Donald B. Freggens, Jr.

As new scientific information becomes available through basic and clinical research, recommended treatments and drug therapies undergo changes. The author and publisher have done everything possible to make this book accurate, up to date, and in accord with accepted standards at the time of publication. The author, editors, and publisher are not responsible for errors or omissions or for consequences from application of the book, and make no warranty, expressed or implied, in regard to the contents of the book. Any practice described in this book should be applied by the reader in accordance with professional standards of care used in regard to the unique circumstances that may apply in each situation. The reader is advised always to check product information (package inserts) for changes and new information regarding dose and contraindications before administering any drug. Caution is especially urged when using new or infrequently ordered drugs.

Library of Congress Cataloging-in-Publication Data

Scanlon, Valerie C., 1946–
   Essentials of anatomy and physiology / Valerie C. Scanlon, Tina
Sanders. — 2nd ed.
      p.   cm.
   Includes bibliographical references and index.
   ISBN 0-8036-7735-9 (alk, paper)
   1. Human anatomy.   2. Human physiology.   I. Sanders, Tina, 1943–
   II. Title.
   [DNLM: 1. Anatomy.   2. Physiology.   QS 4 S283e 1995]
QP345.S36   1995
612—dc20
DNLM/DLC
for Library of Congress   94-20313
                                                            CIP

Preface

All of us associated with the first edition of *Essentials of Anatomy and Physiology,* the author, illustrator, editors, and production staff, extend our thanks to all of you who have adopted our textbook for use in your courses and especially to those who have written to us with suggestions, criticisms, and encouragement. We hope you will find that your concerns have been considered in this second edition, and we thank you for making the second edition possible.

The text has been updated wherever appropriate to include recent research. This will be most apparent in the boxed inserts on clinical applications. Many of the illustrations have been modified or redone entirely, some to add color and others to match more closely the text material. New illustrations also have been added to provide further visual reinforcement of explanations in the text.

Chapter 2 now includes basic material on hydrogen bonds and chemical reactions. Microbiology, which previously was covered in a separate supplement, is now included as Chapter 22: An Introduction to Microbiology and Human Disease. To supplement the glossary we have added Appendix F: Prefixes, Combining Word Roots, and Suffixes Used in Medical Terminology. Appendix G is a list of eponymous terms and their equivalents.

A new ancillary is the computerized test bank, available to adopters of the second edition. This test bank contains more than 1500 questions geared to the review questions in each chapter, and it is fully described, with a demonstration disk, in the *Instructor's Guide.*

As always, your comments and suggestions will be most welcome, and they may be sent to us in care of the publisher: F. A. Davis Company, 1915 Arch Street, Philadelphia, PA 19103.

Valerie C. Scanlon
Dobbs Ferry, New York

Tina Sanders
Castle Creek, New York

To my students, past and present
VCS

To Brooks, for his encouragement
TS

To the Instructor

Teachers of introductory anatomy and physiology courses face a special challenge: we must distill and express the complexities of human structure and function in a simple way, without losing the essence and meaning of the material. That is the goal of this textbook: to make this material readily accessible to students with diverse backgrounds and varying levels of educational preparation.

No prior knowledge of biology or chemistry is assumed, and even the most fundamental terms are defined thoroughly. Essential aspects of anatomy are presented clearly and reinforced with excellent illustrations. Essential aspects of physiology are discussed simply, yet with accuracy and precision. Again, the illustrations complement the text material and foster comprehension on the part of the student. These illustrations were prepared especially for students for whom this is a first course in anatomy and physiology. As you will see, many are full-page images in which detail is readily apparent. All important body parts have been carefully labeled, but the student is not overwhelmed with unnecessary labels. Wherever appropriate, the legends refer students to the text for further description or explanation.

The text has three unifying themes: the relationship between physiology and anatomy, the interrelations among the organ systems, and the relationship of each organ system to homeostasis. Although each type of cell, tissue, organ, or organ system is discussed simply and thoroughly in itself, applicable connections are made to other aspects of the body or to the functioning of the body as a whole. Our goal is to provide your students with the essentials of anatomy and physiology, and in doing so, to help give them an appreciation for the incredible machine that is the human body.

The sequence of chapters is a very traditional one. Cross references are used to remind students of what they have learned from previous chapters. Nevertheless, the textbook is very flexible, and, following the introductory four chapters, the organ systems may be covered in almost any order, depending on the needs of your course.

Each chapter is organized internally from the simple to the more complex, with the anatomy followed by the physiology. The *Instructor's Guide* presents modifications of the topic sequences that may be used, again depending on the needs of your course. Certain more advanced topics may be omitted from each chapter without losing the meaning or flow of the rest of the material, and these are indicated, for each chapter, in the *Instructor's Guide*.

Clinical applications are set apart from the text in boxed inserts. These are often aspects of pathophysiology that are related to the normal anatomy or physiology in the text discussion. Each box presents one particular topic and is referenced at the appropriate point in the text. This material is intended to be an integral part of the chapter but is set apart for ease of reference and to enable you to include or omit as many of these topics as you wish. The use of these boxes also enables students to read the text material without interruption and then to focus on specific aspects of pathophysiology. A comprehensive list of the boxes appears inside the book's front and back covers, and another list at the beginning of each chapter cites the boxes within that chapter.

Tables are utilized as summaries of structure and function, to concisely present a sequence of events, or to present additional material that you may choose to include. Each table is referenced in the text and is intended to facilitate your teaching and to help your students learn.

New terms appear in bold type within the text, and all such terms are fully defined in an extensive glossary, with phonetic pronunciations. Bold type may also be used for emphasis whenever one of these terms is used again in a later chapter.

Each chapter begins with a chapter outline and student objectives to prepare the student for the chapter itself. New terminology and related clinical terms are also listed, with phonetic pronunciations. Each of these terms is fully defined in the glossary, with cross references back to the chapter in which the term is introduced.

At the end of each chapter are a study outline and review questions. The study outline includes all of the essentials of the chapter in a concise outline form. The review questions may be used by the students as a review or self-test. Following each question is a page reference in parentheses. This reference cites the page(s) in the chapter on which the content needed to answer the question correctly can be found. The answers themselves are included in the *Instructor's Guide.*

An important supplementary learning tool for your students is available in the form of a *Student Workbook* that accompanies this text. For each chapter in the textbook, the workbook offers fill-in and matching-column study questions, figure-labeling and figure-coloring exercises, and crossword puzzles based on the chapter's vocabulary list. Also included are comprehensive, multiple-choice chapter tests to provide a thorough review for students. All answers are provided at the end of the workbook.

The instructor's materials for this text include a complete *Instructor's Guide,* a computerized test bank, and a transparency package. The *Instructor's Guide* contains expanded chapter outlines, notes on each chapter's organization and content (useful for modifying the book to your specific teaching needs), topics for class discussions, and answers to the chapter review questions from the textbook. The computerized test bank contains test questions for every chapter of the book, with a total of more than 1500 questions. It uses F. A. Davis's simple but powerful Make-A-Test test-generation software, which allows you to select the questions you wish, modify them if you choose, and even add your own questions. The transparency package offers many clear, sharp, full-color transparencies taken from the textbook's illustrations and tables.

Suggestions and comments from colleagues are always valuable, and yours would be greatly appreciated. When we took on the task of writing and illustrating this textbook, we wanted to make it the most useful book possible for you and your students. Any suggestions that you can give us to help us achieve that goal are most welcome, and they may be sent to us in care of F. A. Davis Company, 1915 Arch Street, Philadelphia, PA 19103.

Valerie C. Scanlon
Dobbs Ferry, New York

Tina Sanders
Castle Creek, New York

To the Student

This is your textbook for your first course in human anatomy and physiology, a subject that is both fascinating and rewarding. That you are taking such a course says something about you: you may simply be curious as to how the human body functions. Or, you may have a personal goal of making a contribution in one of the health-care professions. Whatever your reason, this textbook will help you to be successful in your anatomy and physiology course.

The material is presented simply and concisely, yet with accuracy and precision. The writing style is informal yet clear and specific; it is intended to promote your comprehension and understanding.

Organization of the Textbook

To use this textbook effectively, you should know the purpose of its various parts. Each chapter is organized in the following way:

Chapter Outline—This presents the main topics in the chapter, which correspond to the major headings in the text.

Student Objectives—These summarize what you should know after reading and studying the chapter. These are not questions to be answered, but are rather, with the chapter outline, a preview of the chapter contents.

New Terminology and Related Clinical Terminology—These are some of the new terms you will come across in the chapter. Read through these terms before you read the chapter, but do not attempt to memorize them just yet. When you have finished the chapter, return to the list and see how many terms you can define. All of these terms are fully defined in the glossary.

Study Outline—This is found at the end of the chapter. It is a concise summary of the essentials in the chapter. You may find this outline very useful as a quick review before an exam.

Review Questions—These are also at the end of the chapter. Your instructor may assign some or all of them as homework. If not, the questions may be used as a self-test to evaluate your comprehension of the chapter's content. The page number(s) in parentheses following each question refers you to the page(s) in the chapter on which the content needed to answer the question correctly can be found.

Other Features within Each Chapter

Illustrations—These are an essential part of this textbook. Use them. Look at them and study them carefully, and they will be of great help to you as you learn. They are intended to help you develop your own mental picture of the body and its parts and processes. Each illustration is referenced in the text, so you will know just when to consult it.

Boxes—Discussions of clinical applications are in separate boxes in the text so that you may find and refer to them easily. Your instructor may include all or some of these as required reading. If you are planning a career in the health professions, these boxes are an introduction to pathophysiology, and you will find them interesting and helpful.

Bold Type—This is used whenever a new term is introduced, or when an old term is especially important. The terms in bold type are fully defined in the glossary, which includes phonetic pronunciations.

Tables—This format is used to present material in a very concise form. Some tables are summaries of text material and are very useful for a quick review. Other tables present additional material that complements the text material.

To make the best use of your study time, a Student Workbook is available that will help you to focus your attention on the essentials in each chapter. Also included are comprehensive chapter tests to help you determine which topics you have learned thoroughly and which you may have to review. If your instructor has not made the workbook a required text, you may wish to ask that it be ordered and made available in your bookstore. You will find it very helpful.

Some Final Words of Encouragement

Your success in this course depends to a great extent on you. Try to set aside study time for yourself every day; a little time each day is usually much more productive than trying to cram at the last minute.

Ask questions of yourself as you are studying. What kinds of questions? The simplest ones. If you are studying a part of the body such as an organ, ask yourself: What is its name? Where is it? What is it made of? What does it do? That is: name, location, structure, and function. These are the essentials. If you are studying a process, ask yourself: What is happening here? What is its purpose? That is: What is going on? and what good is it? Again, these are the essentials.

We hope this textbook will contribute to your success. If you have any suggestions or comments, we would very much like to hear them. After all, this book was written for you, to help you achieve your goals in this course and in your education. Please send your suggestions and comments to us in care of F. A. Davis Company, 1915 Arch Street, Philadelphia, PA 19103.

Valerie C. Scanlon
Dobbs Ferry, New York

Tina Sanders
Castle Creek, New York

Acknowledgments

We wish to thank the editors and production staff of the F. A. Davis Company, especially:

- Robert G. Martone, Publisher, Nursing, who oversaw the entire production process.
- Alan Sorkowitz, Nursing Editor, whose many tasks included answer-man, troubleshooter, and problem-solver.
- Susan M. McCoy, Production Editor for the first edition, whose contribution remains a lasting and valued part of the book.
- Crystal McNichol, Production Editor for the second edition, whose thoroughness and diligence, especially on punctuation patrol, add yet another bit of polish to the text.
- Herbert J. Powell, Director of Production, whose efficiency and concern for excellence were matched by his humane deadlines.
- In addition, Tina Sanders wishes especially to thank Dolores Lake Taylor for her consultation on the book's art program.

VCS
TS

Consultants

Kenneth Bynum, PhD
University of North Carolina
Chapel Hill, North Carolina

Barbara Herlihy, PhD
Incarnate Word College
School of Nursing
San Antonio, Texas

Doris A. Rutkowski, BSN
Alvernia School of Practical Nursing
St. Francis Medical Center
Pittsburgh, Pennsylvania

Ann C. Stewart, BA, RNC
Washington Technical College
Marietta, Ohio

Dolores Lake Taylor, MSN, RN
Department of Science
Bucks County Community College
Newtown, Pennsylvania

Carol A. Thomas, BS, MS
Coordinator of Allied Health Science
Santa Fe Community College
Gainesville, Florida

Donna Mazza Wagner, MSEd, RN
Pittsburgh, Pennsylvania

Judith M. Young, RN, BSN
Instructor—Practical Nursing Program
Upper Bucks County Area Vocational-
    Technical School
Perkasie, Pennsylvania

Reviewers

Brenda Berry
Hawkeye Institute of Technology
Waterloo, Iowa

Helen Binda, RN, MEd
Coordinator, LPN Program, Niagara
    County Community College
Department Chairman, Health
    Occupations, Orleans and Niagara
    BOCES
Sanborn, New York

Sylvia A. Blanco, RN, MA
Erwin Technical Center
Tampa, Florida

Nancy Metz Brown
Graff Vocational-Technical Center
Springfield, Missouri

Doris E. Bush, RN, MSN
Formerly Chairperson—Nursing
    Department
Southwest State Technical College
Mobile, Alabama

Margaret K. Butler, RN, BS
Danville School of Practical Nursing
Danville, Kentucky

Ann Carmack
Kansas City Kansas Area Vocational-
    Technical School
Kansas City, Kansas

Evie Chase, RN, BSN, MA
John Adams Community College
San Francisco, California

Marilyn Collins, BSN
Citrus Community College
Glendora, California

Gina Cook, RN, BSN
St. Phillips College
San Antonio, Texas

Margaret Cramer, RN
Virginia Beach Vocational-Technical School
Virginia Beach, Virginia

Judy Datsko, BSN
L.H. Bates Vocational-Technical Institute
Tacoma, Washington

Patrick J. Debold
Concord Career Colleges
Kansas City, Missouri

Serita Dickey, RN, MS
San Jacinto College (North Campus)
School of Nursing
Houston, Texas

Marcelline Eachus, RN
Gloucester County Vocational-Technical
    School
Sewell, New Jersey

Debbie Edwards, RN, BSN
James Martin School of Practical Nursing
Philadelphia, Pennsylvania

Ann Fiala, RN
Clover Park Vocational-Technical Institute
Tacoma, Washington

Gail T. Fox, RN, BSN
Victoria College
Victoria, Texas

William Francis, PhD
South Suburban College
South Holland, Illinois

Sandra L. Freeman, RN, BSEd
Chairperson—Practical Nursing
    Department
Hinds Community College
Jackson, Mississippi

Alice M. Frye, MSN
Augusta Technical Institute
Augusta, Georgia

Nancy Georgeoff
Choffin School of Practical Nursing
Youngstown, Ohio

Donald C. Giersch, AS, BS, MS
Triton College
River Grove, Illinois

Carole Grant
Paris Junior College
Paris, Texas

Julia Haggerty, RN, BSHS, MAEd
Maricopa Skills Center
Phoenix, Arizona

Lee Haroun, MA
Maric College
San Diego, California

Linda A. Howe, RN, MS
Roper Hospital School of Practical Nursing
Charleston, South Carolina

Vivian Hritz
Greater Johnstown Area Vocational-
    Technical School
Johnstown, Pennsylvania

Katie Iverson
Pima Community College
Tucson, Arizona

Linda Jerge, RN, BSN
Erie 1 BOCES
Lancaster, New York

Robert Keck, MS, MHA
Indiana Vocational-Technical College
Indianapolis, Indiana

Grace Kittoe, RN, BS, MEd
Practical Nurse Program of Canton City
    Schools
Canton, Ohio

Susan Lievano, RN, BSN, MS
Vocational and Educational Extension
    Board
School of Nursing
Uniondale, New York

Carolyn Lyon, RN, MSN
Roanoke Memorial Hospital
School of Practical Nursing
Roanoke, Virginia

Florence Maellaro
Sheridan Vocational Center
Hollywood, Florida

Captain Ivy Manning
2076 USARF School
Wilmington, Delaware

Patricia L. Mashburn, BSN, MEd, PhD
Connelley Skill Learning Center
Pittsburgh, Pennsylvania

Sandra Merchant
Raleigh County Vocational-Technical
    Center
Beckley, West Virginia

Ann Montminy, MS, RN
Youville Hospital School of Practical
    Nursing
Cambridge, Massachusetts

Paula Ott, BSN, MEd
Department Head—Health Occupations
    Education
Delta-Ouachita Regional Vocational
    Institute
West Monroe, Louisiana

Theresa Peterson, MA, RN
Oakland Community College
Southfield, Michigan

Alice Phillips, MA, RN
Director of Practical Nursing
Monmouth County Vocational School
    District
Marlboro, New Jersey

Bernice Rudolph, RN, PHN, BS
Casa Loma College
Lakeview Terrace, California

Anna Jane Santosuosso, RN, BS, MEd
Coordinator of Practical Nursing
Quincy College
Quincy, Massachusetts

Sandra J. Scherb, RN, MS
Dakota County Technical College
Rosemont, Minnesota

Jean Seago, BSNEd
Delmar College
Corpus Christi, Texas

Elizabeth Shelton
Richmond Technical Center
Richmond Public School of Practical
    Nursing
Richmond, Virginia

Sharon Van Orden
Atlantic Vocational Center
Coconut Creek, Florida

Bonnie Watts, RN
Minneapolis Technical College
Minneapolis, Minnesota

Contents

Appendices

Chapter 1
Organization and General Plan of the Body

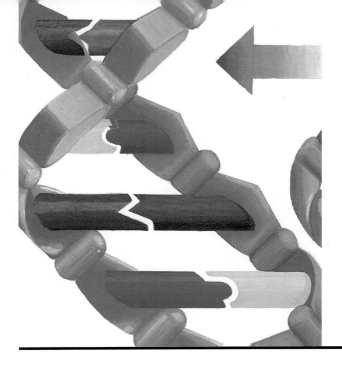

Chapter 1

Chapter Outline

Student Objectives

- Define anatomy, physiology, and pathophysiology. Use an example to explain how they are related.
- Name the levels of organization of the body from simplest to most complex, and explain each.
- Define homeostasis, and use an example to explain.
- Describe the anatomical position.
- State the anatomical terms for the parts of the body.
- Use proper terminology to describe the location of body parts with respect to one another.
- Name the body cavities, their membranes, and some organs within each cavity.
- Describe the possible sections through the body or an organ.
- Explain how and why the abdomen is divided into smaller areas. Be able to name organs in these areas.

Organization and General Plan of the Body

New Terminology

Anatomy (uh–**NAT**–uh–mee)
Body cavity (**BAH**–dee **KAV**–i–tee)
Cell (**SELL**)
Homeostasis (HOH–me–oh–**STAY**–sis)
Inorganic chemicals (**IN**–or–GAN–ik **KEM**–i–kuls)
Meninges (me–**NIN**–jeez)
Organ (**OR**–gan)
Organ system (**OR**–gan **SIS**–tem)
Organic chemicals (or–**GAN**–ik **KEM**–i–kuls)
Pathophysiology (PATH–oh–FIZZ–ee–**AH**–luh–jee)
Pericardial membranes (PER–ee–**KAR**–dee–uhl **MEM**–brains)
Peritoneum—Mesentery (PER–i–toh–**NEE**–um—**MEZ**–en–TER–ee)

Physiology (FIZZ–ee–**AH**–luh–jee)
Plane (**PLAYN**)
Pleural membranes (**PLOOR**–uhl **MEM**–brains)
Section (**SEK**–shun)
Tissue (**TISH**–yoo)

Related Clinical Terminology

Computed tomography (CT) scan (kom–**PEW**–ted toh–**MAH**–grah–fee SKAN)
Diagnosis (DYE–ag–**NO**–sis)
Disease (di–**ZEEZ**)
Magnetic resonance imaging (MRI) (mag–**NET**–ik **REZ**–ah–nanse **IM**–ah–jing)

Terms that appear in **bold type** in the chapter text are defined in the glossary, which begins on p. 549.

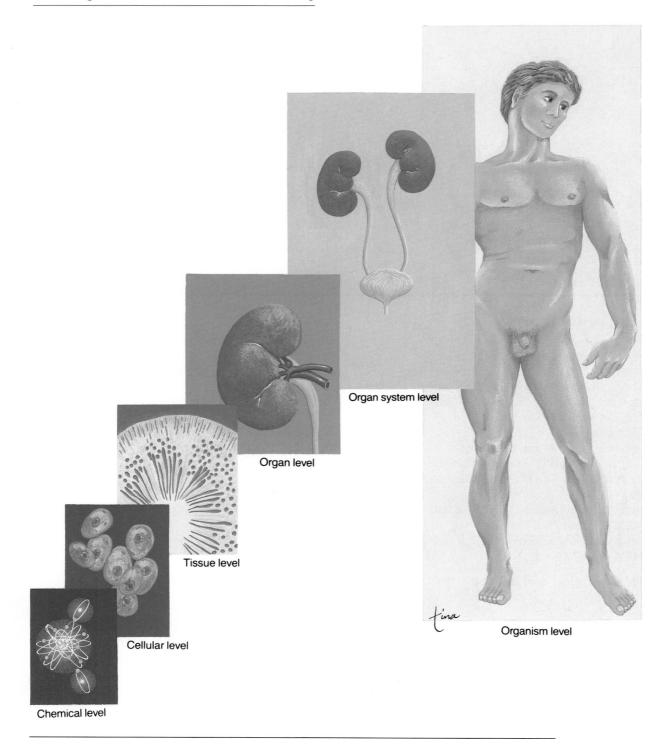

Organ system level

Organ level

Tissue level

Cellular level

Chemical level

Organism level

Figure 1–1   Levels of structural organization of the human body, depicted from the simplest (chemical) to the most complex (organism). The organ shown here is the kidney, and the organ system is the urinary system.

The human body is a precisely structured container of chemical reactions. Have you ever thought of yourself in this way? Probably not, and yet, in the strictly physical sense, that is what each of us is. The body consists of trillions of atoms in specific arrangements and thousands of chemical reactions proceeding in a very orderly manner. That literally describes us, and yet it is clearly not the whole story. The keys to understanding human consciousness and self-awareness are still beyond our grasp. We do not yet know what enables us to study ourselves—no other animals do, as far as we know—but we have accumulated a great deal of knowledge about ourselves. Some of this knowledge makes up the course you are about to take, a course in basic human anatomy and physiology.

Anatomy is the study of body structure, which includes size, shape, composition, and perhaps even coloration. **Physiology** is the study of how the body functions. The physiology of red blood cells, for example, includes what these cells do, how they do it, and how this is related to the functioning of the rest of the body. Physiology is directly related to anatomy. For example, red blood cells contain the mineral iron in molecules of the protein called hemoglobin; this is an aspect of their anatomy. The presence of iron enables red blood cells to carry oxygen, which is their function. All cells in the body must receive oxygen in order to function properly, so the physiology of red blood cells is essential to the physiology of the body as a whole.

Pathophysiology is the study of disorders of functioning, and a knowledge of normal physiology makes such disorders easier to understand. For example, you are probably familiar with the anemia called iron deficiency anemia. With insufficient iron in the diet, there is not enough iron in the hemo-

---

Box 1–1 REPLACING TISSUES AND ORGANS

Blood transfusions are probably the most familiar and frequent form of "replacement parts" for people. Blood is a tissue, and when properly typed and cross-matched (blood types will be discussed in Chapter 11) may safely be given to someone with the same or a compatible blood type.

Organs, however, are much more complex structures. When a patient receives an organ transplant, there is always the possibility of rejection (destruction) of the organ by the recipient's immune system (Chapter 14). With the discovery and use of more effective immune-suppressing medications, however, the success rate for many types of organ transplants has increased. Organs that may be transplanted include corneas, kidneys, the heart, the liver, and the lungs.

The skin is also an organ, but skin transplanted from another person will not survive very long, although such foreign grafts may be used to temporarily cover large areas of damaged skin. Patients with severe burns, for example, will eventually need skin grafts from their own unburned skin to form permanent new skin over the burn sites. It is now possible to "grow" a patient's skin in laboratory culture, so that a small patch of skin may eventually be used to cover a large surface.

Some artificial replacement parts have also been developed. These are made of plastic or metal and are not rejected as foreign by the recipient's immune system. Damaged heart valves, for example, may be replaced by artificial ones, and sections of arteries may be replaced by tubular grafts made of synthetic materials. Artificial joints are available for nearly every joint in the body. Most recently, cochlear implants (a tiny, electronic "ear") have provided some sense of hearing for people with certain types of deafness.

globin of red blood cells, and less oxygen can be transported throughout the body, resulting in the symptoms of the iron deficiency disorder. This example shows the relationship between anatomy, physiology, and pathophysiology.

The purpose of this text is to enable you to gain an understanding of anatomy and physiology with the emphasis on normal structure and function. Many examples of pathophysiology have been included, however, to illustrate the relationship of **disease** to normal physiology and to describe some of the procedures used in the **diagnosis** of disease. Many of the examples are clinical applications that will help you begin to apply what you have learned and demonstrate that your knowledge of anatomy and physiology will become the basis for your further study in the health professions.

LEVELS OF ORGANIZATION

The human body is organized in structural and functional levels of increasing complexity. Each higher level incorporates the structures and functions of the previous level, as you will see. We will begin with the simplest level, which is the chemical level, and proceed to cells, tissues, organs, and organ systems. All of the levels of organization are depicted in Fig. 1–1.

CHEMICALS

The chemicals that make up the body may be divided into two major categories: inorganic and organic. **Inorganic chemicals** are usually simple

Table 1–1   THE ORGAN SYSTEMS

System	Functions	Organs*
Integumentary	• Is a barrier to pathogens and chemicals • Prevents excessive water loss	skin, hair, subcutaneous tissue
Skeletal	• Supports the body • Protects internal organs • Provides a framework to be moved by muscles	bones, ligaments
Muscular	• Moves the skeleton • Produces heat	muscles, tendons
Nervous	• Interprets sensory information • Regulates body functions such as movement by means of electro-chemical impulses	brain, nerves, eyes, ears
Endocrine	• Regulates body functions by means of hormones	thyroid gland, pituitary gland
Circulatory	• Transports oxygen and nutrients to tissues and removes waste products	heart, blood, arteries
Lymphatic	• Returns tissue fluid to the blood • Destroys pathogens that enter the body	spleen, lymph nodes
Respiratory	• Exchanges oxygen and carbon dioxide between the air and blood	lungs, trachea, larynx
Digestive	• Changes food to simple chemicals that can be absorbed and used by the body	stomach, colon, liver
Urinary	• Removes waste products from the blood • Regulates volume and pH of blood	kidneys, urinary bladder, urethra
Reproductive	• Produces eggs or sperm • *In women,* provides a site for the developing embryo-fetus	*Female:* ovaries, uterus *Male:* testes, prostate gland

*These are simply representative organs, not an all-inclusive list.

molecules made of one or two elements other than carbon (with a few exceptions). Examples of inorganic chemicals are water (H_2O); oxygen (O_2); one of the exceptions, carbon dioxide (CO_2); and minerals such as iron (Fe), calcium (Ca), and sodium (Na). **Organic chemicals** are often very complex and always contain the elements carbon and hydrogen. In this category of organic chemicals are carbohydrates, fats, proteins, and nucleic acids. The chemical organization of the body is the subject of Chapter 2.

CELLS

The smallest living units of structure and function are **cells.** There are many different types of cells; each is made of chemicals and carries out specific chemical reactions. Cell structure and function are discussed in Chapter 3.

TISSUES

A **tissue** is a group of cells with similar structure and function. There are four groups of tissues:

Epithelial tissues—cover or line body surfaces; some are capable of producing secretions with specific functions. The outer layer of the skin and sweat glands are examples of epithelial tissues.

Connective tissues—connect and support parts of the body; some transport or store materials. Blood, bone, and adipose tissue are examples of this group.

Muscle tissues—specialized for contraction, which brings about movement. Our skeletal muscles and the heart are examples of muscle tissue.

Nerve tissue—specialized to generate and transmit electrochemical impulses that regulate body functions. The brain and optic nerves are examples of nerve tissue.

The types of tissues in these four groups, as well as their specific functions, are the subject of Chapter 4.

ORGANS

An **organ** is a group of tissues precisely arranged so as to accomplish specific functions. Examples of organs are the kidneys, liver, lungs, and stomach. The stomach is lined with epithelial tissue that secretes gastric juice for digestion. Muscle tissue in the wall of the stomach contracts to mix food with gastric juice and propel it to the small intestine. Nerve tissue carries impulses that increase or decrease the contractions of the stomach (See Box 1–1: Replacing Tissues and Organs).

ORGAN SYSTEMS

An **organ system** is a group of organs that all contribute to a particular function. Examples are the urinary system, digestive system, and respiratory system. In Fig. 1–1 you see the urinary system, which consists of the kidney, ureters, urinary bladder, and urethra. These organs all contribute to the formation and elimination of urine.

As a starting point, Table 1–1 lists the organ systems of the human body with their general functions, and some representative organs (Fig. 1–2). These organ systems make up an individual person, and the balance of this text discusses each system in more detail.

HOMEOSTASIS

A person who is in good health is in a state of **homeostasis.** Homeostasis reflects the ability of the body to maintain relative stability and to function normally despite constant changes. Changes may be external or internal, and the body must respond appropriately.

Eating breakfast, for example, brings about an internal change. Suddenly there is food in the stomach, and something must be done with it. What happens? The food is digested or broken down into simple chemicals that the body can use. The protein in a hard-boiled egg is digested into amino acids, its basic chemical building blocks; these can then be used by the body to produce its own specialized proteins.

An example of an external change is a rise in environmental temperature. On a hot day, the body temperature would also tend to rise. However, body temperature must be kept at its normal level, around 98.6°F (37°C), in order to support normal

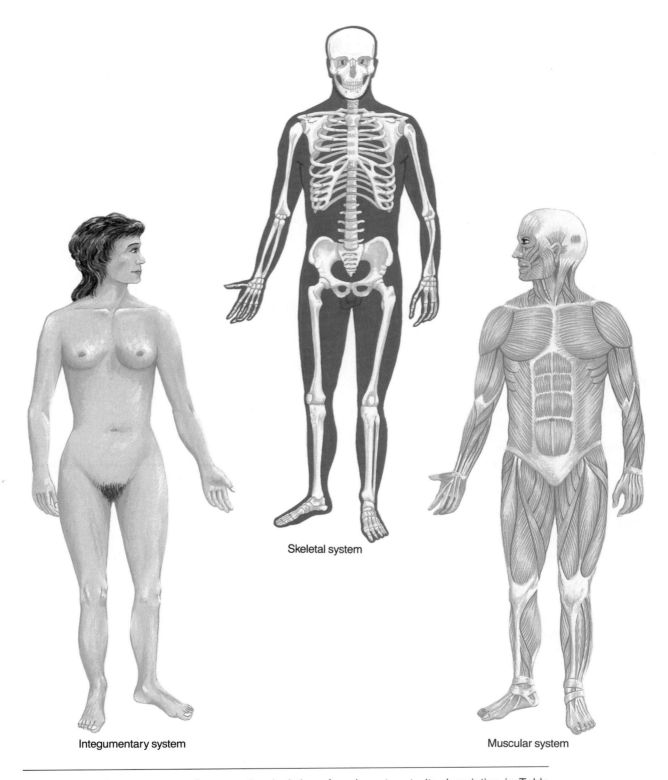

Integumentary system

Skeletal system

Muscular system

Figure 1–2 Organ systems. Compare the depiction of each system to its description in Table 1–1. Try to name at least one organ shown in each system.

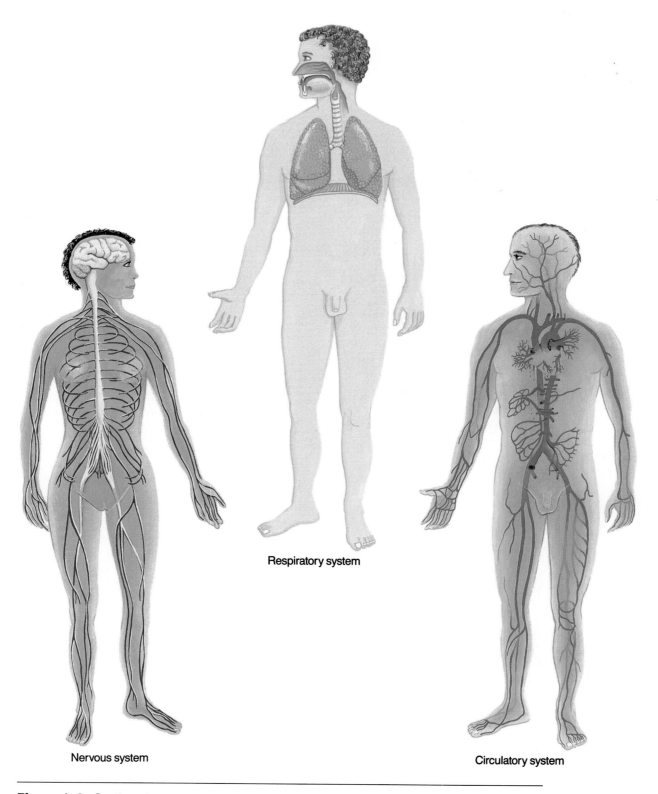

Respiratory system

Nervous system

Circulatory system

Figure 1–2 Continued.

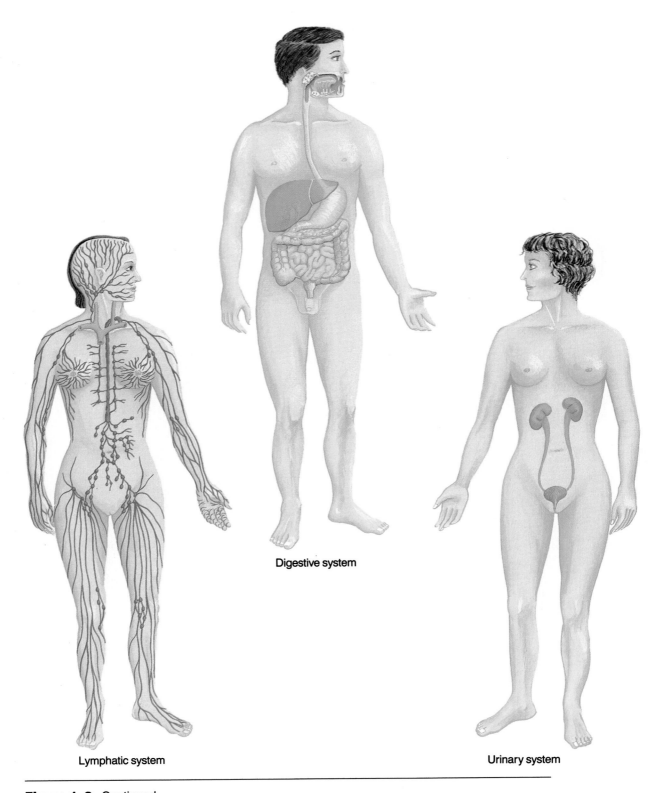

Lymphatic system

Digestive system

Urinary system

Figure 1–2 Continued.

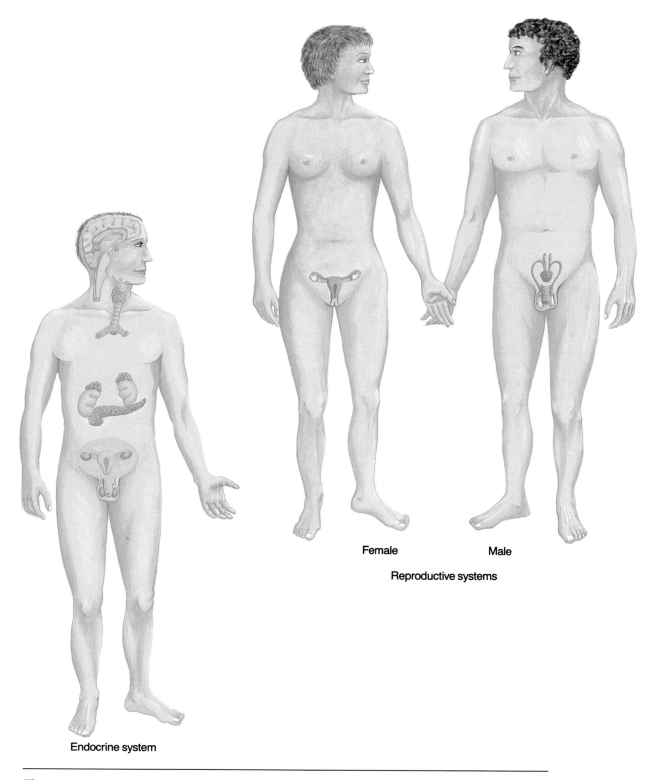

Female   Male

Reproductive systems

Endocrine system

Figure 1–2 Continued.

11

functioning. What happens? One of the body's responses to the external temperature rise is to increase sweating so that excess body heat can be lost by the evaporation of sweat on the surface of the skin. This response, however, may bring about an undesirable internal change, dehydration. What happens? As body water decreases, we feel the sensation of thirst and drink fluids to replace the water lost in sweating.

You can probably think of many other situations in which your body responds to changes and keeps you alive and healthy—this steady-state or equilibrium balancing act is homeostasis. As you continue your study of the human body, keep in mind that the proper functioning of each organ and organ system has a role to perform in maintaining homeostasis.

TERMINOLOGY AND GENERAL PLAN OF THE BODY

As part of your course in anatomy and physiology, you will learn many new words or terms. At times you may feel that you are learning a second language, and indeed you are. Each term has a precise meaning, which is understood by everyone else who has learned the language. Mastering the terminology of your profession is essential to enable you to communicate effectively with your coworkers and your future patients. Although the number of new terms may seem a bit overwhelming at first, you will find that their use soon becomes second nature to you.

The terminology presented in this chapter will be used throughout the text in the discussion of the organ systems. This will help to reinforce the meanings of these terms and will transform these new words into knowledge.

BODY PARTS AND AREAS

Each of the terms listed in Table 1–2 and shown in Fig. 1–3 refers to a specific part or area of the body. For example, "femoral" always refers to the thigh. The femoral artery is a blood vessel that passes through the thigh, and the quadriceps femoris is a large muscle group of the thigh.

Table 1–2   DESCRIPTIVE TERMS FOR BODY PARTS AND AREAS

Term	Definition (Refers to)
Axillary	armpit
Brachial	upper arm
Buccal (oral)	mouth
Cardiac	heart
Cervical	neck
Cranial	head
Cutaneous	skin
Deltoid	shoulder
Femoral	thigh
Frontal	forehead
Gastric	stomach
Gluteal	buttocks
Hepatic	liver
Iliac	hip
Inguinal	groin
Lumbar	small of back
Mammary	breast
Nasal	nose
Occipital	back of head
Orbital	eye
Parietal	crown of head
Patellar	kneecap
Pectoral	chest
Perineal	pelvic floor
Plantar	sole of foot
Popliteal	back of knee
Pulmonary	lungs
Renal	kidney
Sacral	base of spine
Temporal	side of head
Umbilical	naval
Volar	palm

Another example is "pulmonary," which always refers to the lungs, as in pulmonary artery, pulmonary edema, and pulmonary embolism. Although you may not know the exact meaning of each of these terms now, you do know that each has something to do with the lungs.

TERMS OF LOCATION AND POSITION

When describing relative locations, the body is always assumed to be in **anatomical position:** standing upright facing forward, arms at the sides with palms forward, and the feet slightly apart. The

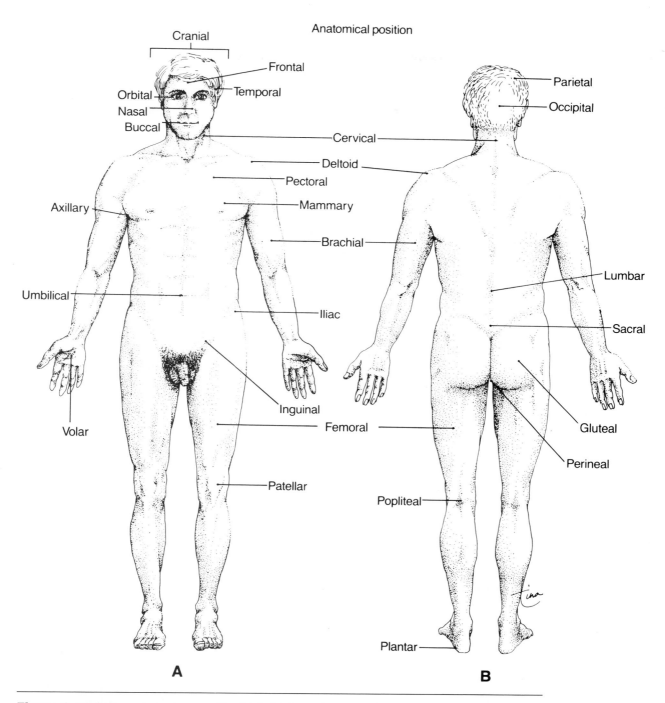

Figure 1–3 Body parts and areas. The body is shown in anatomical position. (**A**), Anterior view. (**B**), Posterior view. (Compare with Table 1–2.)

Table 1–3   TERMS OF LOCATION AND POSITION

Term	Definition	Example
Superior	above, or higher	The heart is superior to the liver.
Inferior	below, or lower	The liver is inferior to the lungs.
Anterior	toward the front	The chest is on the anterior side of the body.
Posterior	toward the back	The lumbar area is posterior to the umbilical area.
Ventral	toward the front	The mammary area is on the ventral side of the body.
Dorsal	toward the back	The buttocks are on the dorsal side of the body.
Medial	toward the midline	The heart is medial to the lungs.
Lateral	away from the midline	The shoulders are lateral to the neck.
Internal	within, or interior to	The brain is internal to the skull.
External	outside, or exterior to	The ribs are external to the lungs.
Superficial	toward the surface	The skin is the most superficial organ.
Deep	within, or interior to	The deep veins of the legs are surrounded by muscles.
Central	the main part	The brain is part of the central nervous system.
Peripheral	extending from the main part	Nerves in the arm are part of the peripheral nervous system.
Proximal	closer to the origin	The knee is proximal to the foot.
Distal	further from the origin	The palm is distal to the elbow.
Parietal	pertaining to the wall of a cavity	The parietal pleura lines the chest cavity.
Visceral	pertaining to the organs within a cavity	The visceral pleura covers the lungs.

terms of location are listed in Table 1–3, with a definition and example for each. As you read each term, find the body parts used as examples in Figs. 1–3 and 1–4. Notice also that these are pairs of terms and that each pair is a set of opposites. This will help you recall the terms and their meanings.

BODY CAVITIES AND THEIR MEMBRANES

The body has two major cavities: the **dorsal cavity** (posterior) and the **ventral cavity** (anterior). Each of these cavities has further subdivisions, which are shown in Fig. 1–4.

Dorsal Cavity

The dorsal cavity consists of the cranial cavity and the vertebral or spinal cavity. The **cranial cavity** is formed by the skull and contains the brain. The **spinal cavity** is formed by the backbone (spine) and contains the spinal cord. The membranes that line these cavities and cover the organs of the central nervous system are called the **meninges.**

Ventral Cavity

The ventral cavity consists of two compartments, the thoracic cavity and the abdominal cavity, which are separated by the diaphragm. The pelvic cavity may be considered a subdivision of the abdominal cavity or as a separate cavity.

Organs in the **thoracic cavity** include the heart and lungs. The membranes of the thoracic cavity are serous membranes called the **pleural membranes.** The parietal pleura lines the chest wall, and the visceral pleura covers the lungs. The heart has

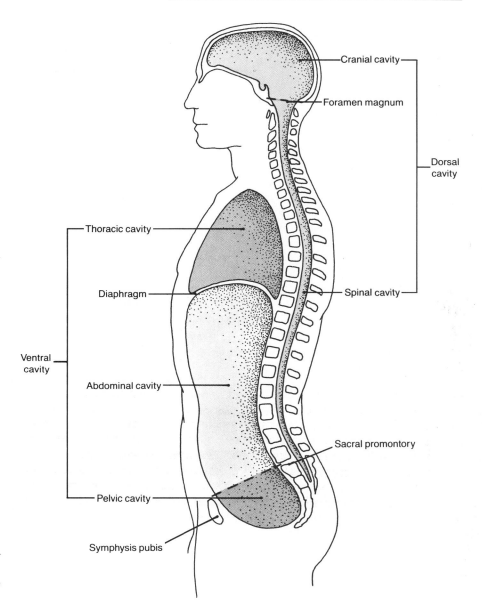

Cranial cavity

Foramen magnum

Dorsal cavity

Thoracic cavity

Diaphragm

Spinal cavity

Ventral cavity

Abdominal cavity

Sacral promontory

Pelvic cavity

Symphysis pubis

Figure 1–4 Body cavities. Shown in lateral view from the left side.

its own set of serous membranes called the pericardial membranes. The parietal pericardium lines the fibrous pericardial sac, and the visceral pericardium covers the heart muscle.

Organs in the **abdominal cavity** include the liver, stomach, and intestines. The membranes of the abdominal cavity are also serous membranes called the peritoneum and mesentery. The **peritoneum** is the membrane which lines the abdominal wall, and the **mesentery** is the membrane folded around and covering the outer surfaces of the abdominal organs.

The **pelvic cavity** is inferior to the abdominal cavity. Although the peritoneum does not line the

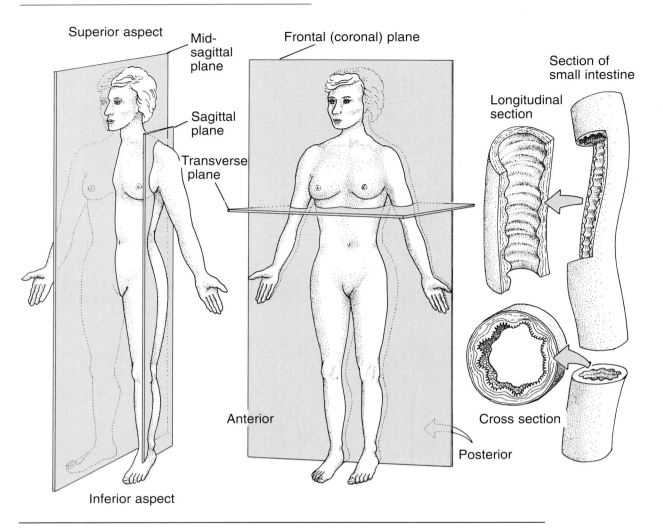

Figure 1–5 Planes and sections of the body and a cross section and longitudinal section of the small intestine. See text for description.

pelvic cavity, it covers the free surfaces of several pelvic organs. Within the pelvic cavity are the urinary bladder and reproductive organs such as the uterus in women and the prostate gland in men.

PLANES AND SECTIONS

When internal anatomy is described, the body or an organ is often cut or **sectioned** in a specific way so as to make particular structures easily visible. A **plane** is an imaginary flat surface that separates two

portions of the body or an organ. These planes and sections are shown in Fig. 1–5 (See Box 1–2: Visualizing the Interior of the Body).

Frontal (coronal) section—a plane from side to side separates the body into front and back portions.

Sagittal section—a plane from front to back separates the body into right and left portions. A midsagittal section creates equal right and left halves.

Transverse section—a horizontal plane separates the body into upper and lower portions.

Box 1–2   VISUALIZING THE INTERIOR OF THE BODY

In the past, the need for exploratory surgery brought with it hospitalization, risk of infection, and discomfort and pain for the patient. Today, however, new technologies and the extensive use of computers permit us to see the interior of the body without surgery.

Computed tomography (CT) scanning uses a narrowly focused x-ray beam that circles rapidly around the body. A detector then measures how much radiation passes through different tissues, and a computer constructs an image of a thin slice through the body. Several images may be made at different levels—each takes only a few seconds—to provide a more complete picture of an organ or part of the body. The images are much more detailed than are those produced by conventional x-rays.

Magnetic resonance imaging (MRI) is another diagnostic tool that is especially useful for visualizing soft tissues, including the brain and spinal cord. Recent refinements have produced images of individual nerve bundles, which had not been possible using any other technique. The patient is placed inside a strong magnetic field, and the tissues are pulsed with radio waves. Since each tissue has different proportions of various atoms, which resonate or respond differently, each tissue emits a characteristic signal. A computer then translates these signals into an image; the entire procedure takes 30 to 45 minutes.

One drawback of the new technologies is their cost; they are expensive. However, the benefits to patients are great: highly detailed images of the body are obtained without the risks of surgery and with virtually no discomfort in the procedures themselves.

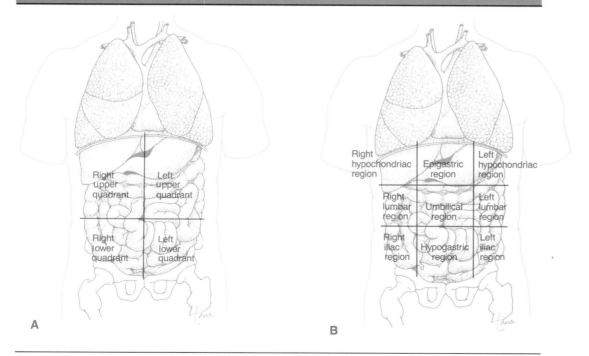

A

B

Figure 1–6   Areas of the abdomen. (**A**), Four quadrants. (**B**), Nine regions.

Cross section—a plane perpendicular to the long axis of an organ. A cross section of the small intestine (which is a tube) would look like a circle with the cavity of the intestine in the center.

Longitudinal section—a plane along the long axis of an organ. A longitudinal section of the intestine is shown in Fig. 1–5, and a frontal section of the femur (thigh bone) would also be a longitudinal section (see Fig. 6–1).

AREAS OF THE ABDOMEN

The abdomen is a large area of the lower trunk of the body. If a patient reported "abdominal pain," the physician or nurse would want to know more precisely where the pain was. In order to do this, the abdomen may be divided into smaller regions or areas, which are shown in Fig. 1–6.

Quadrants—a transverse plane and a midsagittal plane that cross at the umbilicus will divide the abdomen into four quadrants. Clinically, this is probably the division used more frequently. The pain of gall stones might then be described as in the right upper quadrant.

Nine Areas—two transverse planes and two sagittal planes divide the abdomen into nine areas.
Upper areas—above the level of the rib cartilages are the left hypochondriac, epigastric, and right hypochondriac.
Middle areas—the left lumbar, umbilical, and right lumbar.
Lower areas—below the level of the top of the pelvic bone are the left iliac, hypogastric, and right iliac. This division is often used in anatomical studies to describe the location of organs. The liver, for example, is located in the epigastric and right hypochondriac areas.

SUMMARY

As you will see, the terminology presented in this chapter is used throughout the text to describe anatomical structures and in the names of organs and their parts. We will now return to a consideration of the structural organization of the body and to more detailed descriptions of its levels of organization. The first of these, the chemical level, is the subject of the next chapter.

STUDY OUTLINE

Introduction
1. Anatomy—the study of structure.
2. Physiology—the study of function.
3. Pathophysiology—the study of disorders of functioning.

Levels of Organization
1. Chemical—inorganic and organic chemicals make up all matter, both living and non-living.
2. Cells—the smallest living units of the body.
3. Tissues—groups of cells with similar structure and function.
4. Organs—groups of tissues that contribute to specific functions.
5. Organ Systems—groups of organs that work together to perform specific functions (see Table 1–1 and Fig. 1–2).
6. Person—all the organ systems functioning properly.

Homeostasis
1. A state of good health maintained by the normal functioning of the organ systems.
2. The body constantly responds to internal and external changes, yet remains stable.

Terminology and General Plan of the Body
1. Body parts and areas—see Table 1–2 and Fig. 1–3.
2. Terms of location and position—used to describe relationships of position (see Table 1–3 and Figs. 1–3 and 1–4).
3. Body Cavities and their membranes (see Fig. 1–4).
   • Dorsal Cavity—lined with membranes called meninges; consists of the cranial and vertebral cavities.
      ○ Cranial Cavity contains the brain.
      ○ Vertebral Cavity contains the spinal cord.

- Ventral Cavity—the diaphragm separates the thoracic and abdominal cavities; the pelvic cavity is inferior to the abdominal cavity.
  - Thoracic Cavity—contains the lungs and heart.
    - Pleural membranes line the chest wall and cover the lungs.
    - Pericardial membranes surround the heart.
  - Abdominal Cavity—contains many organs including the stomach, liver, and intestines.
    - The peritoneum lines the abdominal cavity; the mesentery covers the abdominal organs.
  - Pelvic Cavity—contains the urinary bladder and reproductive organs.

4. Planes and Sections—cutting the body or an organ in a specific way (see Fig. 1–5).
   - Frontal or Coronal—separates front and back parts.
   - Sagittal—separates right and left parts.
   - Transverse—separates upper and lower parts.
   - Cross—a section perpendicular to the long axis.
   - Longitudinal—a section along the long axis.
5. Areas of the Abdomen—permits easier description of locations:
   - Quadrants—see Fig. 1–6.
   - Nine Areas—see Fig. 1–6.

REVIEW QUESTIONS

1. Explain how the physiology of a bone is related to its anatomy. Explain how the physiology of the hand is related to its anatomy. (p. 5)

2. Describe anatomical position. Why is this knowledge important? (p. 12)

3. Name the organ system with each of the following functions: (p. 6)
   a. Moves the skeleton
   b. Regulates body functions by means of hormones
   c. Covers the body and prevents entry of pathogens
   d. Destroys pathogens that enter the body
   e. Exchanges oxygen and carbon dioxide between the air and blood
4. Name the two major body cavities and their subdivisions. Name the cavity lined by the peritoneum, meninges, parietal pleura. (pp. 14, 16)

5. Name the four quadrants of the abdomen. Name at least one organ in each quadrant. (pp. 17–18)

6. Name the section through the body that would result in each of the following: equal right and left halves, anterior and posterior parts, superior and inferior parts. (pp. 16–18)

7. Review Table 1–2, and try to find each external area on your own body. (pp. 12–13)

8. Define cell. When similar cells work together, what name are they given? (p. 7)

9. Define organ. When a group of organs works together, what name is it given? (p. 7)

10. Define homeostasis. (pp. 7, 12)
    a. Give an example of an external change and explain how the body responds to maintain homeostasis.
    b. Give an example of an internal change and explain how the body responds to maintain homeostasis.

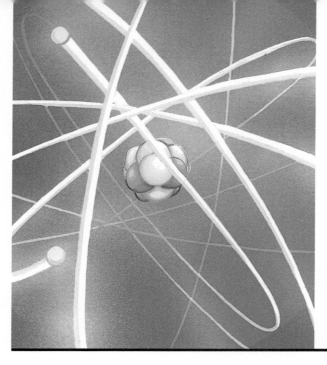

Chapter 2

Student Objectives

- Define the terms element, atom, proton, neutron, electron.
- Describe the formation and purpose of: ionic bonds, covalent bonds, and hydrogen bonds.
- Describe what happens in synthesis and decomposition reactions.
- Explain the importance of water to the functioning of the human body.
- Name and describe the water compartments.
- Explain the roles of oxygen and carbon dioxide in cell respiration.
- State what trace elements are, and name some, with their functions.
- Explain the pH scale. State the normal pH ranges of body fluids.
- Explain how a buffer system limits great changes in pH.
- Describe the functions of monosaccharides, disaccharides, oligosaccharides, and polysaccharides.
- Describe the functions of true fats, phospholipids, and steroids.
- Describe the functions of proteins, and explain how enzymes function as catalysts.
- Describe the functions of DNA, RNA, and ATP.

Some Basic Chemistry

New Terminology

Acid (**ASS**–sid)
Amino acid (ah–**ME**–noh **ASS**–sid)
Atom (**A**–tum)
Base (**BAYSE**)
Buffer system (**BUFF**–er **SIS**–tem)
Carbohydrates (KAR–boh–**HIGH**–drayts)
Catalyst (**KAT**–ah–list)
Cell respiration (SELL RES–pi–**RAY**–shun)
Covalent bond (ko–**VAY**–lent bond)
Dissociation–ionization (dih–SEW–see–**AY**–shun;
    EYE–uh–nih–**ZAY**–shun)
Element (**EL**–uh–ment)
Enzyme (**EN**–zime)
Extracellular fluid (EX–trah–**SELL**–yoo–ler)
Intracellular fluid (IN–trah–**SELL**–yoo–ler)
Ion (**EYE**–on)
Ionic bond (eye–**ON**–ik bond)

Lipids (**LIP**–id)
Matter (**MAT**–ter)
Molecule (**MAHL**–e–kuhl)
Nucleic acids (new–**KLEE**–ik **ASS**–sids)
pH and pH scale (Pee–H SKALE)
Protein (**PRO**–teen)
Salt (**SAWLT**)
Solvent–solution (**SAHL**–vent; suh–**LOO**–shun)
Steroids (**STEER**–oid)
Trace elements (TRAYSE **El**–uh–ments)

Related Clinical Terminology

Acidosis (ASS–i–**DOH**–sis)
Atherosclerosis (ATH–er–oh–skle–**ROH**–sis)
Hypoxia (high–**POCK**–see–ah)
Saturated (**SAT**–uhr–ay–ted) fats
Unsaturated (un–**SAT**–uhr–ay–ted) fats

Terms that appear in **bold type** in the chapter text are defined in the glossary, which begins on p. 549.

When you hear or see the word "chemistry" you may think of test tubes and bunsen burners in a laboratory experiment. However, literally everything in our physical world is made of chemicals. The paper used for this book, which was once the wood of a tree, is made of chemicals. The air we breathe is a mixture of chemicals in the form of gases. Water, lemonade, and diet soda are chemicals in liquid form. Our foods are chemicals, and our bodies are complex arrangements of thousands of chemicals. Recall from Chapter 1 that the simplest level of organization of the body is the chemical level.

This chapter covers some very basic aspects of chemistry as they are related to living organisms, and most especially as they are related to our understanding of the human body.

ELEMENTS

All matter, both living and not living, is made of elements, the simplest chemicals. An element is a substance made of only one type of atom (therefore, an atom is the smallest part of an element). There are 92 naturally occurring elements in the world around us. Examples are hydrogen (H), iron (Fe), oxygen (O), calcium (Ca), nitrogen (N), and carbon (C). In nature, an element does not usually exist by itself but rather combines with the atoms of other elements to form compounds. Examples of some compounds important to our study of the human body are: water (H_2O), in which two atoms of hydrogen combine with one atom of oxygen; carbon dioxide (CO_2), in which an atom of carbon combines with two atoms of oxygen; and glucose ($C_6H_{12}O_6$), in which six carbon atoms and six oxygen atoms combine with 12 hydrogen atoms.

The elements carbon, hydrogen, oxygen, nitrogen, phosphorus, and sulfur are found in all living things. If calcium is included, these seven elements make up approximately 99% of the human body (weight).

More than 20 different elements are found, in varying amounts, in the human body. Some of these are listed in Table 2–1. As you can see, each element has a standard chemical symbol. This is simply

Table 2–1   ELEMENTS IN THE HUMAN BODY

Element	Symbol	Atomic Number*	Percent of the Body by Weight
Hydrogen	H	1	9.5
Carbon	C	6	18.5
Nitrogen	N	7	3.3
Oxygen	O	8	65.0
Fluorine	F	9	Trace
Sodium	Na	11	0.2
Magnesium	Mg	12	0.1
Phosphorus	P	15	1.0
Sulfur	S	16	0.3
Chlorine	Cl	17	0.2
Potassium	K	19	0.4
Calcium	Ca	20	1.5
Manganese	Mn	25	Trace
Iron	Fe	26	Trace
Cobalt	Co	27	Trace
Copper	Cu	29	Trace
Zinc	Zn	30	Trace
Iodine	I	53	Trace

*Atomic number is the number of protons in the nucleus of the atom. It also represents the number of electrons that orbit the nucleus.

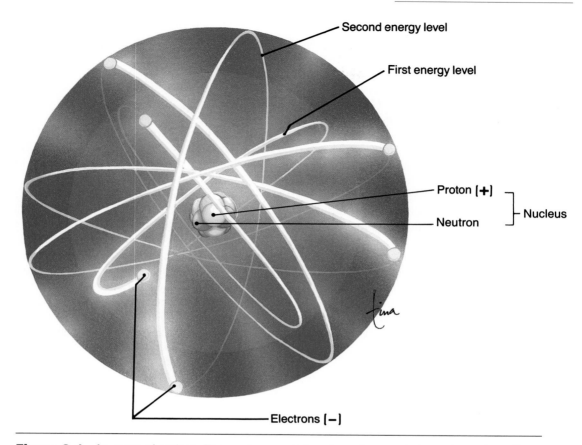

Figure 2–1 An atom of carbon. The nucleus contains six protons and six neutrons (not all are visible here). Six electrons orbit the nucleus, two in the first energy level and four in the second energy level.

the first (and sometimes the second) letter of the element's English or Latin name. You should know the symbols of the elements in this table, since they are used in textbooks, articles, hospital lab reports, and so on. Notice that if a two-letter symbol is used for an element, the second letter is always lower case, not a capital. For example, the symbol for calcium is "Ca," not "CA." "CA" is an abbreviation often used for "cancer."

ATOMS

Atoms are the smallest parts of an element which have the characteristics of that element. An atom

consists of three major subunits or particles: protons, neutrons, and electrons (Fig. 2–1). A **proton** has a positive electrical charge and is found in the nucleus (or center) of the atom. A **neutron** is electrically neutral (has no charge) and is also found in the nucleus. An **electron** has a negative electrical charge and is found outside the nucleus orbiting in what may be called an electron cloud or shell around the nucleus.

The number of protons in an atom gives it its **atomic number.** Protons and neutrons have mass and weight; they give an atom its **atomic weight.** In an atom, the number of protons (+) equals the number of electrons (−); therefore, an atom is electrically neutral. The electrons, however, are important in that they may enable an atom to connect or

bond to other atoms to form **molecules.** A molecule is a combination of atoms (usually of more than one element) which are so tightly bound together that the molecule behaves as a single unit.

Each atom is capable of bonding in only very specific ways. This capability depends on the number and the arrangement of the electrons of the atom. Electrons orbit the nucleus of an atom in shells or **energy levels.** The first, or innermost, energy level can contain a maximum of two electrons and is then considered stable. The second energy level is stable when it contains its maximum of eight electrons. The remaining energy levels, more distant from the nucleus, are also most stable when they contain eight electrons, or a multiple of eight.

A few atoms (elements) are naturally stable, or "uninterested" in reacting, because their outermost energy level already contains the maximum number of electrons. The gases helium and neon are examples of these stable atoms, which do not usually react with other atoms. Most atoms are not stable, however, and tend to gain, lose, or share electrons in order to fill their outermost shell. By doing so, an atom is capable of forming one or more chemical bonds with other atoms. In this way, the atom becomes stable, because its outermost shell of electrons has been filled. It is these reactive atoms that are of interest in our study of anatomy and physiology.

CHEMICAL BONDS

A chemical bond is not a structure, but rather a force or attraction between positive and negative electrical charges that keeps two or more atoms closely associated with each other to form a molecule. By way of comparison, think of gravity. We know that gravity is not a "thing," but rather the force that keeps our feet on the floor. Molecules formed by chemical bonding then have physical characteristics different from those of the atoms of the original elements. For example, the elements hydrogen and oxygen are gases, but atoms of each may chemically bond to form molecules of water, which is a liquid.

The type of chemical bonding depends upon the tendencies of the electrons of atoms involved, as you will see. Three kinds of bonds are very important to the chemistry of the body: ionic bonds, covalent bonds, and hydrogen bonds.

IONIC BONDS

An **ionic bond** involves the loss of one or more electrons by one atom and the gain of the electron(s) by another atom or atoms. Refer to Fig. 2–2 as you read the following.

An atom of sodium (Na) has one electron in its outermost shell, and in order to become stable, it tends to lose that electron. When it does so, the sodium atom has one more proton than it has electrons. Therefore, it now has an electrical charge (or **valence**) of $+1$ and is called a sodium **ion** (Na^+). An atom of chlorine has seven electrons in its outermost shell, and in order to become stable tends to gain one electron. When it does so, the chlorine atom has one more electron than it has protons, and now has a charge (valence) of -1. It is called a chloride ion (Cl^-).

When an atom of sodium loses an electron to an atom of chlorine, their ions have unlike charges (positive and negative) and are thus attracted to one another. The result is the formation of a molecule of sodium chloride: NaCl, or common table salt. The bond that holds these ions together is called an ionic bond.

Another example is the bonding of chlorine to calcium. An atom of calcium has two electrons in its outermost shell and tends to lose those electrons in order to become stable. If two atoms of chlorine each gain one of those electrons, they become chloride ions. The positive and negative ions are then attracted to one another, forming a molecule of calcium chloride, $CaCl_2$, which is also a salt. A **salt** is a molecule made of ions other than hydrogen (H^+) ions or hydroxyl (OH^-) ions.

Ions with positive charges are called **cations.** These include Na^+, Ca^{+2}, K^+, Fe^{+2}, and Mg^{+2}. Ions with negative charges are called **anions,** which include Cl^-, SO_4^{-2} (sulfate), and HCO_3^- (bicarbonate). The types of compounds formed by ionic bonding are salts, acids, and bases. (Acids and bases are discussed later in this chapter.)

In the solid state, ionic bonds are relatively

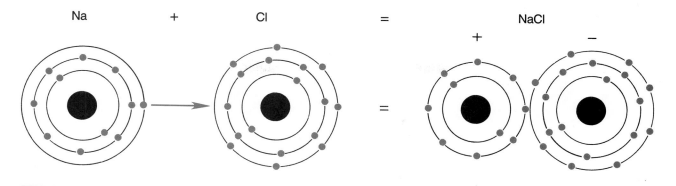

Figure 2–2 Formation of an ionic bond. An atom of sodium loses an electron to an atom of chlorine. The two ions formed have unlike charges, are attracted to one another, and form a molecule of sodium chloride.

strong. Our bones, for example, contain the salt calcium carbonate ($CaCO_3$), which helps give bone its strength. However, in an **aqueous** (water) **solution,** many ionic bonds are weakened. The bonds may become so weak that the bound ions of a molecule separate, creating a solution of free positive and negative ions. For example, if sodium chloride is put in water, it dissolves, then **ionizes.** The water now contains Na^+ ions and Cl^- ions. Ionization, also called **dissociation,** is important to living organisms because once dissociated, the ions are free to take part in other chemical reactions within the body. Cells in the stomach lining produce hydrochloric acid (HCl) and must have Cl^- ions to do so. The chloride in NaCl would not be free to take part in another reaction since it is tightly bound to the sodium atom. However, the Cl^- ions available from ionized NaCl in the cellular water can be used for the **synthesis,** or chemical manufacture, of HCl in the stomach.

COVALENT BONDS

Covalent bonds involve the sharing of electrons between atoms. As shown in Fig. 2–3, an atom of oxygen needs two electrons to become stable. It may share two of its electrons with another atom of oxygen, also sharing two electrons. Together they form a molecule of oxygen gas (O_2), which is the form in which oxygen exists in the atmosphere.

An atom of oxygen may also share two of its electrons with two atoms of hydrogen, each sharing its single electron (see Fig. 2–3). Together they form a molecule of water (H_2O). When writing structural formulas for chemical molecules, a pair of shared electrons is indicated by a single line, as shown in the formula for water; this is a single covalent bond. A double covalent bond is indicated by two lines, as in the formula for oxygen; this represents two pairs of shared electrons.

The element carbon always forms covalent bonds; an atom of carbon has four electrons to share with other atoms. If these four electrons are shared with four atoms of hydrogen, each sharing its one electron, a molecule of methane gas (CH_4) is formed. Carbon may form covalent bonds with other carbons, hydrogen, oxygen, nitrogen, or other elements. Organic compounds such as proteins and carbohydrates are complex and precise arrangements of these atoms covalently bonded to one another. Covalent bonds are relatively strong and are not weakened in an aqueous solution. This is important because the proteins produced by the body, for example, must remain intact in order to function properly in the water of our cells and blood. The functions of organic compounds will be considered later in this chapter.

HYDROGEN BONDS

A hydrogen bond does not involve the sharing or exchange of electrons, but rather is due to a prop-

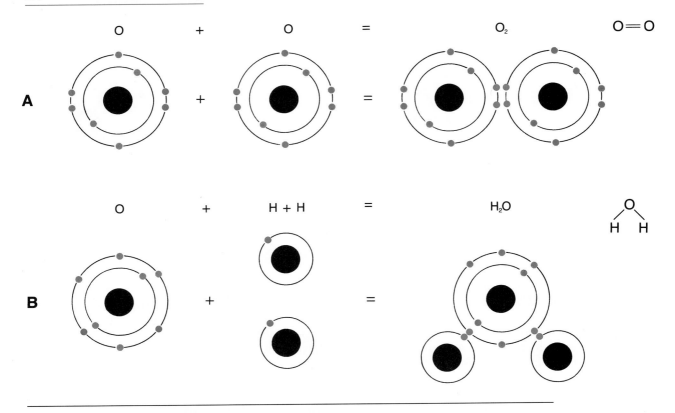

Figure 2–3 Formation of covalent bonds. (**A**), Two atoms of oxygen share two electrons each, forming a molecule of oxygen gas. (**B**), An atom of oxygen shares one electron with each of two hydrogen atoms, each sharing its electron. A molecule of water if formed.

erty of hydrogen atoms. When a hydrogen atom shares its one electron in a covalent bond with another atom, its proton has a slight positive charge and may then be attracted to a nearby oxygen or nitrogen atom, which has a slight negative charge.

Although they are weak bonds, hydrogen bonds are important in several ways. Large organic molecules such as proteins and DNA have very specific functions that depend on their three-dimensional shapes. The shapes of these molecules, so crucial to their proper functioning, are often maintained by hydrogen bonds.

Hydrogen bonds also make water cohesive, that is, each water molecule is attracted to nearby water molecules. Such cohesiveness can be seen if water is dropped onto clean glass; the surface tension created by the hydrogen bonds makes the water form

three-dimensional beads. These bonds are also responsible for the important characteristics of water, which are discussed in a later section.

CHEMICAL REACTIONS

A chemical reaction is a change brought about by the formation or breaking of chemical bonds. Two general types of reactions are synthesis reactions and decomposition reactions.

In a synthesis reaction, bonds are formed to join two or more atoms or molecules to make a new compound. The production of the protein hemoglobin in potential red blood cells is an example of a synthesis reaction. Proteins are synthesized by the bonding of many amino acids, their smaller subunits. Synthesis reactions require energy for the formation of bonds.

In a decomposition reaction, bonds are broken, and a large molecule is changed to two or more smaller ones. One example is the digestion of large molecules of starch to many smaller glucose molecules. Some decomposition reactions release energy; this is described in a later section on cell respiration.

In this and future chapters, keep in mind that the term "reaction" refers to the making or breaking of chemical bonds and thus to changes in the physical and chemical characteristics of the molecules involved.

INORGANIC COMPOUNDS OF IMPORTANCE

Inorganic compounds are usually simple molecules that often consist of only one or two different elements. Despite their simplicity, however, some inorganic compounds are essential to normal structure and functioning of the body.

WATER

Water makes up 60% to 75% of the human body, and is essential to life for several reasons.
1. Water is a **solvent,** that is, many substances (called solutes) can dissolve in water. Nutrients such as glucose are dissolved in blood plasma (which is largely water) to be transported to cells throughout the body. The sense of taste depends upon the solvent ability of saliva; dissolved food stimulates the receptors in taste buds. The excretion of waste products is possible because they are dissolved in the water of urine.
2. Water is a lubricant, which prevents friction where surfaces meet and move. In the digestive tract, mucus is a slippery fluid that permits the smooth passage of food through the intestines. Synovial fluid within joint cavities prevents friction as bones move.

3. Water changes temperature slowly. Water will absorb a great deal of heat before its temperature rises significantly, or it must lose a great deal of heat before its temperature drops significantly. This is one of the factors that helps the body maintain a constant temperature. It is also important for the process of sweating. Excess body heat evaporates sweat on the skin surfaces, rather than overheating the body's cells.

WATER COMPARTMENTS

All water within the body is continually moving, but water is given different names when it is in specific body locations, which are called compartments (Fig. 2–4).

Intracellular fluid (ICF)—the water within cells; about 65% of the total body water
Extracellular fluid (ECF)—all the rest of the water in the body; about 35% of the total. More specific compartments of extracellular fluid include:
  Plasma—water found in blood vessels
  Lymph—water found in lymphatic vessels
  Tissue fluid or interstitial fluid—water found in the small spaces between cells
  Specialized fluids—synovial fluid, cerebrospinal fluid, aqueous humor in the eye, and others

The movement of water between compartments in the body and the functions of the specialized fluids will be discussed in later chapters.

OXYGEN

Oxygen in the form of a gas (O_2) is approximately 21% of the atmosphere, which we inhale. We all know that without oxygen we wouldn't survive very long, but exactly what does it do? Oxygen is important to us because it is essential for a process called cell respiration, in which cells break down simple nutrients such as glucose in order to release energy. The reason we breathe is to obtain oxygen for cell respiration and to exhale the carbon dioxide produced in cell respiration (this will be discussed in the next section). Biologically useful energy that is released by the reactions of cell respiration is trapped in a molecule called adenosine triphos-

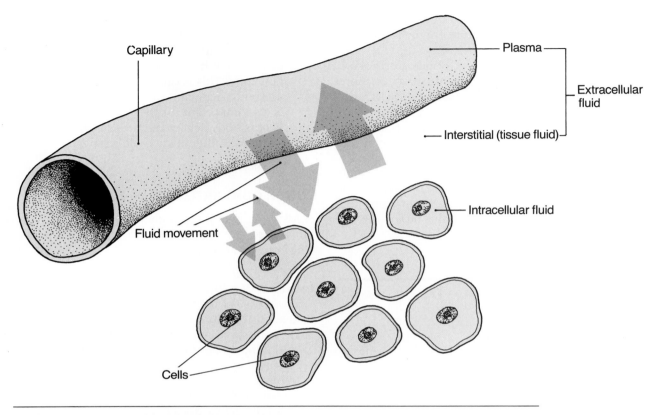

Figure 2–4 Water compartments, showing the names water is given in its different locations and the ways in which water moves between compartments.

phate (ATP). ATP can then be used for cellular processes that require energy.

CARBON DIOXIDE

Carbon dioxide (CO_2) is produced by cells as a waste product of cell respiration. You may ask why a waste product is considered important. Keep in mind that "important" does not always mean "beneficial," but it does mean "significant." If the amount of carbon dioxide in the body fluids increases, it causes these fluids to become too acidic. Therefore, carbon dioxide must be exhaled as rapidly as it is formed to keep the amount in the body within normal limits. Normally this is just what happens, but severe pulmonary diseases such as pneumonia or emphysema decrease gas exchange in the lungs and permit carbon dioxide to accumulate in the blood. When this happens, a person is said to be in a state of **acidosis,** which may seriously disrupt body functioning (See the sections on pH and enzymes later in this chapter; See also Box 2–1: Blood Gases).

CELL RESPIRATION

Cell respiration is the name for energy production within cells and involves both respiratory gases, oxygen and carbon dioxide. There are many chemical reactions involved, but in its simplest form, cell respiration may be summarized by the following equation:

Glucose + O_2 → CO_2 + H_2O + ATP + heat
($C_6H_{12}O_6$)

Box 2–1 BLOOD GASES

A patient is admitted to the emergency room with a possible heart attack, and the doctor in charge orders "blood gases." Another patient hospitalized with pneumonia has "blood gases" monitored at frequent intervals. What are blood gases, and what does measurement of them tell us? The blood gases are oxygen and carbon dioxide, and their levels in arterial blood provide information about the functioning of the respiratory and circulatory systems. Arterial blood normally has a high concentration of oxygen and a low concentration of carbon dioxide. These levels are maintained by gas exchange in the lungs and by the proper circulation of blood.

A pulmonary disease such as pneumonia interferes with efficient gas exchange in the lungs. As a result, blood oxygen concentration may decrease, and blood carbon dioxide concentration may increase. Either of these changes in blood gases may become life-threatening for the patient, so monitoring of blood gases is important. If blood oxygen falls below the normal range, oxygen will be administered; if blood carbon dioxide rises above the normal range, blood pH will be corrected to prevent serious acidosis.

Damage to the heart may also bring about a change in blood gases, especially oxygen. Oxygen is picked up by red blood cells as they circulate through lung capillaries; as red blood cells circulate through the body, they release oxygen to tissues. What keeps the blood circulating or moving? The pumping of the heart.

A mild heart attack, when heart failure is unlikely, is often characterized by a blood oxygen level that is low but still within normal limits. A more severe heart attack that seriously impairs the pumping of the heart will decrease the blood oxygen level to less than normal. This condition is called **hypoxia,** which means that too little oxygen is reaching tissues. When this is determined by measurement of blood gases, appropriate oxygen therapy can be started to correct the hypoxia.

This equation shows us that glucose and oxygen combine to yield carbon dioxide, water, ATP, and heat. Food, represented here by glucose, in the presence of oxygen is broken down into the simpler molecules carbon dioxide and water. The potential energy in the glucose molecule is released in two forms: ATP and heat. Each of the four products of this process has a purpose or significance in the body. The carbon dioxide is a waste product that moves from the cells into the blood to be carried to the lungs and eventually exhaled. The water formed is useful and becomes part of the intracellular fluid. The heat produced contributes to normal body temperature. ATP is used for cell processes such as mitosis, protein synthesis, and muscle contraction, all of which require energy and will be discussed a bit further on in the text.

We will also return to cell respiration in later chapters. For now, the brief description above will suffice to show that eating and breathing are interrelated; both are essential for energy production.

TRACE ELEMENTS

Trace elements are those that are needed by the body in very small amounts. Although they may not be as abundant in the body as are carbon, hydrogen, or oxygen, they are nonetheless essential. Table 2–2 lists some of these trace elements and their functions.

ACIDS, BASES, AND pH

An **acid** may be defined as a substance that increases the concentration of hydrogen ions (H^+) in a water solution. A **base** is a substance that de-

Table 2–2 TRACE ELEMENTS

Element	Function
Calcium	• Provides strength in bones and teeth • Necessary for blood clotting • Necessary for muscle contraction
Phosphorus	• Provides strength in bones and teeth • Part of DNA and RNA • Part of cell membranes
Iron	• Part of hemoglobin in red blood cells; transports oxygen • Part of myoglobin in muscles; stores oxygen • Necessary for cell respiration
Copper	• Necessary for cell respiration
Sodium and potassium	• Necessary for muscle contraction • Necessary for nerve impulse transmission
Sulfur	• Part of some proteins such as insulin and keratin
Cobalt	• Part of vitamin B_{12}
Iodine	• Part of thyroid hormones—thyroxine

creases the concentration of H^+ ions, which in the case of water, has the same effect as increasing the concentration of hydroxyl ions (OH^-).

The acidity or alkalinity (basicity) of a solution is measured on a scale of values called **pH** (parts hydrogen). The values on the **pH scale** range from 0 to 14, with 0 indicating the most acidic level and 14 the most alkaline. A solution with a pH of 7 is neutral because it contains the same number of H^+ ions and OH^- ions. Pure water has a pH of 7. A solution with a higher concentration of H^+ ions than OH^- ions is an acidic solution with a pH below 7. An alkaline solution, therefore, has a higher concentration of OH^- ions than H^+ ions and has a pH above 7.

The pH scale, with the relative concentrations of H^+ ions, and OH^- ions, is shown in Fig. 2–5. A change of one pH unit is a 10-fold change in H^+ ion concentration. This means that a solution with a pH of 4 has 10 times as many H^+ ions as a solution with a pH of 5, and 100 times as many H^+ ions as a solution with a pH of 6. Fig. 2–5 also shows the pH of some body fluids and other familiar solutions. Notice that gastric juice has a pH of 1 and coffee has a pH of 5. This means that gastric juice has 10,000 times as many H^+ ions as does coffee. Although coffee is acidic, it is a weak acid and does not have the corrosive effect of gastric juice, a strong acid.

The cells and internal fluids of the human body have a pH close to neutral. The pH of intracellular fluid is around 6.8, and the normal pH range of blood is 7.35 to 7.45. Fluids such as gastric juice and urine are technically external fluids, since they are in body tracts that open to the environment. The pH of these fluids may be more strongly acidic or alkaline without harm to the body.

The pH of blood, however, must be maintained within its very narrow, slightly alkaline range. A decrease of only one pH unit, which is 10 times as many H^+ ions, would disrupt the chemical reactions of the blood and cause the death of the individual. Normal metabolism tends to make body fluids more acidic, and this tendency to acidosis must be continually corrected. Normal pH of internal fluids is maintained by the kidneys, respiratory system, and buffer systems. Although acid–base balance will be a major topic of Chapter 19, we will briefly mention buffer systems here.

Buffer Systems

A **buffer system** is a chemical or pair of chemicals that minimize changes in pH by reacting with strong acids or strong bases to transform them into substances that will not drastically change pH. Expressed in another way, a buffer may bond to H^+ ions when a body fluid is becoming too acidic, or release H^+ ions when a fluid is becoming too alkaline.

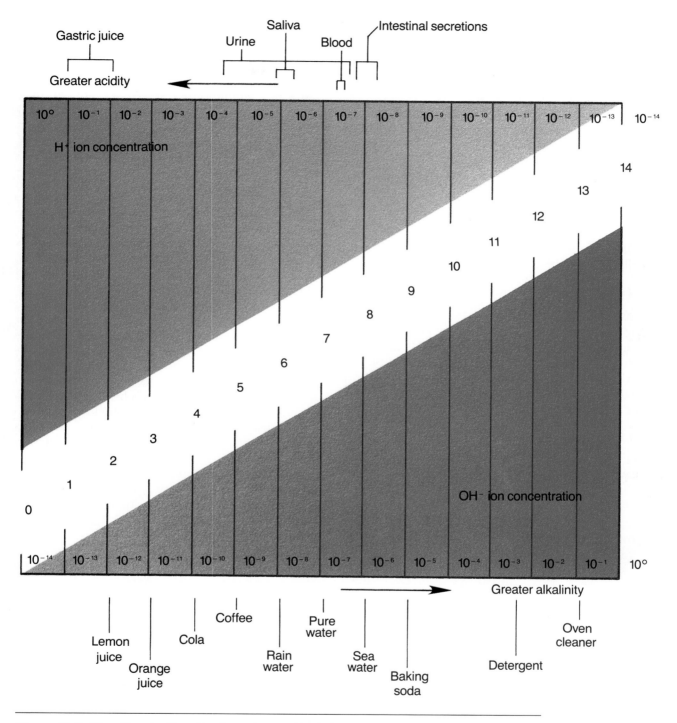

Figure 2–5 The pH scale. The pH values of several body fluids are indicated above the scale. The pH values of some familiar solutions are indicated below the scale.

As a specific example, we will use the bicarbonate buffer system, which consists of carbonic acid (H_2CO_3), a weak acid, and sodium bicarbonate ($NaHCO_3$), a weak base. This pair of chemicals is present in all body fluids but is especially important to buffer blood and tissue fluid.

Carbonic acid ionizes as follows (but remember, because it is a weak acid it does not contribute many H^+ ions to a solution):

$$H_2CO_3 \rightarrow H^+ + HCO_3^-$$

Sodium bicarbonate ionizes as follows:

$$NaHCO_3 \rightarrow Na^+ + HCO_3^-$$

If a strong acid, such as HCl, is added to extracellular fluid, this reaction will occur:

$$HCl + NaHCO_3 \rightarrow NaCl + H_2CO_3$$

What has happened here? Hydrochloric acid, a strong acid that would greatly lower pH, has reacted with sodium bicarbonate. The products of this reaction are NaCl, a salt which has no effect on pH, and H_2CO_3, a weak acid that lowers pH only slightly. This prevents a drastic change in the pH of the extracellular fluid.

If a strong base, such as sodium hydroxide, is added to the extracellular fluid, this reaction will occur:

$$NaOH + H_2CO_3 \rightarrow H_2O + NaHCO_3$$

Sodium hydroxide, a strong base that would greatly raise pH, has reacted with carbonic acid. The products of this reaction are water, which has no effect on pH, and sodium bicarbonate, a weak base that raises pH only slightly. Again, this prevents a drastic change in the pH of the extracellular fluid.

In the body, such reactions take place in less than a second whenever acids or bases are formed that would greatly change pH. Because of the body's tendency to become more acidic, the need to correct acidosis is more frequent. With respect to the bicarbonate buffer system, this means that more $NaHCO_3$ than H_2CO_3 is needed. For this reason, the usual ratio of these buffers is 20:1 ($NaHCO_3$: H_2CO_3).

ORGANIC COMPOUNDS OF IMPORTANCE

Organic compounds all contain covalently bonded carbon and hydrogen atoms and perhaps other elements as well. In the human body there are four major groups of organic compounds: carbohydrates, lipids, proteins, and nucleic acids.

CARBOHYDRATES

A primary function of **carbohydrates** is to serve as sources of energy. All carbohydrates contain carbon, hydrogen, and oxygen and are classified as monosaccharides, disaccharides, oligosaccharides, and polysaccharides. Saccharide means sugar, and the prefix indicates how many are present.

Monosaccharides or single sugar compounds are the simplest sugars. Glucose is a **hexose,** or 6-carbon, sugar with the formula $C_6H_{12}O_6$ (Fig. 2–6). Fructose and galactose also have the same formula, but the physical arrangement of the carbon, hydrogen, and oxygen atoms in each differs from that of glucose. This gives each hexose sugar a different three-dimensional shape. The liver is able to change fructose and galactose to glucose, which is then used by cells in the process of cell respiration to produce ATP.

Another type of monosaccharide is the **pentose,** or 5-carbon, sugar. These are not involved in energy production but rather are structural components of the nucleic acids. Deoxyribose ($C_5H_{10}O_4$) is part of DNA, which is the genetic material of chromosomes. Ribose ($C_5H_{10}O_5$) is part of RNA, which is essential for protein synthesis. We will return to the nucleic acids later in this chapter.

Disaccharides are double sugars, made of two monosaccharides linked together by covalent bonds. Examples are sucrose, lactose, and maltose, which are present in food. They are digested into monosaccharides and then used for energy production.

The prefix "oligo" means "few"; **oligosaccharides** consist of from 3 to 20 monosaccharides. In human cells, oligosaccharides are found on the outer surface of cell membranes. Here they serve as

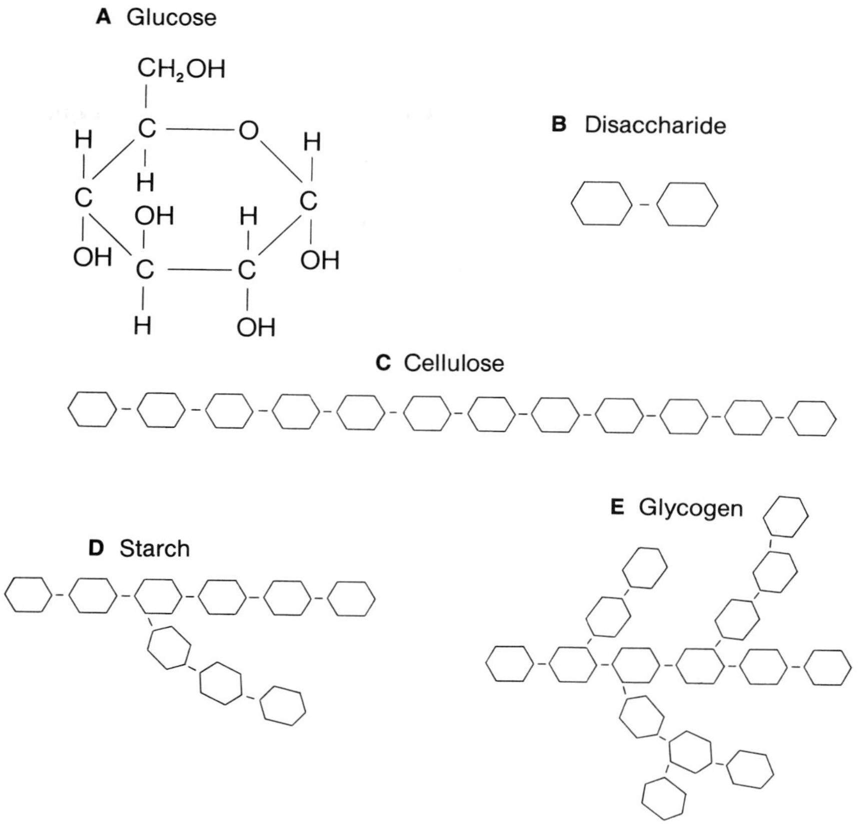

A Glucose

CH_2OH

B Disaccharide

C Cellulose

E Glycogen

D Starch

Figure 2–6 Carbohydrates. **(A)**, Glucose, depicting its structural formula. **(B)**, A disaccharide such as maltose. **(C)**, Cellulose, polysaccharide. **(D)**, Starch, a polysaccharide. **(E)**, Glycogen, a polysaccharide.

antigens, which are chemical markers (or "sign posts") that identify cells. The A, B, and AB blood types, for example, are the result of oligosaccharide antigens on the outer surface of red blood cell membranes. All of our cells have "self" antigens, which identify the cells that belong in an individual. The presence of "self" antigens on our own cells enables the immune system to recognize antigens that are "non-self." Such foreign antigens include bacteria and viruses, and immunity will be a major topic of Chapter 14.

Polysaccharides are made of thousands of glucose molecules, bonded in different ways, resulting in different shapes (see Fig. 2–6). Starches are branched chains of glucose and are produced by plant cells to store energy. We have digestive enzymes that split the bonds of starch molecules, releasing glucose. The glucose is then absorbed and used by cells to produce ATP.

Glycogen, a highly branched chain of glucose molecules, is our own storage form for glucose. After a meal high in carbohydrates, the blood glucose level rises. Excess glucose is then changed to glycogen and stored in the liver and skeletal muscles. When the blood glucose level decreases between meals, the glycogen is converted back to glucose, which is released into the blood. The blood glucose level is kept within normal limits, and cells can take in this glucose to produce energy.

Cellulose is a nearly straight chain of glucose molecules produced by plant cells as part of their cell walls. We have no enzyme to digest the cellulose we consume as part of vegetables and grains, and it passes through the digestive tract unchanged. Another name for dietary cellulose is "fiber," and although we cannot use its glucose for energy, it does have a function. Fiber provides bulk within the cavity of the large intestine. This promotes efficient

peristalsis, the waves of contraction that propel undigested material through the colon. A diet low in fiber does not give the colon much exercise, and the muscle tissue of the colon will contract weakly, just as our skeletal muscles will become flabby without exercise. A diet high in fiber provides exercise for the colon muscle and may help prevent chronic constipation.

The structure and functions of the carbohydrates are summarized in Table 2–3.

LIPIDS

Lipids contain the elements carbon, hydrogen, and oxygen; some also contain phosphorus. In this group of organic compounds are different types of substances with very different functions. We will consider three types: true fats, phospholipids, and steroids (Fig. 2–7).

True fats are made of one molecule of glycerol and one, two, or three fatty acid molecules. If three fatty acid molecules are bonded to a single glycerol, a **triglyceride** is formed (you can usually find this term on the nutrition labels of some highly processed foods; it means that fat has been added). You have undoubtedly heard of saturated and unsaturated fats; the differences between them are discussed in Box 2–2. True fats are a storage form for excess food, that is, they are stored energy. Any type of food consumed in excess of the body's caloric needs will be converted to fat and stored in adipose tissue. Most adipose tissue is subcutaneous, between the skin and muscles. Some organs, however, such as the eyes and kidneys, are enclosed in a layer of fat that acts as a cushion to absorb shock.

Phospholipids are diglycerides with a phosphate group (PO_4) in the third bonding site of glycerol. Although similar in structure to the true fats,

Table 2–3 CARBOHYDRATES

Name	Structure	Function
Monosaccharides—"Single" Sugars		
Glucose	Hexose sugar	• Most important energy source for cells
Fructose and galactose	Hexose sugars	• Converted to glucose by the liver, then used for energy production
Deoxyribose	Pentose sugar	• Part of DNA, the genetic code in the chromosomes of cells
Ribose	Pentose sugar	• Part of RNA, needed for protein synthesis within cells
Disaccharides—"Double" Sugars		
Sucrose, lactose, and maltose	Two hexose sugars	• Present in food, digested to monosaccharides, which are then used for energy production
Oligosaccharides—"Few" Sugars (3–20)		
		• Form "self" antigens on cell membranes; important to permit the immune system to distinguish "self" from foreign antigens (pathogens)
Polysaccharides—"Many" Sugars (Thousands)		
Starches	Branched chains of glucose molecules	• Found in plant foods; digested to monosaccharides and used for energy production
Glycogen	Highly branched chains of glucose molecules	• Storage form for excess glucose in the liver and skeletal muscles
Cellulose	Straight chains of glucose molecules	• Part of plant cell walls; provides fiber to promote peristalsis, especially by the colon

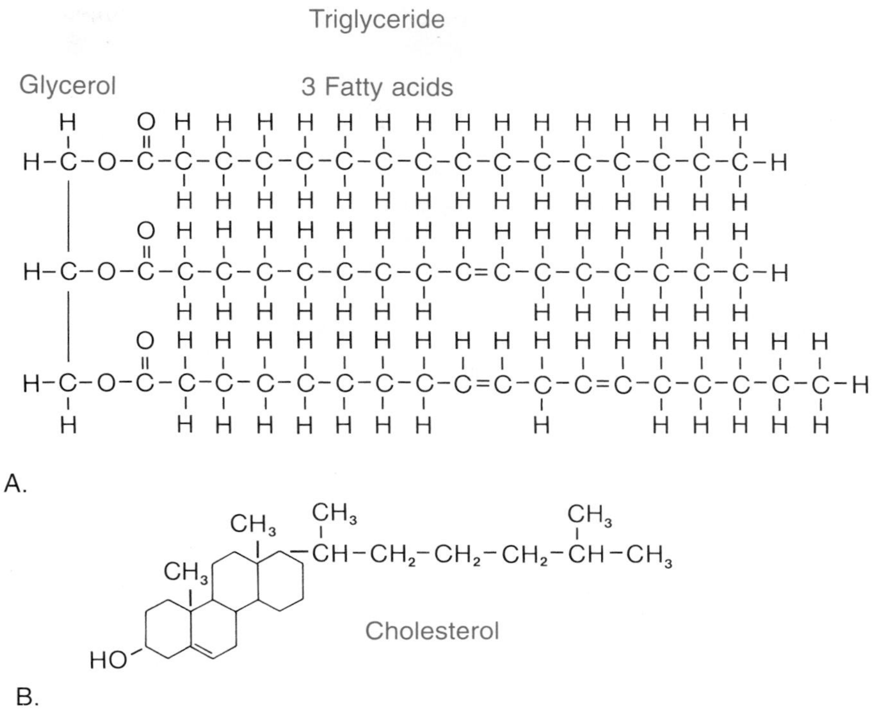

Triglyceride

Glycerol   3 Fatty acids

A.

Cholesterol

B.

Figure 2–7 Lipids. (**A**), A triglyceride made of one glycerol and three fatty acids. (**B**), The steroid cholesterol. The hexagons represent rings of carbons and hydrogens.

phospholipids are not stored energy but rather structural components of cells. Lecithin is a phospholipid that is part of our **cell membranes.** Another example is **myelin,** which forms the myelin sheath around nerve cells and provides electrical insulation for nerve impulse transmission.

The structure of **steroids** is very different from that of the other lipids. **Cholesterol** is an important steroid; it is made of four rings of carbon and hydrogen (not fatty acids and glycerol) and is shown in Fig. 2–7. The liver synthesizes cholesterol, in addition to the cholesterol we eat in food as part of our diet. Cholesterol is another component of cell membranes and is the precursor (raw material) for the synthesis of other steroids. In the ovaries or testes, cholesterol is used to synthesize the steroid hormones estrogen or testosterone, respectively. A form of cholesterol in the skin is changed to vitamin D on exposure to sunlight. Liver cells use cholesterol for the synthesis of bile salts, which emulsify fats in digestion. Despite its link to coronary artery

disease and heart attacks, cholesterol is an essential substance for human beings.

The structure and functions of lipids are summarized in Table 2–4.

PROTEINS

Proteins are made of smaller subunits or building blocks called **amino acids,** which contain the elements carbon, hydrogen, oxygen, nitrogen, and perhaps sulfur. There are about 20 amino acids that make up human proteins. The structure of amino acids is shown in Fig. 2–8. Each amino acid has a central carbon atom covalently bonded to an atom of hydrogen, an amino group (NH_2), and a carboxyl group (COOH). At the fourth bond of the central carbon is the variable portion of the amino acid, represented by R. The R group may be a single hydrogen atom, or a CH_3 group, or a more complex configuration of carbon and hydrogen. This gives each of the 20 amino acids a slightly different physical shape. A bond between two amino acids is

Box 2–2   SATURATED AND UNSATURATED FATS

You have probably heard or read that eating foods high in **unsaturated fats** may help prevent heart disease and that a diet high in **saturated fats** may contribute to heart disease. "Heart disease" actually refers to **atherosclerosis** of the coronary arteries, those that supply the heart muscle with oxygen. Atherosclerosis is the deposition of cholesterol and other substances in the walls of these arteries, leading to obstruction or clot formation and a heart attack. These abnormal deposits of cholesterol are more likely to occur when blood cholesterol levels are high, and diet does have an effect on blood cholesterol.

The true fats may be divided into saturated and unsaturated fats. Refer to Fig. 2–7 and notice that one of the fatty acids has single covalent bonds between all its carbon atoms. Each of these carbons is then bonded to the maximum number of hydrogens; this is a saturated fatty acid. The other fatty acids shown have one or more (poly) double covalent bonds between their carbons and less than the maximum number of hydrogens; these are unsaturated fatty acids. At room temperature, saturated fats are often in solid form, while unsaturated fats are often (not always) in liquid form.

Saturated fats tend to be found in animal foods such as beef, pork, eggs, and cheese, but palm oil and coconut oil are also saturated. Unsaturated fats are found in other plant oils such as corn oil, sunflower oil, and safflower oil, but certain fish oils are also unsaturated.

The breakdown products of saturated fats are used by the liver to synthesize cholesterol, which raises the blood cholesterol level. This is not true for the unsaturated fats; they do not contribute to cholesterol formation. This is why a diet high in unsaturated fats may help prevent atherosclerosis. Unsaturated fats, especially polyunsaturated fats, may actually help lower blood cholesterol by decreasing the formation of cholesterol by the liver.

There are other contributing factors to coronary artery disease, such as heredity, smoking, being overweight, and lack of exercise. Diet alone cannot prevent atherosclerosis. However, a diet low in total fat and high in polyunsaturated fats is a good start.

called a **peptide bond,** and a short chain of amino acids linked together by peptide bonds is a **polypeptide.**

A protein may consist of from 50 to thousands of amino acids. The sequence of the amino acids is specific and unique for each protein. This unique sequence determines the protein's characteristic three-dimensional shape, which in turn determines its function. Our body proteins have many functions; some of these are listed in Table 2–5 and will be mentioned again in later chapters. However, one very important function of proteins will be discussed further here: the role of proteins as enzymes.

Enzymes

Enzymes are **catalysts,** which means that they speed up chemical reactions without the need for an external source of energy such as heat. The many reactions that take place within the body are catalyzed by specific enzymes; all of these reactions must take place at body temperature.

The way in which enzymes function as catalysts is called the **Active Site Theory,** which is based on the shape of the enzyme and the shapes of the reacting molecules, called **substrates.** A simple reaction is depicted in Fig. 2–9. Notice that the enzyme has a specific shape, as do the substrate

Table 2–4 LIPIDS

Name	Structure	Function
True fats	A triglyceride consists of three fatty acid molecules bonded to a glycerol molecule (some are monoglycerides or diglycerides)	• Storage form for excess food molecules in subcutaneous tissue • Cushion organs such as the eyes and kidneys
Phospholipids	Diglycerides with a phosphate group bonded to the glycerol molecule	• Part of cell membranes (lecithin) • Form the myelin sheath to provide electrical insulation for neurons
Steroids (cholesterol)	Four carbon–hydrogen rings	• Part of cell membranes • Converted to vitamin D in the skin on exposure to UV rays of the sun • Converted by the liver to bile salts, which emulsify fats during digestion • Precursor for the steroid hormones such as estrogen in women (ovaries) or testosterone in men (testes)

molecules. The active site of the enzyme is the part that matches the shapes of the substrates. The substrates must "fit" into the active site of the enzyme, and temporary bonds may form between the enzyme and the substrate. This is called the enzyme–substrate complex. In this case, two sub-

Table 2–5 FUNCTIONS OF PROTEINS

Type of Protein	Function
Structural proteins	• Form pores and receptor sites in cell membranes • Keratin—part of skin and hair • Collagen—part of tendons and ligaments
Hormones	• Insulin—enables cells to take in glucose; lowers blood glucose level • Growth hormone—increases protein synthesis and cell division
Hemoglobin	• Enables red blood cells to carry oxygen
Antibodies	• Produced by lymphocytes (white blood cells); label pathogens for destruction
Myosin and actin	• Muscle structure and contraction
Enzymes	• Catalyze reactions

strate molecules are thus brought close together so that chemical bonds are formed between them, creating a new compound. The product of the reaction, the new compound, is then released, leaving the enzyme itself unchanged and able to catalyze another reaction of the same type.

Each enzyme is specific in that it will catalyze only one type of reaction. An enzyme that digests the protein in food, for example, has the proper shape for that reaction but cannot digest starches. For starch digestion, another enzyme with a differently shaped active site is needed. Thousands of chemical reactions take place within the body, and therefore we have thousands of enzymes, each with its own shape and active site.

The ability of enzymes to function may be limited or destroyed by changes in the intracellular or extracellular fluids in which they are found. Changes in pH and temperature are especially crucial. Recall that the pH of intracellular fluid is approximately 6.8, and that a decrease in pH means that more H^+ ions are present. If pH decreases significantly, the excess H^+ ions will react with the active sites of cellular enzymes, change their shapes, and prevent them from catalyzing reactions. This is why a state of acidosis may cause the death of cells—the cells' enzymes are unable to function properly.

With respect to temperature, most human enzymes have their optimum functioning in the nor-

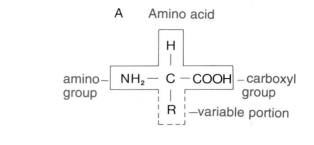

A   Amino acid

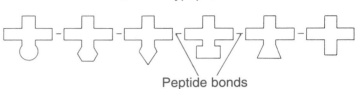

B   Polypeptide

Peptide bonds

Figure 2–8 Amino acids. (**A**), The structural formula of an amino acid. The "R" represents the variable portion of the molecule. (**B**), A polypeptide. Several amino acids, represented by different shapes, are linked by peptide bonds.

mal range of body temperature: 97° to 99°F (36° to 38°C). A temperature of 106°F, a high fever, may break the chemical bonds that maintain the shapes of enzymes. If an enzyme loses its shape, it is said to be **denatured,** and a denatured enzyme is unable to function as a catalyst.

NUCLEIC ACIDS

DNA and RNA

The **nucleic acids, DNA** (deoxyribonucleic acid) and **RNA** (ribonucleic acid), are large molecules made of smaller subunits called nucleotides. A **nu-** **cleotide** consists of a pentose sugar, a phosphate group, and one of several nitrogenous bases. In DNA nucleotides, the sugar is deoxyribose, and the bases are adenine, guanine, cytosine, or thymine. In RNA nucleotides, the sugar is ribose, and the bases are adenine, guanine, cytosine, or uracil. Small segments of DNA and RNA molecules are shown in Fig. 2–10.

Notice that DNA looks somewhat like a twisted ladder; this is two strands of nucleotides called a double helix (coil). Alternating phosphate and sugar molecules form the uprights of the ladder, and pairs of nitrogenous bases form the rungs. The size of the bases and the number of hydrogen

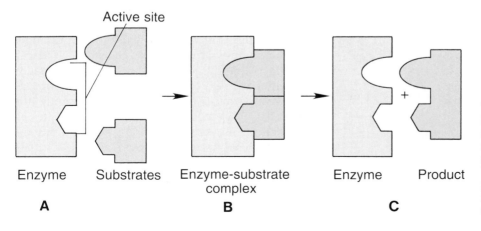

Active site

Enzyme   Substrates
A

Enzyme-substrate
complex
B

Enzyme   Product
C

Figure 2–9 Active site theory, as shown in a synthesis reaction. (**A**), The enzyme and substrates of this reaction. (**B**), The enzyme-substrate complex. (**C**), The product of the reaction and the intact enzyme.

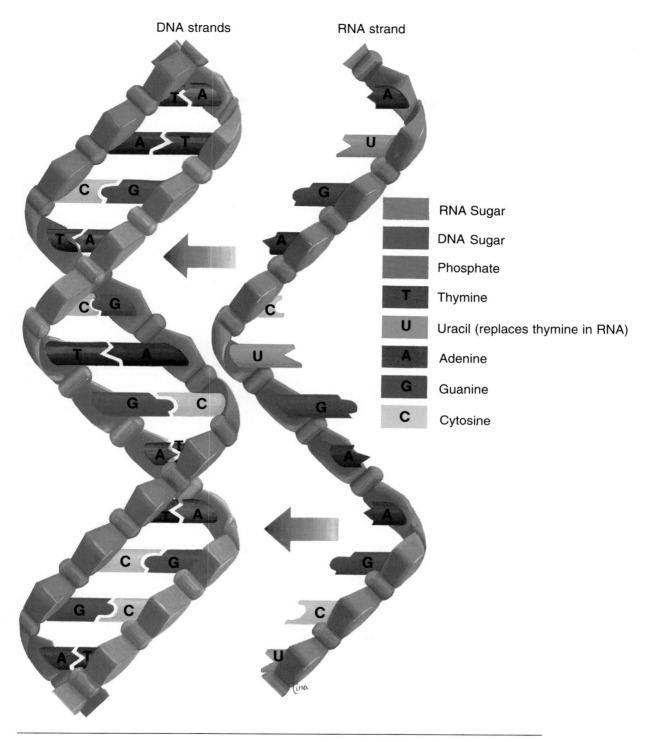

Figure 2–10 DNA and RNA. A small portion of each molecule is shown, with each part of a nucleotide represented by a different color. Note the complementary base pairing of DNA (A-T and G-C). When RNA is synthesized, it is a complementary copy of half the DNA molecule (with U in place of T).

Table 2–6   NUCLEIC ACIDS

Name	Structure	Function
DNA (deoxyribonucleic acid)	A double helix of nucleotides; adenine paired with thymine and guanine paired with cytosine	• Found in the chromosomes in the nucleus of a cell • Is the genetic code for hereditary characteristics
RNA (ribonucleic acid)	A single strand of nucleotides; adenine, guanine, cytosine, and uracil	• Copies the genetic code of DNA to direct protein synthesis in the cytoplasm of cells
ATP (adenosine triphosphate)	A single adenine nucleotide with three phosphate groups	• An energy-transfering molecule • Formed when cell respiration releases energy from food molecules • Used for energy-requiring cellular processes

bonds each can form the complementary base pairing of the nucleic acids. In DNA, adenine is always paired with thymine, and guanine is always paired with cytosine.

DNA makes up the chromosomes of cells, and is, therefore, the **genetic code** for hereditary characteristics. The sequence of bases in the DNA strands is actually a code for the many kinds of proteins our cells produce. The functioning of DNA will be covered in more detail in the next chapter.

RNA is a single strand of nucleotides (see Fig. 2–10), with uracil nucleotides in place of thymine nucleotides. RNA is synthesized from DNA in the nucleus of a cell but carries out its function in the cytoplasm. This function is protein synthesis, which will also be discussed in the following chapter.

ATP

ATP (adenosine triphosphate) is a specialized nucleotide that consists of the base adenine, the sugar ribose, and three phosphate groups. Mention has already been made of ATP as a product of cell respiration that contains biologically useful energy. ATP is one of several "energy transfer" molecules within cells, transferring the potential energy in food molecules to cell processes. When a molecule of glucose is broken down into carbon dioxide and water with the release of energy, some of this en-

ergy is used by the cell to synthesize ATP. Present in cells are molecules of ADP (adenosine diphosphate) and phosphate. The energy released from glucose is used to loosely bond a third phosphate to ADP, forming ATP. When the bond of this third phosphate is again broken and energy is released, ATP then becomes the energy source for cell processes such as mitosis.

All cells have enzymes that can remove the third phosphate group from ATP to release its energy, forming ADP and phosphate. As cell respiration continues, ATP is resynthesized from ADP and phosphate. ATP formation to trap energy from food and breakdown to release energy for cell processes is a continuing cycle in cells.

The structure and functions of the nucleic acids are summarized in Table 2–6.

SUMMARY

All the chemicals we have just described are considered to be non-living, even though they are essential parts of all living organisms. The cells of our bodies are precise arrangements of these non-living chemicals and yet are considered living matter. The cellular level, therefore, is the next level of organization we will examine.

STUDY OUTLINE

Elements
1. Elements are the simplest chemicals, which make up all matter.
2. Carbon, hydrogen, oxygen, nitrogen, phosphorus, sulfur, and calcium make up 99% of the human body.
3. Elements combine in many ways to form molecules.

Atoms (see Fig. 2–1)
1. Atoms are the smallest part of an element which still retain the characteristics of the element.
2. Atoms consist of positively and negatively charged particles and neutral (or uncharged) particles.
   - Protons have a positive charge and are found in the nucleus of the atom.
   - Neutrons have no charge and are found in the nucleus of the atom.
   - Electrons have a negative charge and orbit the nucleus.
3. The number and arrangement of electrons give an atom its bonding capabilities.

Chemical Bonds
1. An ionic bond involves the loss of electrons by one atom and the gain of these electrons by another atom: ions are formed which attract one another (see Fig. 2–2).
   - Cations are ions with positive charges: Na^+, Ca^{+2}.
   - Anions are ions with negative charges: Cl^-, HCO_3^-.
   - Salts, acids, and bases are formed by ionic bonding.
   - In water, many ionic bonds break; dissociation releases ions for other reactions.
2. A covalent bond involves the sharing of electrons between two atoms (see Fig. 2–3).
   - Oxygen gas (O_2) and water (H_2O) are covalently bonded molecules.
   - Carbon always forms covalent bonds; these are the basis for the organic compounds.
   - Covalent bonds are not weakened in an aqueous solution.
3. A hydrogen bond is the attraction of a covalently bonded hydrogen to a nearby oxygen or nitrogen atom.
   - The three-dimensional shape of proteins and nucleic acids is maintained by hydrogen bonds.
   - Water is cohesive because of hydrogen bonds.

Chemical Reactions
1. A change brought about by the formation or breaking of chemical bonds.
2. Synthesis—bonds are formed to join two or more molecules.
3. Decomposition—bonds are broken within a molecule.

Inorganic Compounds of Importance
1. Water—makes up 60% to 75% of the body.
   - Solvent—for transport of nutrients in the blood and excretion of wastes in urine.
   - Lubricant—mucus in the digestive tract.
   - Prevents sudden changes in body temperature; absorbs body heat in evaporation of sweat.
   - Water compartments—the locations of water within the body (see Fig. 2–4).
     - Intracellular—within cells; 65% of total body water.
     - Extracellular—35% of total body water
       - Plasma—in blood vessels.
       - Lymph—in lymphatic vessels.
       - Tissue Fluid—in tissue spaces between cells.
2. Oxygen—21% of the atmosphere.
   - Essential for cell respiration: the breakdown of food molecules to release energy.
3. Carbon Dioxide
   - Produced as a waste product of cell respiration.
   - Must be exhaled; excess CO_2 causes acidosis.
4. Cell Respiration—the energy–producing processes of cells.
   - Glucose + $O_2 \rightarrow CO_2 + H_2O$ + ATP + heat
   - This is why we breathe: to take in oxygen to break down food; to exhale the CO_2 produced.

5. Trace Elements—needed in small amounts (see Table 2–2).
6. Acids, Bases, and pH
   - The pH scale ranges from 0 to 14; 7 is neutral; below 7 is acidic; above 7 is alkaline.
   - An acid increases the H^+ ion concentration of a solution; a base decreases the H^+ ion concentration (or increases the OH^- ion concentration) (see Fig. 2–5).
   - The pH of cells is about 6.8. The pH range of blood is 7.35 to 7.45.
   - Buffer systems maintain normal pH by reacting with strong acids or strong bases to change them to substances that do not greatly change pH.
   - The bicarbonate buffer system consists of H_2CO_3 and $NaHCO_3$.

Organic Compounds of Importance

1. Carbohydrates (see Table 2–3 and Fig. 2–6).
   - Monosaccharides are simple sugars. Glucose, a hexose sugar ($C_6H_{12}O_6$), is the primary energy source for cell respiration.
     - Pentose sugars are part of the nucleic acids DNA and RNA.
   - Disaccharides are made of two hexose sugars. Sucrose, lactose, and maltose are digested to monosaccharides and used for cell respiration.
   - Oligosaccharides consists of from 3 to 20 monosaccharides; they are antigens on the cell membrane that identify cells as "self."
   - Polysaccharides are made of thousands of glucose molecules.
     - Starches are plant products broken down in digestion to glucose.
     - Glycogen is the form in which our bodies store glucose in the liver and muscles.
     - Cellulose, the fiber portion of plant cells, cannot be digested but promotes efficient peristalsis in the colon.
2. Lipids (see Table 2–4 and Fig. 2–7).
   - True fats are made of fatty acids and glycerol; a storage form for energy in adipose tissue. The eyes and kidneys are cushioned by fat.
   - Phospholipids are part of cell membranes. An example is myelin, which provides electrical insulation for nerve cells.
   - Steroids consist of four rings of carbon and hydrogen. Cholesterol, produced by the liver and consumed in food, is the basic steroid from which the body manufactures others: steroid hormones, vitamin D, and bile salts.
3. Proteins
   - Amino acids are the subunits of proteins; 20 amino acids make up human proteins. Peptide bonds join amino acids to one another (see Fig. 2–8).
   - A protein consists of from 50 to thousands of amino acids in a specific sequence.
   - Protein functions—see Table 2–5.
   - Enzymes are catalysts, which speed up reactions without additional energy. The Active Site Theory is based on the shapes of the enzyme and the substrate molecules: these must "fit" (see Fig. 2–9). The enzyme remains unchanged after the product of the reaction is released. Each enzyme is specific for one type of reaction. The functioning of enzymes may be disrupted by changes in pH or body temperature, which change the shape of the active sites of enzymes.
4. Nucleic Acids (see Table 2–6 and Fig. 2–10).
   - Nucleotides are the subunits of nucleic acids. A nucleotide consists of a pentose sugar, a phosphate group, and a nitrogenous base.
   - DNA is a double strand of nucleotides, coiled into a double helix, with complimentary base pairing: A–T and G–C. DNA makes up the chromosomes of cells and is the genetic code for the synthesis of proteins.
   - RNA is a single strand of nucleotides, synthesized from DNA, with U in place of T. RNA functions in protein synthesis.
   - ATP is a nucleotide which is specialized to trap and release energy. Energy released from food in cell respiration is used to synthesize ATP from ADP + P. When cells need energy, ATP is broken down to ADP + P, and the energy is released for cell processes.

REVIEW QUESTIONS

1. State the chemical symbol for each of the following elements: sodium, potassium, iron, calcium, oxygen, carbon, hydrogen, copper, chlorine. (p. 22)

2. Explain, in terms of their electrons, how an atom of sodium and an atom of chlorine form a molecule of sodium chloride. (pp. 24–25)

3. Explain, in terms of their electrons, how an atom of carbon and two atoms of oxygen form a molecule of carbon dioxide. (pp. 25–26)

4. Name the subunits (smaller molecules) of which each of the following is made: DNA, glycogen, a true fat, a protein. (pp. 33–35, 38)

5. State precisely where in the body each of these fluids is found: plasma, intracellular water, lymph, tissue fluid. (p. 27)

6. Explain the importance of the fact that water changes temperature slowly. (p. 27)

7. Describe two ways the solvent ability of water is important to the body. (p. 27)

8. Name the organic molecule with each of the following functions: (pp. 34, 37, 40)
   a. The genetic code in chromosomes
   b. "Self" antigens in our cell membranes
   c. The storage form for glucose in the liver
   d. The storage form for excess food in adipose tissue
   e. The precursor molecule for the steroid hormones
   f. The undigested part of food that promotes peristalsis
   g. The sugars that are part of the nucleic acids

9. State the simple equation of cell respiration. (p. 28)

10. State the role or function of each of the following in cell respiration: CO_2, glucose, O_2, heat, ATP. (pp. 28–29)

11. State a specific function of each of the following in the human body: Ca, Fe, Na, I, Co. (p. 30)

12. Explain, in terms of relative concentrations of H^+ ions and OH^- ions, each of the following: acid, base, neutral substance. (pp. 29–30)

13. State the normal pH range of blood. (p. 30)

14. Complete the following equation, and state how each of the products affects pH: $HCl + NaHCO_3 \rightarrow$ _____ + _____. (p. 32)

15. Explain the Active Site Theory of enzyme functioning. (pp. 36–38)

16. Explain the difference between a synthesis reaction and a decomposition reaction (pp. 26–27).

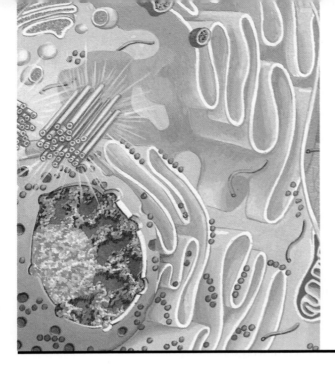

Chapter 3

Chapter Outline

Student Objectives

- Name the organic molecules that make up cell membranes and state their functions.
- State the function of the nucleus and chromosomes.
- Describe the functions of the cell organelles.
- Define each of these cellular transport mechanisms and give an example of the role of each in the body: diffusion, osmosis, facilitated diffusion, active transport, filtration, phagocytosis, pinocytosis.
- Describe the triplet code of DNA.
- Explain how the triplet code of DNA is translated in the synthesis of proteins.
- Describe what happens in mitosis and in meiosis.
- Use examples to explain the importance of mitosis.
- Explain the importance of meiosis.

Cells

New Terminology

Active transport (**AK**–tiv **TRANS**–port)
Aerobic (air–**ROH**–bik)
Cell membrane (SELL **MEM**–brain)
Chromosomes (**KROH**–muh–sohms)
Cytoplasm (**SIGH**–toh–plazm)
Diffusion (di–**FEW**–zhun)
Diploid number (**DIH**–ployd)
Filtration (fill–**TRAY**–shun)
Gametes (**GAM**–eets)
Gene (**JEEN**)
Haploid number (**HA**–ployd)
Meiosis (my–**OH**–sis)
Mitochondria (MY–to–**CHON**–dree–ah)
Mitosis (my–**TOH**–sis)
Nucleus (**NEW**–klee–us)
Organelles (OR–gan–**ELLS**)
Osmosis (ahs–**MOH**–sis)

Pinocytosis (PIN–oh–sigh–**TOH**–sis)
Phagocytosis (FAG–oh–sigh–**TOH**–sis)
Selectively permeable (se–**LEK**–tiv–lee **PER**–me–uh–buhl)
Theory (**THEER**–ree)

Related Clinical Terminology

Benign (bee–**NINE**)
Carcinogen (kar–**SIN**–oh–jen)
Chemotherapy (KEE–moh–**THER**–uh–pee)
Genetic disease (je–**NET**–ik di–**ZEEZ**)
Hypertonic (HIGH–per–**TOHN**–ik)
Hypotonic (HIGH–po–**TOHN**–ik)
Isotonic (EYE–so–**TOHN**–ik)
Malignant (muh–**LIG**–nunt)
Metastasis (muh–**TASS**–tuh–sis)
Mutation (mew–**TAY**–shun)

Terms that appear in **bold type** in the chapter text are defined in the glossary, which begins on p. 549.

All living organisms are made of cells and cell products. This simple statement, called the Cell Theory, was first proposed over 150 years ago. You may think of a **theory** as a guess or hypothesis, and sometimes this is so. A theory, however, is actually the best explanation of all the available evidence. All of the evidence science has gathered so far supports the validity of the Cell Theory.

Cells are the smallest living subunits of a multi-cellular organism such as a human being. A cell is a complex arrangement of the chemicals discussed in the previous chapter, is living, and carries out specific activities. Microorganisms, such as amoebas and bacteria, are single cells which function independently. Human cells, however, must work together and function interdependently. Homeostasis depends upon the contributions of all of the different kinds of cells.

Human cells vary in size, shape, and function. Most human cells are so small they can only be seen with the aid of a microscope and are measured in units called **microns** (1 micron = 1/25,000 of an inch—See Appendix 1: Units of Measure). One exception is the human ovum or egg cell, which is about 1 millimeter in diameter, just visible to the unaided eye. Some nerve cells, although microscopic in diameter, may be quite long. Those in our arms and legs, for example, are at least 2 feet (60 cm) long.

With respect to shape, human cells vary greatly. Some are round or spherical, others rectangular, still others irregular. White blood cells even change shape as they move.

Cell functions also vary, and since our cells do not act independently, we will cover specialized cell functions in Chapter 4. This chapter will be concerned with the basic structure of cells and the cellular activities common to all our cells.

CELL STRUCTURE

Despite their many differences, human cells have several similar structural features: a cell membrane, cytoplasm and cell organelles, and a nucleus. Red blood cells are an exception since they have no nuclei when mature. The cell membrane forms the outer boundary of the cell and surrounds the cytoplasm, organelles, and nucleus.

CELL MEMBRANE

Also called the **plasma membrane,** the **cell membrane** is made of phospholipids, cholesterol, and proteins. The arrangement of these organic molecules is shown in Fig. 3–1. The phospholipids permit lipid-soluble materials to easily enter or leave the cell by diffusion through the cell membrane. The presence of cholesterol decreases the fluidity of the membrane, thus making it more stable. The proteins have several functions: Some form **pores** or openings to permit passage of materials; others are **enzymes** that also help substances enter the cell. Still other proteins, with oligosaccharides on their outer surface, are **antigens,** markers that identify the cells of an individual as "self." Yet another group of proteins serves as **receptor sites** for hormones. Many hormones bring about their specific effects by first bonding to a particular receptor on the cell membrane. This bonding then triggers chemical reactions within the cell membrane or the interior of the cell.

Although the cell membrane is the outer boundary of the cell, it should already be apparent to you that it is not a static or wall-like boundary, but rather an active, dynamic one. The cell membrane is **selectively permeable,** that is, certain substances are permitted to pass through and others are not. These mechanisms of cellular transport will be covered later in this chapter.

NUCLEUS

With the exception of mature red blood cells, all human cells have a nucleus. The **nucleus** floats in the cytoplasm and is bounded by a double-layered **nuclear membrane** with many pores. It contains one or more nucleoli and the chromosomes of the cell (Fig. 3–2).

A **nucleolus** is a small sphere made of DNA, RNA, and protein. The nucleoli form a type of RNA called ribosomal RNA, which becomes part of ribosomes (a cell organelle) and is involved in protein synthesis.

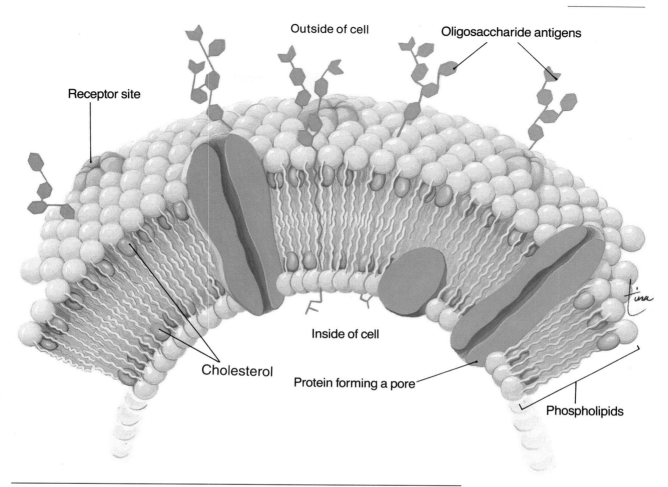

Figure 3–1 The cell (plasma) membrane depicting the types of molecules present.

The nucleus is the control center of the cell because it contains the chromosomes. The 46 **chromosomes** of a human cell are usually not visible; they are long threads called **chromatin.** When a cell divides, however, the chromatin coils extensively into visible chromosomes. Chromosomes are made of DNA and protein. Remember from our earlier discussion that the DNA is the genetic code for the characteristics and activities of the cell. Although the DNA in the nucleus of each cell contains all of the genetic information for all human traits, only a small number of genes (a **gene** is the genetic code for one protein) are actually active in a partic-ular cell. These active genes are the codes for the proteins necessary for the specific cell type. How the genetic code in chromosomes is translated into proteins will be covered in a later section.

CYTOPLASM AND CELL ORGANELLES

Cytoplasm is a watery solution of minerals, gases, and organic molecules that is found between the cell membrane and the nucleus. Chemical reactions take place within the cytoplasm, and the cell organelles are found here. Cell **organelles** are intracellular structures, often bounded by their own

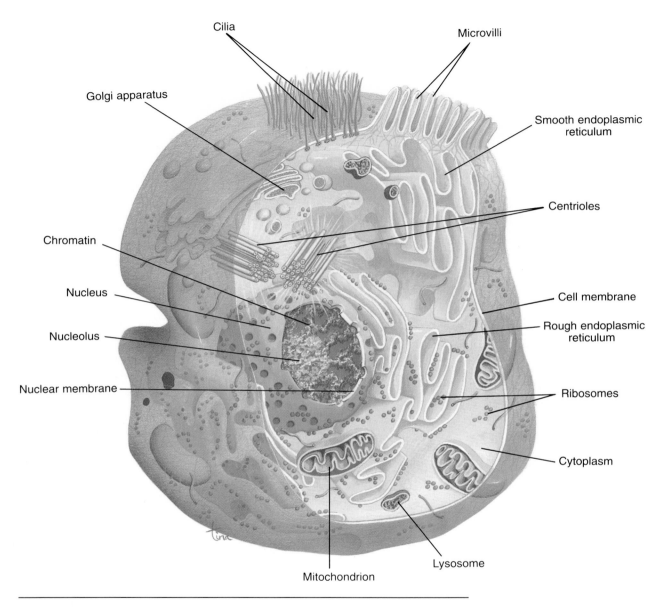

Figure 3–2 Generalized human cell depicting the structural components. See text and Table 3–1 for descriptions.

membranes, that have specific roles in cellular functioning. They are also shown in Fig. 3–2.

The **endoplasmic reticulum** (ER) is an extensive network of membranous tubules that extend from the nuclear membrane to the cell membrane.

Rough ER has numerous ribosomes on its surface, while smooth ER has no ribosomes at all. As a network of interconnected tunnels, the ER serves as a passageway for the transport of the materials necessary for cell function within the cell. These include proteins synthesized by the ribosomes on the rough ER, and lipids synthesized by the smooth ER.

Ribosomes are very small structures made of protein and ribosomal RNA. Some are found on the surface of rough ER, while others float freely within the cytoplasm. Ribosomes are the site of protein synthesis.

The **Golgi apparatus** is a series of flat, membranous sacs, somewhat like a stack of saucers. Carbohydrates are synthesized within the Golgi apparatus, and are packaged, along with other materials, for secretion from the cell. To secrete a substance, small sacs of the Golgi membrane break off and fuse with the cell membrane, releasing the substance to the exterior of the cell.

Mitochondria are oval or spherical organelles within the cytoplasm, bounded by a double membrane. The inner membrane has folds called cristae. Within the mitochondria, the **aerobic** (oxygen-requiring) reactions of cell respiration take place. Therefore, mitochondria are the site of ATP (and hence energy) production. Cells that require large amounts of ATP, such as muscle cells, have many mitochondria to meet their need for energy.

Lysosomes are single membrane structures within the cytoplasm that contain digestive enzymes. When certain white blood cells engulf bacteria, the bacteria are digested and destroyed by these lysosomal enzymes. Worn-out cell parts and dead cells are also digested by these enzymes, which contributes to the process of inflammation in damaged tissues and is necessary before tissue repair can begin.

Centrioles are a pair of rod-shaped structures perpendicular to one another, located just outside the nucleus. Their function is to organize the spindle fibers during cell division.

Cilia and **flagella** are mobile thread-like projections through the cell membrane. **Cilia** serve the function of sweeping materials across the cell surface. They are usually shorter than flagella, and an individual cell has many of them. Cells lining the fallopian tubes, for example, have cilia to sweep the egg cell toward the uterus. The only human cell with a **flagellum** is the sperm cell. The flagellum provides **motility,** or movement, for the sperm cell.

The functions of the cell organelles are summarized in Table 3–1.

Table 3–1 FUNCTIONS OF CELL ORGANELLES

Organelle	Function(s)
Endoplasmic reticulum (ER)	• Passageway for transport of materials within the cell • Synthesis of lipids
Ribosomes	• Site of protein synthesis
Golgi apparatus	• Synthesis of carbohydrates • Packaging of materials for secretion from the cell
Mitochondria	• Site of aerobic cell respiration—ATP production
Lysosomes	• Contain enzymes to digest ingested material or damaged tissue
Centrioles	• Organize the spindle fibers during cell division
Cilia	• Sweep materials across the cell surface
Flagellum	• Enables a cell to move

CELLULAR TRANSPORT MECHANISMS

Living cells constantly interact with the blood or tissue fluid around them, taking in some substances and secreting or excreting others. There are several mechanisms of transport that enable cells to move materials into or out of the cell: diffusion, osmosis, facilitated diffusion, active transport, filtration, phagocytosis, and pinocytosis. Some of these take place without the expenditure of energy by the cells. But others *do* require energy, often in the form of ATP. Each of these mechanisms is described below and an example is included to show how each is important to the body.

DIFFUSION

Diffusion is the movement of molecules from an area of greater concentration to an area of lesser concentration (that is, with or along a **concentration gradient**). Diffusion occurs because molecules have free energy, that is, they are always in

Sugar cube in water   Sugar dissolving   Equilibrium

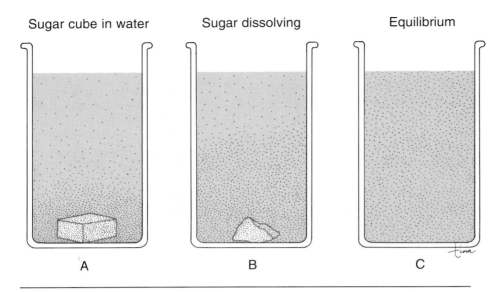

A   B   C

Figure 3–3 Diffusion of sugar in water. (**A**), Sugar cube. (**B**), Partial dissolving and diffusion of sugar molecule. (**C**), Sugar molecules distributed evenly throughout the water (equilibrium).

motion. The molecules in a solid move very slowly; those in a liquid move faster, and those in a gas move faster still, as when ice absorbs heat energy, melts, and then evaporates. In Fig. 3–3, a sugar cube in a glass of water is shown. As the sugar dissolves, the sugar molecules collide with one another. These collisions spread out the sugar molecules until they are evenly dispersed among the water molecules. The molecules are still moving, but as some go to the top others go to the bottom, and so on. Thus, an equilibrium (or steady-state balance) is reached.

Within the body, the gases oxygen and carbon dioxide move by diffusion. In the lungs, for example, there is a high concentration of oxygen in the alveoli (air sacs) and a low concentration of oxygen in the blood in the surrounding pulmonary capillaries. The opposite is true for carbon dioxide: A low concentration in the air in the alveoli and a high concentration in the blood in the pulmonary capillaries. These gases diffuse in opposite directions, each moving from where there is more to where there is less. Oxygen diffuses from the air to the blood to be circulated throughout the body. Carbon dioxide diffuses from the blood to the air to be exhaled.

OSMOSIS

Osmosis may be simply defined as the diffusion of water through a selectively permeable membrane or barrier. That is, water will move from an area with more water present to an area with less water. Another way to say this is that water will naturally tend to move to an area where there is more dissolved material, such as salt or sugar. If a 2% salt solution and a 6% salt solution are separated by a membrane allowing water but not salt to pass through it, water will diffuse from the 2% salt solution to the 6% salt solution. The result is that the 2% solution will become more concentrated, and the 6% solution will become more dilute.

In the body, the cells lining the small intestine absorb water from digested food by osmosis. These cells have first absorbed salts, have become more "salty," and water follows salt into the cells. The process of osmosis also takes place in the kidneys, which reabsorb large amounts of water (many gallons each day) to prevent its loss in urine. In Box 3–1: Terminology of Solutions is some terminology we use when discussing solutions and the effects of various solutions on cells.

Box 3–1 TERMINOLOGY OF SOLUTIONS

Human cells or other body fluids contain many dissolved substances (called **solutes**) such as salts, sugars, acids, and bases. The concentration of solutes in a fluid creates the **osmotic pressure** of the solution, which in turn determines the movement of water through membranes.

As an example here, we will use sodium chloride (NaCl). Human cells have a NaCl concentration of 0.9%. With human cells as a reference point, the relative NaCl concentrations of other solutions may be described with the following terms:

Isotonic—a solution with the same salt concentration as in cells.
   The blood plasma is isotonic to red blood cells.
Hypotonic—a solution with a lower salt concentration than in cells.
   Distilled water (0% salt) is hypotonic to human cells.
Hypertonic—a solution with a higher salt concentration than in cells.
   Sea water (3% salt) is hypertonic to human cells.
Refer now to the diagrams below of red blood cells (RBCs) in each of these different types of solutions, and note the effect of each on osmosis:

- When RBCs are in plasma, water moves into and out of them at equal rates, and the cells remain normal in size and water content.
- If RBCs are placed in distilled water, more water will enter the cells than leaves, and the cells will swell and eventually burst.
- If RBCs are placed in seawater, more water will leave the cells then enters, and the cells will shrivel and die.

This knowledge of osmotic pressure is used when replacement fluids are needed for a patient who has become dehydrated. Isotonic solutions are usually used; normal saline and Ringer's solution are examples. These will provide rehydration without causing osmotic damage to cells or extensive shifts of fluid between the blood and tissues.

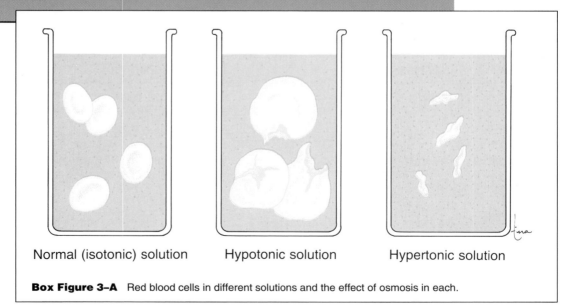

Normal (isotonic) solution   Hypotonic solution   Hypertonic solution

Box Figure 3–A Red blood cells in different solutions and the effect of osmosis in each.

FACILITATED DIFFUSION

The word facilitate means to help or assist. In **facilitated diffusion,** molecules move through a membrane from an area of greater concentration to an area of lesser concentration, but they need some help to do this.

In the body, our cells must take in glucose to use for ATP production. Glucose, however, will not diffuse through most cell membranes by itself, even if there is more outside the cell than inside. Diffusion of glucose into most cells requires **carrier enzymes,** proteins that are part of the cell membrane. Glucose bonds to the carrier enzymes, and by doing so becomes soluble in the phospholipids of the cell membrane. The glucose-carrier molecule diffuses through the membrane, and glucose is released to the interior of the cell.

ACTIVE TRANSPORT

Active transport requires the energy of ATP to move molecules from an area of lesser concentration to an area of greater concentration. Notice that this is the opposite of diffusion, in which the free energy of molecules causes them to move to where there are fewer of them. Active transport is therefore said to be movement against a concentration gradient.

In the body, nerve cells and muscle cells have "sodium pumps" to move sodium ions (Na^+) out of the cells. Sodium ions are more abundant outside the cells, and they constantly diffuse into the cell, their area of lesser concentration. Without the sodium pumps to return them outside, the incoming sodium ions would bring about an unwanted nerve impulse or muscle contraction. Nerve and muscle cells constantly produce ATP to keep their sodium pumps working and prevent spontaneous impulses.

Another example of active transport is the absorption of glucose and amino acids by the cells lining the small intestine. The cells use ATP to absorb these nutrients from digested food, even when their intracellular concentration becomes greater than their extracellular concentration.

FILTRATION

The process of **filtration** also requires energy, but the energy needed does not come directly from

Table 3–2   CELLULAR TRANSPORT MECHANISMS

Mechanism	Definition	Example in the Body
Diffusion	Movement of molecules from an area of greater concentration to an area of lesser concentration.	Exchange of gases in the lungs or body tissues.
Osmosis	The diffusion of water.	Absorption of water by the small intestine or kidneys.
Facilitated diffusion	Carrier enzymes move molecules across cell membranes.	Intake of glucose by most cells.
Active transport	Movement of molecules from an area of lesser concentration to an area of greater concentration (requires ATP).	Absorption of amino acids and glucose from food by the cells of the small intestine.
Filtration	Movement of water and dissolved substances from an area of higher pressure to an area of lower pressure (blood pressure).	Formation of tissue fluid; the first step in the formation of urine.
Phagocytosis	A moving cell engulfs something.	White blood cells engulf bacteria.
Pinocytosis	A stationary cell engulfs something.	Cells of the kidney tubules reabsorb small proteins.

ATP. It is the energy of mechanical pressure. Filtration means that water and dissolved materials are forced through a membrane from an area of higher pressure to an area of lower pressure.

In the body, **blood pressure** is created by the pumping of the heart. Filtration occurs when blood flows through capillaries, whose walls are only one cell thick and very permeable. The blood pressure in capillaries is higher than the pressure of the surrounding tissue fluid. In capillaries throughout the body, blood pressure forces plasma and dissolved materials through the capillary membranes into the surrounding tissue spaces. This creates more tissue fluid and is how cells receive glucose, amino acids, and other nutrients. Blood pressure in the capillaries of the kidneys also brings about filtration, which is the first step in the formation of urine.

PHAGOCYTOSIS AND PINOCYTOSIS

These two processes are similar in that both involve a cell engulfing something. An example of **phagocytosis** is a white blood cell engulfing bacteria. The white blood cell flows around the bacterium, taking it in and eventually digesting it.

Other cells that are stationary may take in small molecules that become adsorbed or attached to their membranes. The cells of the kidney tubules reabsorb small proteins by **pinocytosis,** so that the protein is not lost in urine.

Table 3–2 summarizes the cellular transport mechanisms.

THE GENETIC CODE AND PROTEIN SYNTHESIS

The structure of DNA, RNA, and protein was described in Chapter 2 but will be reviewed briefly here.

DNA AND THE GENETIC CODE

DNA is a double strand of nucleotides in the form of a **double helix,** very much like a spiral ladder. The rungs of the ladder are made of the four ni-

trogenous bases, always found in complementary pairs: adenine with thymine (A–T) and guanine with cytosine (G–C). Although DNA contains just these four bases, the bases may be arranged in many different sequences (reading up or down the ladder). It is the sequence of bases that is the **genetic code.** The DNA of our 46 chromosomes is estimated to contain about 6 billion base pairs, which make up as many as 50,000 to 100,000 genes.

Recall that a **gene** is the genetic code for one protein, and a protein is a specific sequence of amino acids. Therefore, a gene, or segment of DNA, is the code for the sequence of amino acids in a particular protein.

The code for a single amino acid consists of three bases in the DNA molecule; this **triplet** of bases may be called a **codon** (Fig. 3–4). There is a triplet of bases in the DNA for each amino acid in the protein. If a protein consists of 100 amino acids, the gene for that protein would consist of 100 triplets, or 300 bases. Some of the triplets will be the same, since the same amino acid may be present in several places within the protein. Also part of the gene are other triplets that start and stop the process of making the protein, rather like punctuation marks.

RNA AND PROTEIN SYNTHESIS

The transcription of the genetic code in DNA into proteins requires the other nucleic acid, **RNA.** DNA is found in the chromosomes in the nucleus of the cell, but protein synthesis takes place on the ribosomes in the cytoplasm. **Messenger RNA (mRNA)** is the intermediary molecule between these two sites.

When a protein is to be made, the segment of DNA that is its gene uncoils, and the hydrogen bonds between the base pairs break (see Fig. 3–4). Within the nucleus are RNA nucleotides (A,C,G,U) and enzymes to construct a single strand of nucleotides that is a complementary copy of half the DNA gene (with uracil in place of thymine). This copy of the gene is mRNA, which then separates from the DNA. The gene coils back into the double helix, and the mRNA leaves the nucleus, enters the cytoplasm, and becomes attached to ribosomes.

As the copy of the gene, mRNA is a series of triplets of bases; each triplet is the code for one amino

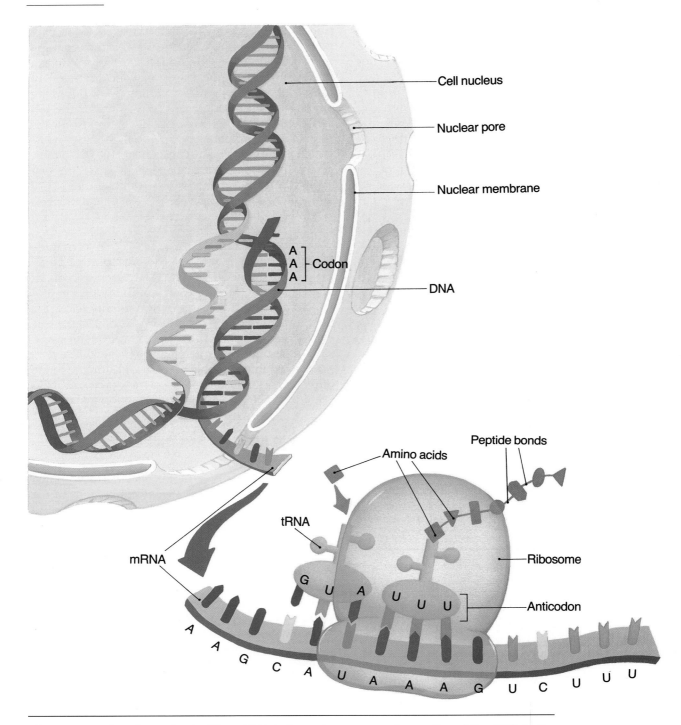

Figure 3–4 Protein synthesis. The mRNA is formed as a copy of a portion of the DNA in the nucleus of a cell. In the cytoplasm, the mRNA becomes attached to ribosomes. See text for further description.

acid. Another type of RNA, called **transfer RNA (tRNA)**, is also found in the cytoplasm. Each tRNA molecule has an **anticodon,** a triplet complementary to a triplet on the mRNA. The tRNA molecules pick up specific amino acids (which have come from protein in our food) and bring them to their proper triplets on the mRNA. The ribosomes contain enzymes to catalyze the formation of **peptide bonds** between the amino acids. When an amino acid has been brought to each triplet on the mRNA, and all peptide bonds have been formed, the protein is finished.

The protein then leaves the ribosomes and may be transported by the ER to where it is needed in the cell, or it may be packaged by the Golgi apparatus for secretion from the cell. A summary of the process of protein synthesis is found in Table 3–3.

Thus, the expression of the genetic code may be described by the following sequence:

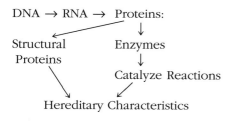

Table 3–3   PROTEIN SYNTHESIS

Molecule or Organelle	Functions
DNA	· A double strand (helix) of nucleotides that is the genetic code in the chromosomes of cells · A gene is the sequence of bases (segment of DNA) that is the code for one protein
mRNA (messenger RNA)	· A single strand of nucleotides formed as a complementary copy of a gene in the DNA · Now contains the triplet code: three bases is the code for one amino acid · Leaves the DNA in the nucleus, enters the cytoplasm of the cell, and becomes attached to ribosomes
Ribosomes	· The cell organelles that are the site of protein synthesis · Attaches the mRNA molecule · Contain enzymes to form peptide bonds between amino acids
tRNA (transfer RNA)	· Picks up amino acids (from food) in the cytoplasm and transports them to their proper sites (triplets) along the mRNA molecule

Box 3–2   GENETIC DISEASE— SICKLE-CELL ANEMIA

A **genetic disease** is a hereditary disorder, one that may be passed from generation to generation. Although there are hundreds of genetic diseases, they all have the same basis: a mistake in DNA. Since DNA makes up the chromosomes that are found in egg and sperm, this mistake may be passed from parents to children.

Sickle-cell anemia is the most common genetic disorder among people of African descent and affects the hemoglobin in red blood cells. Normal hemoglobin, called hemoglobin A (HbA), is a protein made of two identical alpha chains (141 amino acids each) and two identical beta chains (146 amino acids each). In sickle-cell hemoglobin (HbS), the 6th amino acid in each beta chain is incorrect; valine is present instead of the glutamic acid found in HbA. This difference seems minor— only two incorrect amino acids out of a total of over 500—but the consequences for the person are very serious.

HbS has a great tendency to crystallize when oxygen levels are low, as is true in capillaries. When HbS crystallizes, the red blood cells are deformed into crescents (sickles) and other irregular shapes. These irregular, rigid red blood cells clog and

Box 3–2   GENETIC DISEASE— SICKLE-CELL ANEMIA (Continued)

rupture capillaries, causing internal bleeding and severe pain. These red blood cells are also fragile, have a short life span, and break up easily, leading to anemia and hypoxia (lack of oxygen). In the past, children with sickle-cell anemia often died before the age of 20 years. Treatment of this disease has improved greatly, but it is still incurable.

What has happened to cause the formation of HbS rather than HbA? Since hemoglobin is a protein, the genetic code for it is in the DNA (of chromosome #11). One amino acid in the beta chains is incorrect, therefore, one triplet in the DNA gene for the beta chain must be, and is, incorrect. This mistake is copied by mRNA in the cells of the red bone marrow, and HbS is synthesized in red blood cells.

Sickle-cell anemia is a recessive genetic disease, which means that a person with one gene for HbS and one gene for HbA will have "sickle-cell trait." Such a person, sometimes called a carrier, usually will not have the severe effects of sickle-cell anemia, but may pass the gene for HbS to children. It is estimated that 9% of African-Americans have sickle-cell trait and about 1% have sickle-cell anemia.

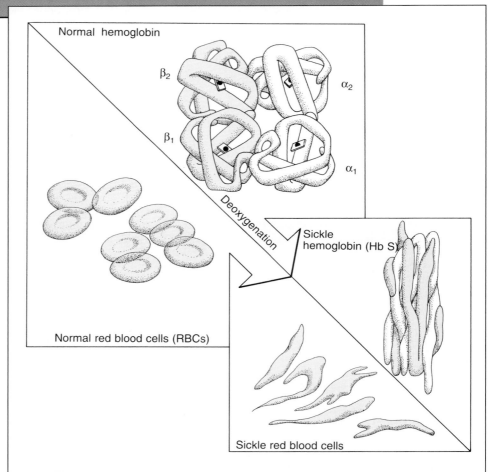

Box Figure 3–B Structure of hemoglobin A and sickle cell hemoglobin and their effect on red blood cells.

Each of us is the sum total of our genetic characteristics. Blood type, hair color, muscle proteins, nerve cells, and thousands of other aspects of our structure and functioning have their basis in the genetic code of DNA.

If there is a "mistake" in the DNA, that is, incorrect bases or triplets of bases, this mistake will be copied by the mRNA. The result is the formation of a malfunctioning or nonfunctioning protein. This is called a **genetic** or **hereditary disease,** and a specific example is described in Box 3–2: Genetic Disease—Sickle-Cell Anemia.

CELL DIVISION

Cell division is the process by which a cell reproduces itself. There are two types of cell division, mitosis and meiosis. Although both types involve cell reproduction, their purposes are very different.

MITOSIS

Each of us began life as one cell, a fertilized egg. Each of us now consists of billions of cells produced by the process of mitosis. In **mitosis,** one cell with the **diploid number** of chromosomes (the usual number, 46 for people) divides into two identical cells, each with the diploid number of chromosomes. This production of identical cells is necessary for the growth of the organism and for repair of tissues.

Before mitosis can take place, a cell must have two complete sets of chromosomes, since each new cell must have the diploid number. The process of **DNA replication** enables each chromosome to make a copy of itself. The time during which this takes place is called **interphase,** the time between mitotic divisions. Although interphase is sometimes referred to as the resting stage, resting means "not dividing," not "inactive." The cell is quite actively producing a second set of chromosomes and storing energy in ATP.

The stages of mitosis are **prophase, metaphase, anaphase,** and **telophase.** What happens in each of these stages is described in Table 3–4. As you read the events of each stage, refer to Fig. 3–5,

Table 3–4   STAGES OF MITOSIS

Stage	Events
Prophase	1. The chromosomes coil up and become visible as short rods. Each chromosome is really 2 chromatids (original DNA plus its copy) still attached at a region called the centromere. 2. The nuclear membrane disappears. 3. The centrioles move toward opposite poles of the cell and organize the spindle fibers, which extend across the equator of the cell.
Metaphase	1. The pairs of chromatids line up along the equator of the cell. The centromere of each pair is attached to a spindle fiber. 2. The centromeres now divide.
Anaphase	1. Each chromatid is now considered a separate chromosome; there are two complete and separate sets. 2. The spindle fibers contract and pull the chromosomes, one set toward each pole of the cell.
Telophase	1. The sets of chromosomes reach the poles of the cell and become indistinct as their DNA uncoils to form chromatin. 2. A nuclear membrane reforms around each set of chromosomes.
Cytokinesis	1. The cytoplasm divides; new cell membrane is formed.

which depicts mitosis in a cell with a diploid number of four.

As mentioned above, mitosis is essential for repair of tissues, to replace damaged or dead cells. Some examples may help illustrate this. In several areas of the body, mitosis takes place constantly. These sites include the epidermis of the skin, the stomach lining, and the red bone marrow. For each of these sites, there is a specific reason why this constant mitosis is necessary.

What happens to the surface of the skin? The dead, outer cells are worn off by contact with the

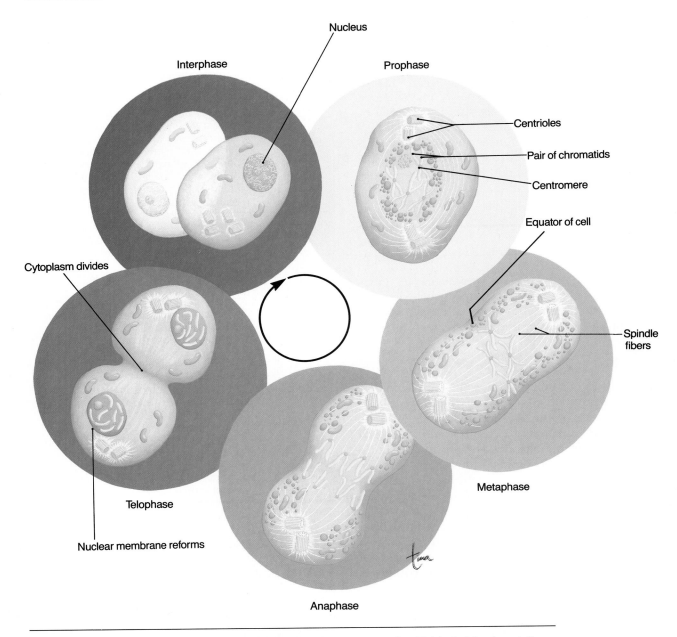

Figure 3–5 States of mitosis in a cell with the diploid number of four. See Table 3–3 for descriptions.

environment. Mitosis in the lower living layer replaces these cells, and the epidermis maintains its normal thickness.

The stomach lining, although internal, is also constantly worn away. Gastric juice, especially the hydrochloric acid, is very damaging to cells. Rapid mitosis replaces damaged cells and keeps the stomach lining intact.

Box 3–3   ABNORMAL CELLULAR FUNCTIONING—CANCER

There are more than 100 different types of **cancer,** all of which are characterized by abnormal cellular functioning. Normally, our cells undergo mitosis only when necessary and stop when appropriate. A cut in the skin, for example, is repaired by mitosis, usually without formation of excess tissue. The new cells fill in the damaged area, and mitosis slows when the cells make contact with surrounding cells. This is called contact inhibition, which limits the new tissue to just what is needed. **Malignant** (cancer) cells, however, are characterized by uncontrolled cell division. Our cells are genetically programmed to have particular life spans and to divide or die. One gene is known to act as a brake on cell division; another gene enables cells to live indefinitely, beyond their normal life spans, and to keep dividing. Any imbalance in the activity of these genes may lead to abnormal cell division. Such cells are not inhibited by contact with other cells, keep dividing, and tend to spread.

A malignant tumor begins in a primary site such as the colon, then may spread or metastasize. Often the malignant cells are carried by the lymph or blood to other organs such as the liver, where secondary tumors develop. **Metastasis** is characteristic only of malignant cells; **benign** tumors do not metastasize but remain localized in their primary site.

What causes normal cells to become malignant? At present, we have only partial answers. A malignant cell is created by a **mutation,** a genetic change that brings about abnormal cell functions or responses. Environmental substances that cause mutations are called **carcinogens.** One example is the tar found in cigarette smoke, which is definitely linked to lung cancer. Ultraviolet light may also cause mutations, especially in skin that is overexposed to sunlight. For a few specific kinds of cancer, the trigger is believed to be infection with certain viruses that cause cellular mutations. Carriers of hepatitis B virus, for example, are more likely to develop primary liver cancer than are people who have never been exposed to this virus. Recent research has discovered two genes, one on chromosome 2 and the other on chromosome 3, that causes a certain form of colon cancer. Both of these genes are the codes for proteins that correct the "mistakes" that may occur when the new DNA is synthesized. When these proteins do not function properly, the mistakes (mutations) in the DNA lead to the synthesis of yet other faulty proteins that impair the functioning of the cell and predispose it to becoming malignant.

Once cells have become malignant, their functioning cannot return to normal. Therefore, the treatments for cancer are directed at removing or destroying the abnormal cells. Surgery to remove tumors, radiation to destroy cells, and **chemotherapy** to stop cell division or interfere with other aspects of cell metabolism are all aspects of cancer treatment.

One of the functions of red bone marrow is the production of red blood cells. Since red blood cells have a life span of only about 120 days, new ones are needed to replace the older ones that die. Very rapid mitosis in the red bone marrow produces ap- proximately 2 million new red blood cells every second.

It is also important to be aware of the areas of the body where mitosis cannot take place. In an adult, muscle cells and neurons (nerve cells) cannot reproduce themselves. If they die, their functions are also lost. Someone whose spinal cord has been severed will have paralysis and loss of sensation below the level of the injury. The spinal cord neurons cannot undergo mitosis to replace the ones that were lost, and such an injury is permanent.

The heart is made of cardiac muscle cells, which are also incapable of mitosis. A heart attack (myocardial infarction) means that a portion of cardiac muscle dies because of lack of oxygen. These cells cannot be replaced, and the heart will be a less effective pump. If a large enough area of the heart muscle dies, the heart attack may be fatal (See Box 3–3: Abnormal Cellular Functioning—Cancer).

MEIOSIS

Meiosis is a more complex process of cell division that results in the formation of **gametes,** which are egg and sperm cells. In meiosis, one cell with the diploid number of chromosomes divides twice to form four cells, each with the **haploid number** (half the usual number) of chromosomes. In women, meiosis takes place in the ovaries and is called **oogenesis.** In men, meiosis takes place in the testes and is called **spermatogenesis.** The differences between oogenesis and spermatogenesis will be discussed in Chapter 20, the Reproductive System.

The egg and sperm cells produced by meiosis have the haploid number of chromosomes, which is 23 for humans. Meiosis is sometimes called reduction division because the division process reduces the chromosome number in egg or sperm. Then, during **fertilization** in which the egg unites with the sperm, the 23 chromosomes of the sperm plus the 23 chromosomes of the egg will restore the diploid number of 46 in the fertilized egg. Thus the proper chromosome number is maintained in the cells of the new individual.

SUMMARY

As mentioned at the beginning of this chapter, human cells work closely together and function interdependently. Each type of human cell makes a contribution to the body as a whole. Usually, however, cells do not function as individuals, but rather in groups. Groups of cells with similar structure and function form a tissue, which is the next level of organization.

STUDY OUTLINE

Human cells vary in size, shape, and function. Our cells function interdependently to maintain homeostasis.
Cell Structure—the major parts of a cell are the cell membrane, nucleus (except mature RBCs), cytoplasm, and cell organelles
1. Cell Membrane—the selectively permeable boundary of the cell (see Fig. 3–1).
   • Phospholipids permit diffusion of lipid-soluble materials.
   • Cholesterol provides stability.
   • Proteins form pores, carrier enzymes, "self" antigens, and receptor sites for hormones.
2. Nucleus—the control center of the cell; has a double-layer membrane.

   • Nucleolus—forms ribosomal RNA.
   • Chromosomes—made of DNA and protein; DNA is the genetic code for the structure and functioning of the cell. A gene is a segment of DNA that is the code for one protein. Human cells have 46 chromosomes.
3. Cytoplasm—a watery solution of minerals, gases, and organic molecules; contains the cell organelles; site for many chemical reactions.
4. Cell Organelles—intracellular structures with specific functions (see Table 3–1 and Fig. 3–2).

Cellular Transport Mechanisms—the processes by which cells take in or secrete or excrete

materials through the selectively permeable cell membrane

1. Diffusion—movement of molecules from an area of greater concentration to an area of lesser concentration; occurs because molecules have free energy: they are constantly in motion. Oxygen and carbon dioxide are exchanged by diffusion in the lungs.
2. Osmosis—the diffusion of water. Water diffuses to an area of less water, that is, to an area of more dissolved material. The small intestine absorbs water from digested food by osmosis. Isotonic, hypertonic, hypotonic (see Box 3–1).
3. Facilitated Diffusion—carrier enzymes that are part of the cell membrane permit cells to take in materials that would not diffuse in by themselves. Most cells take in glucose by facilitated diffusion.
4. Active Transport—a cell uses ATP to move substances from an area of lesser concentration to an area of greater concentration. Nerve cells and muscle cells have sodium pumps to return Na$^+$ ions to the exterior of the cells; this prevents spontaneous impulses. Cells of the small intestine absorb glucose and amino acids from digested food by active transport.
5. Filtration—pressure forces water and dissolved materials through a membrane from an area of higher pressure to an area of lower pressure. Tissue fluid is formed by filtration: blood pressure forces plasma and dissolved nutrients out of capillaries and into tissues. Blood pressure in the kidney capillaries creates filtration that is the first step in the formation of urine.
6. Phagocytosis—a moving cell engulfs something; white blood cells phagocytize bacteria to destroy them.
7. Pinocytosis—a stationary cell engulfs small molecules; kidney tubule cells reabsorb small proteins by pinocytosis.

The Genetic Code and Protein Synthesis (see Fig. 3–4 and Table 3–3)

1. DNA and the Genetic Code
   - DNA is a double helix with complementary base pairing: A–T and G–C.
   - The sequence of bases in the DNA is the genetic code for proteins.
   - The triplet code: three bases (a codon) is the code for one amino acid.
   - A gene consists of all the triplets that code for a single protein.
2. RNA and Protein Synthesis
   - mRNA is a complementary copy of the sequence of bases in a gene (DNA).
   - mRNA moves from the nucleus to the ribosomes in the cytoplasm.
   - tRNA molecules (in the cytoplasm) have anticodons for the triplets on the mRNA.
   - tRNA molecules bring amino acids to their proper triplets on the mRNA.
   - Ribosomes contain enzymes to form peptide bonds between the amino acids.
3. Expression of the Genetic Code
   - DNA → RNA → Proteins (structural and enzymes) → Hereditary Characteristics.
   - A genetic disease is a "mistake" in the DNA, which is copied by mRNA and results in a malfunctioning protein.

Cell Division

1. Mitosis—one cell with the diploid number of chromosomes divides once to form two cells, each with the diploid number of chromosomes (46 for humans).
   - DNA replication forms two sets of chromosomes during Interphase.
   - Stages of mitosis (see Fig. 3–5 and Table 3–4): prophase, metaphase, anaphase, and telophase. Cytokinesis is the division of the cytoplasm following telophase.
   - Mitosis is essential for growth and for repair and replacement of damaged cells.
   - Adult nerve and muscle cells cannot divide; their loss may involve permanent loss of function.
2. Meiosis—one cell with the diploid number of chromosomes divides twice to form four cells, each with the haploid number of chromosomes (23 for humans).
   - Oogenesis in the ovaries forms egg cells.
   - Spermatogenesis in the testes forms sperm cells.
   - Fertilization of an egg by a sperm restores the diploid number in the fertilized egg.

REVIEW QUESTIONS

1. State the functions of the organic molecules of cell membranes: cholesterol, proteins, phospholipids. (p. 46)

2. Describe the function of each of these cell organelles: mitochondria, lysosomes, Golgi apparatus, ribosomes, endoplasmic reticulum. (pp. 47–49)

3. Explain why the nucleus is the control center of the cell. (p. 47)

4. What part of the cell membrane is necessary for facilitated diffusion? Describe one way this process is important within the body. (p. 52)

5. What provides the energy for filtration? Describe one way this process is important within the body. (pp. 52–53)

6. What provides the energy for diffusion? Describe one way this process is important within the body. (pp. 49–50)

7. What provides the energy for active transport? Describe one way this process is important within the body. (p. 52)

8. Define osmosis, and describe one way this process is important within the body. (p. 50)

9. Explain the difference between hypertonic and hypotonic, using human cells as a reference point. (p. 51)

10. In what way are phagocytosis and pinocytosis similar? Describe one way each process is important within the body. (p. 53)

11. How many chromosomes does a human cell have? What are these chromosomes made of? (p. 53)

12. Name the stage of mitosis in which each of the following takes place: (p. 57)
    a. the two sets of chromosomes are pulled toward opposite poles of the cell
    b. the chromosomes become visible as short rods
    c. a nuclear membrane reforms around each complete set of chromosomes
    d. the pairs of chromatids line up along the equator of the cell
    e. the centrioles organize the spindle fibers
    f. cytokinesis takes place after this stage

13. Describe two specific ways mitosis is important within the body. Explain why meiosis is important. (pp. 57–60)

14. Compare mitosis and meiosis in terms of: (pp. 57, 60)
    a. number of divisions
    b. number of cells formed
    c. chromosome number of the cell formed

15. Explain the triplet code of DNA. Name the molecule that copies the triplet code of DNA. Name the organelle that is the site of protein synthesis. What other function does this organelle have in protein formation? (pp. 53–55)

Chapter 4
Tissues and Membranes

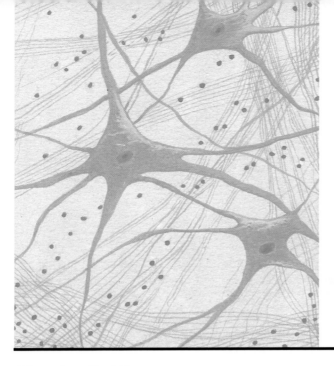

Chapter 4

Student Objectives

- Describe the general characteristics of each of the four major categories of tissues.
- Describe the functions of the types of epithelial tissues with respect to the organs in which they are found.
- Describe the functions of the connective tissues, and relate them to the functioning of the body or a specific organ system.
- Explain the differences, in terms of location and function, among skeletal muscle, smooth muscle, and cardiac muscle.
- Name the three parts of a neuron and state the function of each. Name the organs made of nerve tissue.
- Describe the locations of the pleural membranes, the pericardial membranes, and the peritoneum–mesentery. State the function of serous fluid in each of these locations.
- State the locations of mucous membranes and the functions of mucus.
- Name some membranes made of connective tissue.
- Explain the difference between exocrine and endocrine glands, and give an example of each.

Tissues and Membranes

New Terminology

Absorption (ab–**ZORB**–shun)
Bone (**BOWNE**)
Cartilage (**KAR**–ti–lidj)
Chondrocytes (**KON**–droh–sites)
Collagen (**KAH**–lah–jen)
Connective tissue (kah–**NEK**–tiv **TISH**–yoo)
Elastin (eh–**LAS**–tin)
Endocrine gland (**EN**–doh–krin GLAND)
Epithelial tissue (EP–i–**THEE**–lee–uhl **TISH**–yoo)
Exocrine gland (**ECK**–so–krin GLAND)
Hemopoietic (HEE–moh–poy–**ET**–ik)
Matrix (**MAY**–tricks)

Microvilli (MY–kro–**VILL**–eye)
Mucous membrane (**MEW**–kuss **MEM**–brain)
Muscle tissue (**MUSS**–uhl **TISH**–yoo)
Myocardium (MY–oh–**KAR**–dee–um)
Nerve tissue (NERV **TISH**–yoo)
Neuron (**NYOOR**–on)
Neurotransmitter (NYOOR–oh–**TRANS**–mih–ter)
Osteocytes (**AHS**–tee–oh–SITES)
Plasma (**PLAZ**–mah)
Secretion (see–**KREE**–shun)
Serous membrane (**SEER**–us **MEM**–brain)
Synapse (**SIN**–aps)

Terms that appear in **bold type** in the chapter text are defined in the glossary, which begins on p. 549.

A **tissue** is a group of cells with similar structure and function. The tissue then contributes to the functioning of the organs in which it is found. You may recall that in Chapter 1 the four major groups of tissues were named and very briefly described. These four groups are epithelial, connective, muscle, and nerve tissue.

This chapter presents more detailed descriptions of the tissues in these four categories. For each tissue, its functions are related to the organs of which it is a part. Also in this chapter is a discussion of **membranes,** which are sheets of tissues. As you might expect, each type of membrane has its specific locations and functions.

EPITHELIAL TISSUE

Epithelial tissues are found on surfaces as either coverings (outer surfaces) or linings (inner sur-

faces). Since they have no capillaries of their own, epithelial tissues receive oxygen and nutrients from the connective tissue beneath them. Many epithelial tissues are capable of secretion and may be called glandular epithelium, or more simply, **glands.**

Classification of the epithelial tissues is based on the type of cell of which the tissue is made, its characteristic shape, and the number of layers of cells. There are three distinctive shapes: **squamous** cells are flat, **cuboidal** cells are cube-shaped, and **columnar** cells are tall and narrow. **"Simple"** is the term for a single layer of cells, and **"stratified"** means that many layers of cells are present (Fig. 4–1 and Table 4–1).

SIMPLE SQUAMOUS EPITHELIUM

Simple squamous epithelium is a single layer of flat cells (Fig. 4–2). These cells are very thin and very smooth—these are important physical characteristics. The alveoli (air sacs) of the lungs are sim-

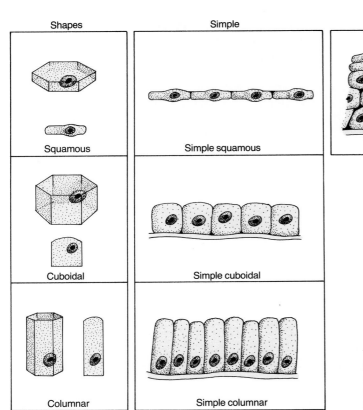

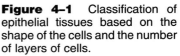

Figure 4–1 Classification of epithelial tissues based on the shape of the cells and the number of layers of cells.

Table 4–1   TYPES OF EPITHELIAL TISSUE

Type	Structure	Location and Function
Simple Squamous	One layer of flat cells	• Alveoli of the lungs—thin to permit diffusion of gases • Capillaries—thin to permit exchanges of materials; smooth to prevent abnormal blood clotting
Stratified Squamous	Many layers of cells; surface cells flat; lower cells rounded; lower layer undergoes mitosis	• Epidermis—surface cells are dead; a barrier to pathogens • Lining of esophagus, vagina—surface cells are living; a barrier to pathogens
Transitional	Many layers of cells; surface cells change from rounded to flat	• Lining of urinary bladder—permits expansion without tearing the lining
Cuboidal	One layer of cube-shaped cells	• Thyroid gland—secretes thyroxine • Salivary glands—secrete saliva
Columnar	One layer of column-shaped cells	• Lining of stomach—secretes gastric juice • Lining of small intestine—secretes enzymes and absorbs end products of digestion (microvilli present)
Ciliated	One layer of columnar cells with cilia on their free surfaces	• Lining of trachea—sweeps mucus and dust to the pharynx • Lining of fallopian tube—sweeps ovum toward uterus

ple squamous epithelium. The thinness of the cells permits the diffusion of gases between the air and blood.

Another location of this tissue is capillaries, the smallest blood vessels. Capillary walls are only one cell thick, which permits the exchange of gases, nutrients, and waste products between the blood and tissue fluid. The interior surface of capillaries is also very smooth (and these cells continue as the lining of the arteries, veins, and heart); this is important because it prevents abnormal blood clotting within blood vessels.

STRATIFIED SQUAMOUS EPITHELIUM

Stratified squamous epithelium consists of many layers of mostly flat cells, although lower cells are rounded. Mitosis takes place in the lowest layer to continually produce new cells to replace those worn off the surface (see Fig. 4–2). This type of epithelium makes up the epidermis of the skin; here the surface cells are dead. Stratified squamous epithelium also lines the oral cavity, esophagus, and, in women, the vagina. In these locations the surface cells are living and make up the mucous membranes of these organs. In all its body locations, this tissue is a barrier to microorganisms because the cells of which it is made are very close together. The more specialized functions of the epidermis will be covered in the next chapter.

TRANSITIONAL EPITHELIUM

Transitional epithelium is a type of stratified epithelium in which the surface cells change shape from round to squamous. The urinary bladder is lined with transitional epithelium. When the bladder is empty, the surface cells are rounded (see Fig. 4–2). As the bladder fills, these cells become flat-

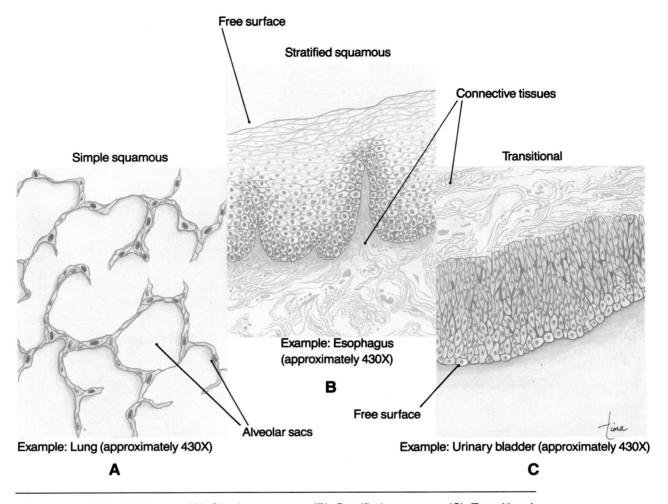

Figure 4–2 Epithelial tissues. (**A**), Simple squamous. (**B**), Stratified squamous. (**C**), Transitional.

tened. Transitional epithelium enables the bladder to fill and stretch without tearing the lining.

SIMPLE CUBOIDAL EPITHELIUM

Simple cuboidal epithelium is a single layer of cube-shaped cells (Fig. 4–3). This type of tissue makes up the functional units of the thyroid gland and salivary glands; these are examples of **glandular epithelium.** In these glands the cuboidal cells are arranged in small spheres and **secrete** into the cavity formed by the sphere. In the thyroid gland, the cuboidal epithelium secretes the thyroid

hormones; thyroxine is an example. In the salivary glands the cuboidal cells secrete saliva.

SIMPLE COLUMNAR EPITHELIUM

Columnar cells are taller than they are wide and are specialized for secretion and absorption. The stomach lining is made of **columnar epithelium** that secretes gastric juice for digestion. The lining of the small intestine (see Fig. 4–3) secretes digestive enzymes, but these cells also **absorb** the end products of digestion. In order to absorb efficiently, the columnar cells of the small intestine have **mi-**

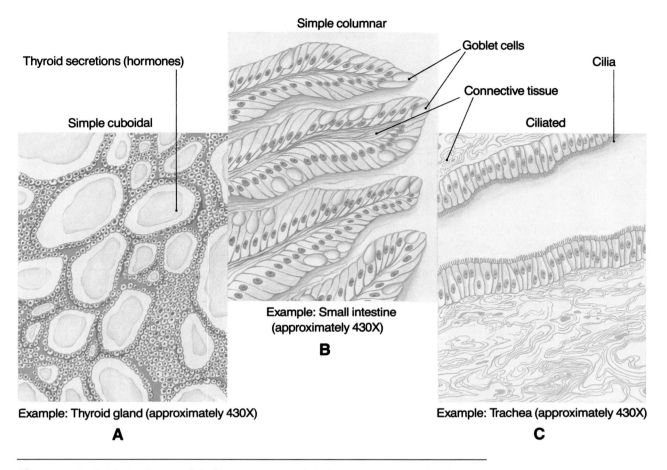

Simple columnar

Goblet cells

Cilia

Thyroid secretions (hormones)

Connective tissue

Simple cuboidal

Ciliated

Example: Small intestine
(approximately 430X)

B

Example: Thyroid gland (approximately 430X)

A

Example: Trachea (approximately 430X)

C

Figure 4–3 Epithelial tissues. (**A**), Simple cuboidal, (**B**), Simple columnar. (**C**), Ciliated.

crovilli, which are folds of the cell membrane on their free surfaces. These microscopic folds greatly increase the surface area for absorption.

Yet another type of columnar cell is the **goblet cell,** which is a unicellular gland. Goblet cells secrete **mucus** and are found in the lining of the intestines and parts of the respiratory tract such as the trachea.

CILIATED EPITHELIUM

Ciliated epithelium consists of columnar cells that have **cilia** on their free surfaces (see Fig. 4–3). Recall from Chapter 3 that the function of cilia is to sweep materials across the cell surface. Ciliated ep-

ithelium lines the nasal cavities, larynx, trachea, and large bronchial tubes. The cilia sweep mucus, with trapped dust and bacteria, toward the pharynx to be swallowed. Bacteria are then destroyed by the hydrochloric acid in the stomach.

Another location of ciliated epithelium in women is the lining of the fallopian tubes. The cilia here sweep the ovum, which has no means of self-locomotion, toward the uterus.

GLANDS

Glands are cells or organs that **secrete** something, that is, produce a substance that has a function either at that site or at a more distant site.

Box 4–1   CYSTIC FIBROSIS

   Cystic fibrosis (CF) is a genetic disorder of certain exocrine glands including the salivary glands, sweat glands, the pancreas, and the mucus glands of the respiratory tract. The genetic mistake in CF involves the gene for a tunnel-like protein that helps cells remove chloride ions. When this protein is abnormal, the cells produce a thick mucus rather than a thin, more watery mucus.

   In the pancreas, thick mucus clogs the ducts and prevents pancreatic enzymes from reaching the small intestine, thus impairing digestion, especially of fats. The most serious effects of CF are in the lungs, where the thick mucus obstructs the small bronchial tubes. Air cannot get to and from the alveoli of the lungs, leading to greatly diminished gas exchange. The mucus also traps bacteria from the air, and the CF patient is subject to frequent episodes of pneumonia.

   CF is one of several disorders believed to be correctable by gene therapy. One current research trial involves the use of non-pathogenic viruses to transfer a normal gene into the cells that line the respiratory airways. Because it involves human subjects, this kind of work proceeds slowly, and it may be several years before conclusive results are obtained as to whether gene therapy offers a cure for CF.

   One possible new treatment is an enzyme (to be inhaled in the form of a fine mist) that will break up the thick mucus to a thinner form that may be more easily coughed up. Although this would not be a cure, it may help prolong the lives of people with CF until a cure becomes possible.

Unicellular Glands

   Unicellular means "one cell." Goblet cells are an example of unicellular glands. As mentioned earlier, goblet cells are found in the lining of the respiratory and digestive tracts. Their secretion is mucus.

Multicellular Glands

   Most glands are made of many similar cells. **Multicellular** glands may be divided into two major groups: exocrine glands and endocrine glands.

   Exocrine glands have **ducts** (tubes) to take the secretion away from the gland to the site of its function. Salivary glands, for example, secrete saliva that is carried by ducts to the oral cavity. Sweat glands secrete sweat that is transported by ducts to the skin surface, where it can be evaporated by excess body heat (see also Box 4–1: Cystic Fibrosis).

   Endocrine glands are ductless glands. The secretions of endocrine glands are a group of chemicals called **hormones,** which enter capillaries and are circulated throughout the body. Hormones then bring about specific effects in their target organs. These will be covered in more detail in Chapter 10. Examples of endocrine glands are the thyroid gland, adrenal glands, and pituitary gland.

   The pancreas is an organ that functions as both an exocrine and an endocrine gland. The exocrine portions secrete digestive enzymes that are carried by ducts to the duodenum, their site of action. The endocrine portions of the pancreas, called Islets of Langerhans, secrete the hormones insulin and glucagon directly into the blood.

CONNECTIVE TISSUE

   There are several kinds of **connective tissue,** some of which may at first seem more different than

Table 4–2 TYPES OF CONNECTIVE TISSUE

Type	Structure	Location and Function
Blood	Plasma (matrix) and red blood cells, white blood cells, and platelets	Within blood vessels: • *Plasma*—transports materials • *RBCs*—carry oxygen • *WBCs*—destroy pathogens • *Platelets*—prevent blood loss
Areolar (Loose)	Fibroblasts and a matrix of tissue fluid, collagen, and elastin fibers	Subcutaneous • Connects skin to muscles; WBCs destroy pathogens Mucous membranes (digestive, respiratory, urinary, reproductive tracts) • WBCs destroy pathogens
Adipose	Adipocytes that store fat (little matrix)	Subcutaneous • Stores excess energy Around eyes and kidneys • Cushions
Fibrous	Mostly collagen fibers (matrix) with few fibroblasts	Tendons and ligaments (regular) • Strong to withstand forces of movement of joints Dermis (irregular) • The strong inner layer of the skin
Elastic	Mostly elastin fibers (matrix) with few fibroblasts	Walls of large arteries • Helps maintain blood pressure Around alveoli in lungs • Promotes normal exhalation
Bone	Osteocytes in a matrix of calcium salts and collagen	Bones: • Support the body • Protect internal organs from mechanical injury • Store excess calcium • Contain and protect red bone marrow
Cartilage	Chondrocytes in a flexible protein matrix	Wall of trachea • Keeps airway open On joint surfaces of bones • Smooth to prevent friction Tip of nose and outer ear • Support Between vertebrae • Absorb shock

alike. The types of connective tissue include areolar, adipose, fibrous, and elastic tissue as well as blood, bone, and cartilage (Table 4–2). A characteristic that all connective tissues have in common is the presence of a matrix in addition to cells. The **matrix** is a structural network or solution of nonliving intercellular material. Each connective tissue has its own specific kind of matrix. The matrix of blood, for example, is blood plasma, which is mostly water. The matrix of bone is made primarily of calcium salts, which are hard and strong. As each type of connective tissue is described below, mention will be made of the types of cells present as well as the kind of matrix.

BLOOD

Although **blood** is the subject of Chapter 11, a brief description will be given here. The matrix of blood is **plasma,** which is 52% to 62% of the total blood volume in the body. The water portion of plasma contains dissolved salts, nutrients, and waste products. As you might expect, one of the primary functions of plasma is transport of these materials within the body.

The cells of blood are red blood cells, white blood cells, and platelets, which are actually fragments of cells. These are shown in Fig. 4–4. The blood-forming or **hemopoietic tissues** are the red bone marrow and lymphatic tissue, which includes the spleen and the lymph nodes. Red bone marrow produces red blood cells, the five types of white blood cells, and the platelets. Two kinds of white blood cells are also produced in lymphatic tissue.

The blood cells make up 38% to 48% of the total blood, and each type of cell has its specific function. **Red blood cells** (RBCs) carry oxygen bonded to their hemoglobin. **White blood cells** (WBCs) destroy pathogens and provide us with immunity to some diseases. **Platelets** prevent blood loss; the process of blood clotting involves platelets.

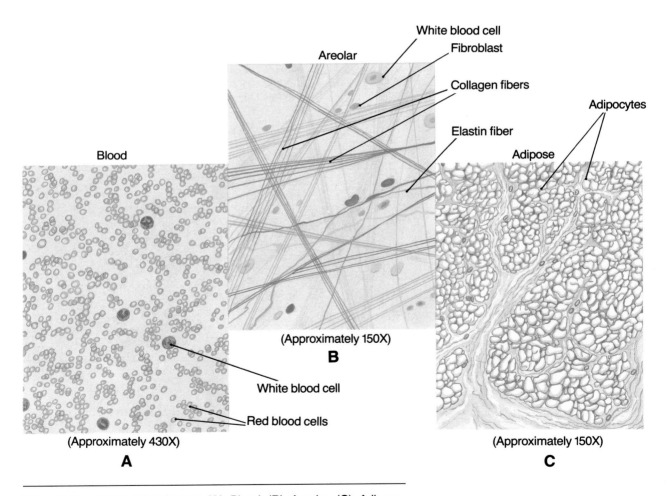

Figure 4–4   Connective tissues. (**A**), Blood. (**B**), Areolar. (**C**), Adipose.

AREOLAR CONNECTIVE TISSUE

The cells of **areolar (or loose) connective tissue** are called **fibroblasts,** which produce protein fibers. **Collagen** fibers are very strong; **elastin** fibers are elastic, that is, able to return to their original length, or recoil, after being stretched. These protein fibers and tissue fluid make up the matrix, or non-living portion, of areolar connective tissue (see Fig. 4–4). Also within the matrix are many white blood cells, which are capable of self-locomotion. Their importance here is related to the locations of areolar connective tissue.

Areolar tissue is found beneath the dermis of the skin and beneath the epithelial tissue of all the body systems that have openings to the environment. Recall that one function of white blood cells is to destroy pathogens. How do pathogens enter the body? Many do so through breaks in the skin. Bacteria and viruses also enter with the air we breathe and the food we eat, and some may get through the epithelial linings of the respiratory and digestive tracts. Areolar connective tissue with its many white blood cells is strategically placed to intercept pathogens before they get to the blood and circulate throughout the body.

ADIPOSE TISSUE

The cells of **adipose tissue** are called **adipocytes** and are specialized to store fat in microscopic droplets. True fats are the chemical form of long-term energy storage. Excess nutrients have calories that are not wasted but are converted to fat to be stored for use when food intake decreases. The amount of matrix in adipose tissue is small and consists of tissue fluid and a few collagen fibers (see Fig. 4–4).

Most fat is stored subcutaneously in the areolar

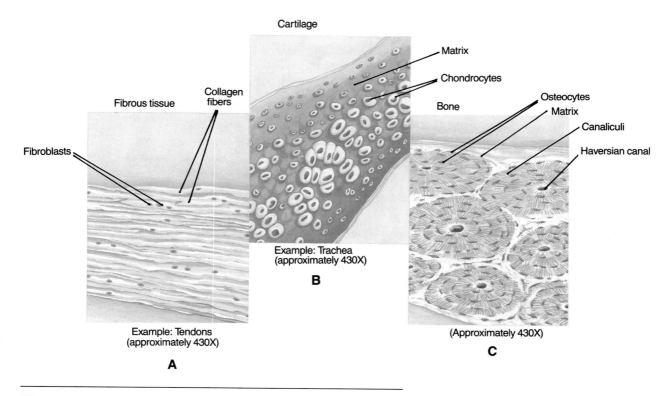

Figure 4–5 Connective tissues. (**A**), Fibrous. (**B**), Cartilage. (**C**), Bone.

connective tissue between the dermis and the muscles. This layer varies in thickness among individuals; the more excess calories consumed, the thicker the layer. As was mentioned in Chapter 2, adipose tissue cushions organs such as the eyes and kidneys.

FIBROUS CONNECTIVE TISSUE

Fibrous connective tissue consists mainly of parallel (regular) collagen fibers with few fibroblasts scattered among them (Fig. 4–5). This parallel arrangement of collagen provides great strength, yet is flexible. The locations of this tissue are related to the need for flexible strength. The outer walls of arteries are reinforced with fibrous connective tissue, because the blood in these vessels is under high pressure. The strong outer wall prevents rupture of the artery (see also Box 4–2: Vitamin C and Collagen). Tendons and ligaments are made of fibrous connective tissue. Tendons connect muscle to bone; ligaments connect bone to bone. When the skeleton is moved, these structures must be able to withstand the great mechanical forces exerted upon them.

Fibrous connective tissue has a relatively poor blood supply, which makes repair a slow process. If you have ever had a severely sprained ankle (which means the ligaments have been overly stretched), you know that complete healing may take several months.

An irregular type of fibrous connective tissue forms the dermis of the skin and the fascia (membranes) around muscles. Although the collagen fibers here are not parallel to one another, the tissue is still strong. The dermis is different from other fibrous connective tissue in that it has a good blood supply (see also Box 4–3: Cosmetic Collagen).

ELASTIC CONNECTIVE TISSUE

As its name tells us, **elastic connective tissue** is primarily elastin fibers. One of its locations is the walls of large arteries. These vessels are stretched when the heart contracts and pumps blood, then recoil when the heart relaxes; this is important to maintain normal blood pressure.

Elastic connective tissue is also found surrounding the alveoli of the lungs. The elastic fibers are stretched during inhalation, then recoil during exhalation to squeeze air out of the lungs. If you pay

Box 4–2 VITAMIN C AND COLLAGEN

In the last two decades, many claims have been made for the effectiveness of vitamin C in preventing or treating the common cold and even cancer. For none of these claims is there any conclusive evidence, but one function of vitamin C is known with certainty, and that is its role in the synthesis of collagen.

Imagine the protein collagen as a ladder with three uprights and rungs that connect adjacent uprights. Vitamin C is essential for forming the "rungs," without which the uprights will not stay together as a strong unit. Collagen formed in the absence of vitamin C is weak, and the effects of weak collagen are dramatically seen in the disease called scurvy.

In 1753 James Lind, a Scottish surgeon, recommended to the British Navy that lime juice be taken on long voyages to prevent scurvy among the sailors. Scurvy is characterized by bleeding gums and loss of teeth, poor healing of wounds, fractures, and bleeding in the skin, joints, and elsewhere in the body. The lime juice did prevent this potentially fatal disease, as did consumption of fresh fruits and vegetables, although at the time no one knew why. Vitamin C was finally isolated in the laboratory in 1928.

Box 4–3 COSMETIC COLLAGEN

Collagen is the protein that makes tendons, ligaments, and other connective tissues strong. In 1981, the Food and Drug Administration (FDA) approved the use of cattle collagen by injection for cosmetic purposes, to minimize wrinkles and scars. Indeed, collagen injected below the skin will flatten out deep facial wrinkles and make them less prominent, and more than half a million people in the United States have had this seemingly simple cosmetic surgery.

There are, however, drawbacks. Injected collagen lasts only a few months; the injections must be repeated several times a year, and they are expensive. Some people experience allergic reactions to the cattle collagen, which is perceived by the immune system as foreign tissue. More seriously, an autoimmune response may be triggered in some individuals, and the immune system may begin to destroy the person's own connective tissue.

In an effort to avoid these problems, some cosmetic surgeons now use the person's own collagen and fat, which may be extracted from the thigh, hip, or abdomen. This technique is still relatively new, and long-term consequences and outcomes have yet to be evaluated. We might remember that for many years the use of silicone injections had been considered safe. Silicone injections are now banned by the FDA, since we now know that they carry significant risk of serious tissue damage.

attention to your breathing for a few moments, you will notice that normal exhalation does not require "work" or energy. This is because of the normal elasticity of the lungs.

BONE

The prefix that designates bone is "osteo," so bone cells are called **osteocytes.** The matrix of **bone** is made of calcium salts and collagen and is strong, hard, and not flexible. In the shafts of long bones such as the femur, the osteocytes, matrix, and blood vessels are in very precise arrangements called **haversian systems** (see Fig. 4–5). Bone has a good blood supply, which enables it to serve as a storage site for calcium and to repair itself relatively rapidly after a simple fracture. Some bones, such as the sternum (breastbone) and pelvic bone, contain red bone marrow, one of the hemopoietic tissues that produces blood cells.

Other functions of bone tissue are related to the strength of bone matrix. The skeleton supports the body, and some bones protect internal organs from mechanical injury. A more complete discussion of bone will be found in Chapter 6.

CARTILAGE

The protein matrix of **cartilage** differs from that of bone in that it is firm, yet smooth and flexible. Cartilage is found on the joint surfaces of bones, where its smooth surface helps prevent friction. The tip of the nose and external ear are supported by flexible cartilage. The wall of the trachea, the airway to the lungs, contains rings of cartilage to maintain an open air passageway. Discs of cartilage are found between the bony vertebrae of the spine. Here the cartilage absorbs shock and permits movement.

Within the cartilage matrix are the **chondrocytes,** or cartilage cells (see Fig. 4–5). There are no capillaries within the cartilage matrix, so these cells are nourished by diffusion through the matrix, a slow process. This becomes clinically important when cartilage is damaged, for repair will take place very slowly or not at all. Athletes sometimes damage cartilage within the knee joint. Such damaged cartilage is usually surgically removed in order to preserve as much joint mobility as possible.

MUSCLE TISSUE

Muscle tissue is specialized for contraction. When muscle cells contract, they shorten and bring about some type of movement. There are three types of muscle tissue: skeletal, smooth, and cardiac (Table 4–3). The movements each can produce have very different purposes.

SKELETAL MUSCLE

Skeletal muscle also may be called **striated** muscle or **voluntary** muscle. Each name describes a particular aspect of this tissue, as you will see. The skeletal muscle cells are cylindrical, have several nuclei each, and appear striated, or striped (Fig. 4–6). The striations are the result of the precise arrangement of the contracting proteins within the cells.

Skeletal muscle tissue makes up the muscles that are attached to bones. These muscles are supplied with motor nerves, and thus move the skeleton.

They also produce a significant amount of body heat. Each muscle cell has its own motor nerve ending. The nerve impulses which can then travel to the muscles are essential to cause contraction. Although we do not have to consciously plan all our movements, the nerve impulses for them originate in the cerebrum, the "thinking" part of the brain.

Let us return to the three names for this tissue: "skeletal" describes its location, "striated" describes its appearance, and "voluntary" describes how it functions. The skeletal muscles and their functioning are the subject of Chapter 7.

SMOOTH MUSCLE

Smooth muscle also may be called **visceral** muscle or **involuntary** muscle. The cells of smooth muscle have tapered ends, a single nucleus, and no striations (see Fig. 4–6). Although nerve impulses do bring about contractions, this is not something most of us can control. The term "visceral" refers to internal organs, many of which contain smooth muscle. The functions of smooth muscle are actually functions of the organs in which the muscle is found.

In the stomach and intestines, smooth muscle contracts in waves called peristalsis to propel food through the digestive tract.

In the walls of arteries and veins, smooth muscle constricts or dilates the vessels to maintain normal

Table 4–3   TYPES OF MUSCLE TISSUE

Type	Structure	Location and Function	Effect of Nerve Impulses
Skeletal	Large cylindrical cells with striations and several nuclei each	Attached to bones • Moves the skeleton and produces heat	Essential to cause contraction (voluntary)
Smooth	Small tapered cells with no striations and one nucleus each	Walls of arteries • Maintains blood pressure Walls of stomach and intestines • Peristalsis Iris of eye • Regulates size of pupil	Bring about contraction or regulate the rate of contraction (involuntary)
Cardiac	Branched cells with faint striations and one nucleus each	Walls of the chambers of the heart • Pumps blood	Regulate only the rate of contraction

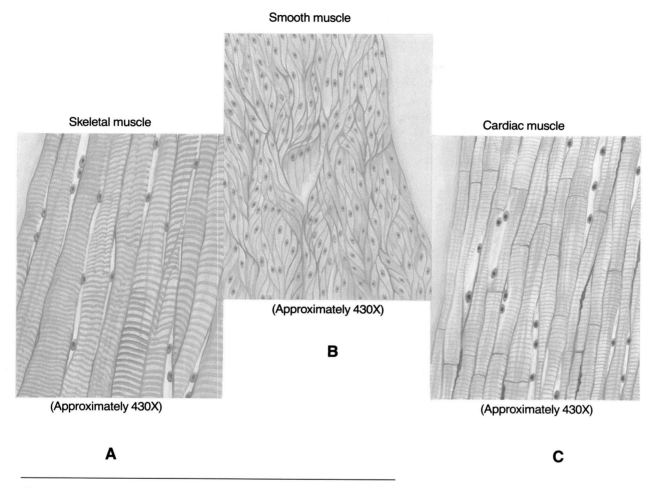

Smooth muscle

Skeletal muscle

Cardiac muscle

(Approximately 430X)

B

(Approximately 430X)

(Approximately 430X)

A

C

Figure 4–6 Muscle tissues. (**A**), Skeletal. (**B**), Smooth. (**C**), Cardiac.

blood pressure. The iris of the eye has two sets of smooth muscle fibers to constrict or dilate the pupil, which regulates the amount of light that strikes the retina.

Other functions of smooth muscle will be mentioned in later chapters. This is an important tissue that you will come across again and again in our study of the human body.

CARDIAC MUSCLE

The cells of **cardiac muscle** are shown in Fig. 4–6. They are branched, have one nucleus each,

and have faint striations. Cardiac muscle, called the **myocardium,** forms the walls of the chambers of the heart. Its function, therefore, is the function of the heart, to pump blood. The contractions of the myocardium create blood pressure and keep blood circulating throughout the body, so that the blood can carry out its many functions.

Cardiac muscle cells have the ability to contract by themselves. Thus the heart maintains its own beat. The role of nerve impulses is to increase or decrease the heart rate, depending upon whatever is needed by the body in a particular situation. We will return to the heart in Chapter 12.

NERVE TISSUE

Nerve tissue consists of nerve cells called **neurons** and some specialized cells found only in the nervous system. The nervous system has two divisions: the central nervous system (CNS) and the peripheral nervous system (PNS). The brain and spinal cord are the organs of the CNS. They are made of neurons and specialized cells called neuroglia. The CNS and the neuroglia are discussed in detail in Chapter 8. The PNS consists of all the nerves that emerge from the CNS and supply the rest of the body. These nerves are made of neurons and specialized cells called Schwann cells. The Schwann cells form the myelin sheath to electrically insulate neurons (Table 4–4).

Neurons are capable of generating and transmitting electrochemical impulses. There are many different kinds of neurons, but they all have the same basic structure (Fig. 4–7). The **cell body** contains the nucleus and is essential for the continuing life of the neuron. An **axon** is a process (cellular extension) that carries impulses away from the cell body; a neuron has only one axon. **Dendrites** are processes that carry impulses toward the cell body; a neuron may have several dendrites. A nerve impulse along the cell membrane of a neuron is electrical, but where neurons meet there is a small space called a **synapse,** which an electrical impulse cannot cross. At a synapse, between the axon of one neuron and the dendrite or cell body of the next neuron, impulse transmission depends upon chemicals called **neurotransmitters.** Each of these aspects of nerve tissue will be covered in more detail in Chapter 8.

Nerve tissue makes up the brain, spinal cord, and peripheral nerves. As you can imagine, each of these organs has very specific functions. For now, we will just summarize the functions of nerve tissue. These functions include sensation, movement, the rapid regulation of body functions such as heart rate and breathing, and the organization of information for learning and memory.

MEMBRANES

Membranes are sheets of tissue that cover or line surfaces or separate organs or parts (lobes) of organs from one another. Many membranes produce secretions that have specific functions. The two major categories of membranes are epithelial membranes and connective tissue membranes.

Table 4–4   NERVE TISSUE

Part	Structure	Function
Neuron (nerve cell)		
Cell body	• Contains the nucleus	• Regulates the functioning of the neuron
Axon	• Cellular process (extension)	• Carries impulses away from the cell body
Dendrites	• Cellular process (extension)	• Carry impulses toward the cell body
Synapse	• Space between axon of one neuron and the dendrite or cell body of the next neuron	• Transmits impulses from one neuron to others
Neurotransmitters	• Chemicals released by axons	• Transmit impulses across synapses
Neuroglia	• Specialized cells in the central nervous system	• Form myelin sheaths and other functions
Schwann cells	• Specialized cells in the peripheral nervous system	• Form the myelin sheaths around neurons

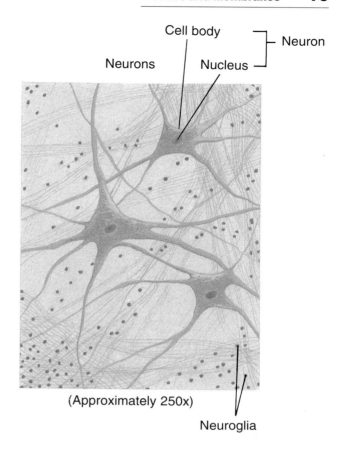

(Approximately 250x)

Figure 4–7 Nerve tissue of the central nervous system (CNS).

EPITHELIAL MEMBRANES

There are two types of epithelial membranes, serous and mucous. Each type is found in specific locations within the body and secretes a fluid. These fluids are called serous fluid and mucus.

Serous Membranes

Serous membranes are sheets of simple squamous epithelium that line some closed body cavities and cover the organs in these cavities (Fig. 4–8). The **pleural membranes** are the serous membranes of the thoracic cavity. The parietal pleura lines the chest wall and the visceral pleura covers the lungs. (Notice that "line" means "on the inside" and "cover" means "on the outside." These terms cannot be used interchangeably, because each indicates a different location.) The pleural membranes secrete **serous fluid,** which prevents friction between them as the lungs expand and recoil during breathing.

The heart, in the thoracic cavity between the lungs, has its own set of serous membranes. The parietal **pericardium** lines the fibrous pericardium (a connective tissue membrane), and the visceral **pericardium,** or epicardium, is on the surface of the heart muscle. Serous fluid is produced to prevent friction as the heart beats.

In the abdominal cavity, the **peritoneum** is the serous membrane that lines the cavity. The **mesentery,** or visceral peritoneum, is folded over and covers the abdominal organs. Here, the serous fluid prevents friction as the stomach and intestines contract and slide against other organs (see also Fig. 16–4).

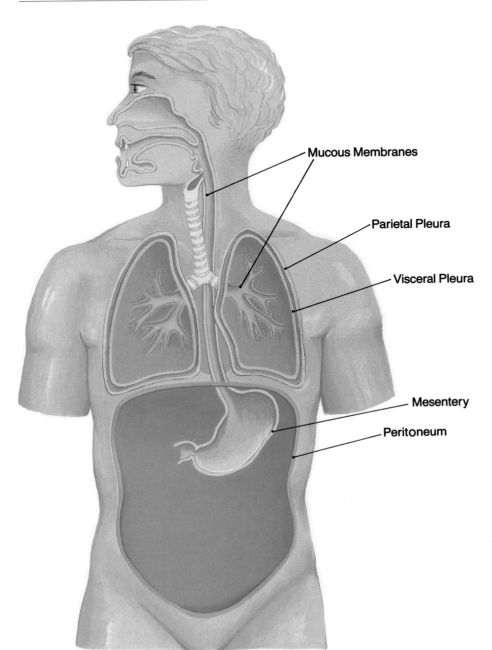

Mucous Membranes

Parietal Pleura

Visceral Pleura

Mesentery

Peritoneum

Figure 4–8 Epithelial membranes. Mucous membranes line body tracts that open to the environment. Serous membranes are found within closed body cavities such as the thoracic and abdominal cavities.

Mucous Membranes

Mucous membranes line the body tracts (systems) that have openings to the environment. These are the respiratory, digestive, urinary, and repro-ductive tracts. The epithelium of a mucous membrane **(mucosa)** varies with the different organs involved. The mucosa of the esophagus and of the vagina is stratified squamous epithelium; the mu-

Table 4–5 CONNECTIVE TISSUE MEMBRANES

Membrane	Location and Function
Superficial Fascia	• Between the skin and muscles; adipose tissue stores fat
Periosteum	• Covers each bone; contains blood vessels that enter the bone • Anchors tendons and ligaments
Perichondrium	• Covers cartilage; contains capillaries, the only blood supply for cartilage
Synovial	• Lines joint cavities; secretes synovial fluid to prevent friction when joints move
Deep Fascia	• Covers each skeletal muscle; anchors tendons
Meninges	• Cover the brain and spinal cord; contain cerebrospinal fluid
Fibrous Pericardium	• Forms a sac around the heart

cosa of the trachea is ciliated epithelium; the mucosa of the stomach is columnar epithelium.

The **mucus** secreted by these membranes keeps the lining epithelial cells wet. Remember that these are living cells, and if they dry out, they will die. In the digestive tract, mucus also lubricates the surface to permit the smooth passage of food. In the respiratory tract the mucus traps dust and bacteria, which are then swept to the pharynx by ciliated epithelium.

CONNECTIVE TISSUE MEMBRANES

Many membranes are made of connective tissue. Since these will be covered with the organ systems of which they are a part, their locations and functions are summarized in Table 4–5.

SUMMARY

The tissues and membranes described in this chapter are more complex than the individual cells of which they are made. However, we have only reached an intermediate level with respect to the structural and functional complexity of the body as a whole. The following chapters are concerned with the organ systems, the most complex level. In the descriptions of the organs of these systems, you will find mention of the tissues and their contributions to each organ and organ system.

STUDY OUTLINE

A Tissue is a group of cells with similar structure and function. The four main groups of tissues are: epithelial, connective, muscle, and nerve.
Epithelial Tissue—found on surfaces; have no capillaries; some are capable of secretion; classified as to shape of cells and number of layers of cells (see Table 4–1 and Figs. 4–1, 4–2, and 4–3)
1. Simple Squamous—one layer of flat cells; thin and smooth. Sites: alveoli (to permit diffusion of gases); capillaries (to permit exchanges between blood and tissues).
2. Stratified Squamous—many layers of mostly flat cells; mitosis takes place in lowest layer. Sites: epidermis, where surface cells are dead (a barrier to pathogens); lining of mouth; esophagus; and vagina (a barrier to pathogens).
3. Transitional—stratified, yet surface cells are rounded and flatten when stretched. Site: urinary bladder (to permit expansion without tearing the lining).
4. Simple Cuboidal—one layer of cube-shaped cells. Sites: thyroid gland (to secrete thyroid hormones); salivary glands (to secrete saliva).
5. Simple Columnar—one layer of column-shaped cells. Sites: stomach lining (to secrete gastric juice); small intestinal lining (to secrete digestive

enzymes and absorb nutrients—microvilli increase surface area for absorption).
6. Ciliated—columnar cells with cilia on free surfaces. Sites: trachea (to sweep mucus and bacteria to the pharynx); fallopian tubes (to sweep ovum to uterus).
7. Glands—Epithelial tissues that produce secretions.
   • Unicellular—one-celled glands. Goblet cells secrete mucus in the respiratory and digestive tracts.
   • Multicellular—many-celled glands.
      ○ Exocrine glands have ducts; salivary glands secrete saliva into ducts that carry it to the oral cavity.
      ○ Endocrine glands secrete hormones directly into capillaries (no ducts); thyroid gland secretes thyroxine.

Connective Tissue—all have a non-living intercellular matrix and specialized cells (see Table 4–2 and Figs. 4–4 and 4–5)
1. Blood—the matrix is plasma, mostly water; transports materials in the blood. Red blood cells carry oxygen; white blood cells destroy pathogens and provide immunity; platelets prevent blood loss, as in clotting.
2. Areolar (loose)—cells are fibroblasts, which produce protein fibers: collagen is strong, elastin is elastic; the matrix is collagen, elastin, and tissue fluid. White blood cells are also present. Sites: below the dermis and below the epithelium of tracts that open to the environment (to destroy pathogens that enter the body).
3. Adipose—cells are adipocytes that store fat; little matrix. Sites: between the skin and muscles (to store energy); around the eyes and kidneys (to cushion).
4. Fibrous—mostly matrix, strong collagen fibers; cells are fibroblasts. Regular fibrous sites: tendons (to connect muscle to bone); ligaments (to connect bone to bone). Irregular fibrous sites: dermis of the skin and the fascia around muscles.
5. Elastic—mostly matrix, elastin fibers. Sites: walls of large arteries (to maintain blood pressure); around alveoli (to promote normal exhalation).

6. Bone—cells are osteocytes; matrix is calcium salts and collagen, strong and not flexible. Sites: bones of the skeleton (to support the body and protect internal organs from mechanical injury).
7. Cartilage—cells are chondrocytes; protein matrix is firm yet flexible; no capillaries in matrix. Sites: joint surfaces of bones (to prevent friction); tip of nose and external ear (to support); wall of trachea (to keep air passage open); discs between vertebrae (to absorb shock).

Muscle Tissue—specialized to contract and bring about movement (see Table 4–3 and Fig. 4–6)
1. Skeletal—also called striated or voluntary muscle. Cells are cylindrical, have several nuclei, and have striations. Each cell has a motor nerve ending; nerve impulses are essential to cause contraction. Site: skeletal muscles attached to bones (to move the skeleton and produce heat).
2. Smooth—also called visceral or involuntary muscle. Cells have tapered ends, one nucleus each, and no striations. Contraction is not under voluntary control. Sites: stomach and intestines to produce peristalsis; walls of arteries and veins (to maintain blood pressure); iris (to constrict or dilate pupil).
3. Cardiac—cells are branched, have one nucleus each, and faint striations. Site: walls of chambers of the heart (to pump blood; nerve impulses regulate the rate of contraction).

Nerve Tissue—neurons are specialized to generate and transmit impulses (see Table 4–4 and Fig. 4–7)
1. Cell body contains the nucleus; axon carries impulses away from the cell body; dendrites carry impulses toward the cell body.
2. A synapse is the space between two neurons; a neurotransmitter carries the impulse across a synapse.
3. Specialized cells in nerve tissue are neuroglia in the CNS and Schwann cells in the PNS.
4. Sites: brain; spinal cord; and peripheral nerves (to provide sensation, movement, regulation of body functions, learning, and memory).

Membranes—sheets of tissue on surfaces, or separating organs or lobes

1. Epithelial Membranes (see Fig. 4–8)
   - Serous membranes—in closed body cavities; the serous fluid prevents friction between the two layers of the serous membrane.
     ◦ Thoracic cavity—partial pleura lines chest wall; visceral pleura covers the lungs.
     ◦ Pericardial sac—parietal pericardium lines fibrous pericardium; visceral pericardium (epicardium) covers the heart muscle.
     ◦ Abdominal cavity—peritoneum lines the abdominal cavity; mesentery covers the abdominal organs.
   - Mucous membranes—line body tracts that open to the environment: respiratory, digestive, urinary, and reproductive. Mucus keeps the living epithelium wet; provides lubrication in the digestive tract; traps dust and bacteria in the respiratory tract.
2. Connective Tissue Membranes—see Table 4–5.

REVIEW QUESTIONS

1. Explain the importance of each tissue in its location: (pp. 66–67, 69, 75)
   a. simple squamous epithelium in the alveoli of the lungs
   b. ciliated epithelium in the trachea
   c. cartilage in the trachea

2. Explain the importance of each tissue in its location: (pp. 74–75)
   a. bone tissue in bones
   b. cartilage on the joint surfaces of bones
   c. fibrous connective tissue in ligaments

3. State the functions of red blood cells, white blood cells, and platelets. (p. 72)

4. Name two organs made primarily of nerve tissue, and state the general functions of nerve tissue. (p. 78)

5. State the location and function of cardiac muscle. (p. 77)

6. Explain the importance of each of these tissues in the small intestine: smooth muscle, columnar epithelium. (pp. 68, 76)

7. State the precise location of each of the following membranes: (p. 79)
   a. peritoneum
   b. visceral pericardium
   c. parietal pleura

8. State the function of: (pp. 72, 79, 81)
   a. serous fluid
   b. mucus
   c. blood plasma

9. State two functions of skeletal muscles. (p. 76)

10. Name three body tracts lined with mucous membranes. (p. 80)

11. Explain how endocrine glands differ from exocrine glands. (p. 70)

12. State the function of adipose tissue: (pp. 71, 74)
    a. around the eyes
    b. between the skin and muscles

13. State the location of: (p. 81)
    a. meninges
    b. synovial membranes

14. State the important physical characteristics of collagen and elastin, and name the cells that produce these protein fibers (pp. 73).

Chapter 5
The Integumentary System

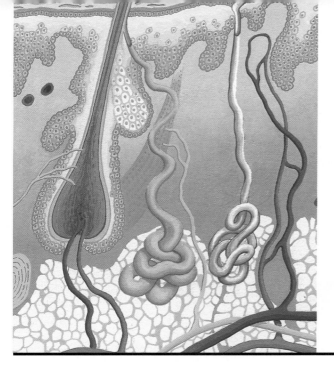

Chapter 5

Student Objectives

- Name the two major layers of the skin and the tissue of which each is made.
- State the locations and describe the functions of the stratum corneum and stratum germinativum.
- Describe the function of melanocytes and melanin.
- Describe the functions of hair and nails.
- Name the cutaneous senses and explain their importance.
- Describe the functions of the secretions of sebaceous glands, ceruminous glands, and eccrine sweat glands.
- Describe how the arterioles in the dermis respond to heat, cold, and stress.
- Name the tissues that make up the subcutaneous tissue, and describe their functions.

The Integumentary System

New Terminology

Arterioles (ar–**TEER**–ee–ohls)
Ceruminous gland/Cerumen (suh–**ROO**–mi–nus
    GLAND/suh–**ROO**–men)
Dermis (**DER**–miss)
Eccrine sweat gland (**ECK**–rin SWET GLAND)
Epidermis (EP–i–**DER**–miss)
Hair follicle (HAIR **FAH**–li–kull)
Keratin (**KER**–uh–tin)
Melanin (**MEL**–uh–nin)
Melanocyte (muh–**LAN**–o–site)
Nail follicle (NAIL **FAH**–li–kull)
Papillary layer (**PAP**–i–LAR–ee LAY–er)
Receptors (ree–**SEP**–turs)
Sebaceous gland/sebum (suh–**BAY**–shus GLAND/
    SEE–bum)
Stratum corneum (**STRA**–tum KOR–**NEE**–um)
Stratum germinativum (**STRA**–tum JER–min–ah–
    TEE–vum)
Subcutaneous tissue (SUB–kew–**TAY**–nee–us
    TISH–yoo)
Vasoconstriction (VAY–so–kon–**STRICK**–shun)
Vasodilation (VAY–so–dye–**LAY**–shun)

Related Clinical Terminology

Acne (**ACK**–nee)
Alopecia (AL–oh–**PEE**–she–ah)
Biopsy (**BYE**–op–see)
Carcinoma (KAR–sin–**OH**–mah)
Circulatory shock (**SIR**–kew–lah–TOR–ee
    SHAHCK)
Contusion (kon–**TOO**–zhun)
Decubitis ulcer (dee–**KEW**–bi–tuss **UL**–ser)
Dehydration (DEE–high–**DRAY**–shun)
Dermatology (DER–muh–**TAH**–luh–gee)
Eczema (**ECK**–zuh–mah)
Erythema (ER–i–**THEE**–mah)
Histamine (**HISS**–tah–meen)
Hives (**HIGH**–VZ)
Inflammation (IN–fluh–**MAY**–shun)
Melanoma (MEL–ah–**NO**–mah)
Nevus (**NEE**–vus)
Pruritus (proo–**RYE**–tus)
Septicemia (SEP–tih–**SEE**–mee–ah)

Terms that appear in **bold type** in the chapter text are defined in the glossary, which begins on p. 549.

The **integumentary system** consists of the skin, its accessory structures such as hair and sweat glands, and the subcutaneous tissue below the skin. The **skin** is made of several different tissue types and is considered an organ. Since the skin covers the surface of the body, one of its functions is readily apparent: it separates the body from the external environment and prevents the entry of many harmful substances. The **subcutaneous tissue** directly underneath the skin connects it to the muscles and has other functions as well.

THE SKIN

The two major layers of the skin are the outer **epidermis** and the inner **dermis.** Each of these layers is made of different tissues and has very different functions.

EPIDERMIS

The **epidermis** is made of stratified squamous epithelial tissue and is thickest on the palms and soles. Although the epidermis may be further subdivided into four or five sub-layers, two of these are of greatest importance: the innermost layer, the stratum germinativum, and the outermost layer, the stratum corneum (Fig. 5–1).

Stratum Germinativum

The **stratum germinativum** is the inner epidermal layer in which **mitosis** takes place. New cells are continually being produced, pushing the older cells toward the skin surface. These cells produce the protein **keratin,** and as they get further away from the capillaries in the dermis, they die. As dead cells are worn off the skin's surface, they are replaced by cells from within.

Stratum Corneum

The **stratum corneum,** the outermost epidermal layer, consists of many layers of dead cells; all that is left is their **keratin.** The protein keratin is relatively waterproof and prevents evaporation of

body water. Also of importance, keratin prevents the entry of water. Without a waterproof stratum corneum, it would be impossible to swim in a pool or even take a shower without damaging our cells.

The stratum corneum is also a barrier to pathogens and chemicals. Most bacteria and other microorganisms cannot penetrate unbroken skin. Most chemicals, unless they are corrosive, will not get through unbroken skin to the living tissue within. One painful exception is the sap of poison ivy. This resin does penetrate the skin and initiates an allergic reaction in susceptible people. The inflammatory response that characterizes allergies causes blisters and severe itching. The importance of the stratum corneum becomes especially apparent when it is lost (see Box 5–1: Burns).

Certain changes in the epidermis are undoubtedly familiar to you. When first wearing new shoes, for example, the skin of the foot may be subjected to friction. This will separate layers of the epidermis, or separate the epidermis from the dermis, and tissue fluid may collect, causing a **blister.** If the skin is subjected to pressure, the rate of mitosis in the stratum germinativum will increase and create a thicker epidermis; we call this a **callus.** Although calluses are more common on the palms and soles, they may occur on any part of the skin.

Melanocytes

Another type of cell found in the lower epidermis is the melanocyte. **Melanocytes** produce another protein, a pigment called **melanin.** People of the same size have approximately the same number of melanocytes. In people with dark skin, the melanocytes continuously produce large amounts of melanin. The melanocytes of light-skinned people produce less melanin. The activity of melanocytes is genetically regulated; skin color is one of our hereditary characteristics.

In all people, melanin production is increased by exposure of the skin to ultraviolet rays, which are part of sunlight. As more melanin is produced, it is taken in by the epidermal cells as they are pushed toward the surface. This gives the skin a darker color, which prevents further exposure of the living stratum germinativum to ultraviolet rays. People with dark skin already have good protection against

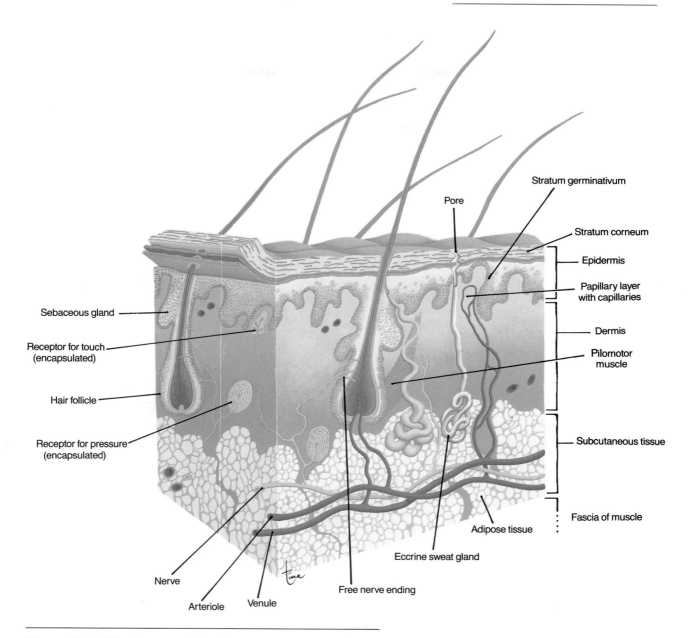

Figure 5–1   Skin. Structure of the skin and subcutaneous tissue.

the damaging effects of ultraviolet rays; people with light skin do not (see Box 5–2: Preventing Skin Cancer: Common Sense and Sunscreens). The functions of the epidermis are summarized in Table 5–1.

DERMIS

The **dermis** is made of an irregular type of fibrous connective tissue. Fibroblasts produce both collagen and elastin fibers. Recall that **collagen** fi-

Box 5–1 BURNS

Burns of the skin may be caused by flames, hot water or steam, sunlight, electricity, or corrosive chemicals. The severity of burns ranges from minor to fatal, and the classification of burns is based on the extent of damage.

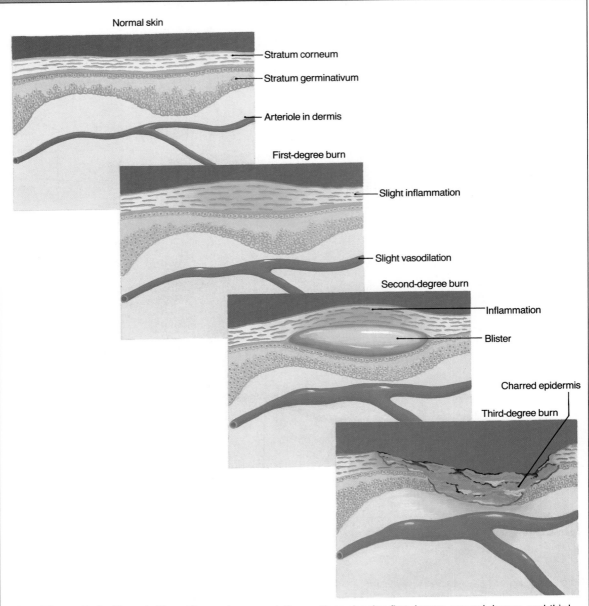

Box Figure 5–A Normal skin section and representative sections showing first-degree, second-degree, and third-degree burns.

Box 5–1   BURNS (Continued)

FIRST-DEGREE BURN—only the superficial epidermis is burned, and is painful but not blistered. Light-colored skin will appear red due to localized **vasodilation** in the damaged area. Vasodilation is part of the inflammatory response that brings more blood to the injured site.

SECOND-DEGREE BURN—deeper layers of the epidermis are affected. Another aspect of **inflammation** is that damaged cells release **histamine,** which makes capillaries more permeable. More plasma leaves these capillaries and becomes tissue fluid, which collects at the burn site, creating blisters. The burned skin is often very painful.

THIRD-DEGREE BURN—the entire epidermis is charred or burned away, and the burn may extend into the dermis or subcutaneous tissue. Often such a burn is not painful at first, if the receptors in the dermis have been destroyed.

Extensive third-degree burns are potentially life-threatening because of the loss of the stratum corneum. Without this natural barrier, living tissue is exposed to the environment and is susceptible to infection and dehydration.

Bacterial infection is a serious problem for burn patients; the pathogens may get into the blood **(septicemia)** and quickly spread throughout the body. Dehydration may also be fatal if medical intervention does not interrupt and correct the following sequence: tissue fluid evaporates from the burned surface, and more plasma is pulled out of capillaries into the tissue spaces. As more plasma is lost, blood volume and blood pressure decrease. This is called **circulatory shock;** eventually the heart simply does not have enough blood to pump, and heart failure is the cause of death. To prevent these serious consequences, third-degree burns are covered with donor skin or artificial skin until skin grafts of the patient's own skin can be put in place.

bers are strong, and **elastin** fibers are able to recoil after being stretched. Strength and elasticity are two characteristics of the dermis. With increasing age, however, the deterioration of the elastin fibers causes the skin to lose its elasticity. We can all look forward to at least a few wrinkles as we get older.

The uneven junction of the dermis with the epidermis is called the **papillary layer** (see Fig. 5–1). Capillaries are abundant here to nourish not only the dermis but also the stratum germinativum. This epidermal layer has no capillaries of its own and depends on the blood supply in the dermis for oxygen and nutrients.

Within the dermis are the accessory skin structures: hair and nail follicles, sensory receptors, and several types of glands. Some of these project through the epidermis to the skin surface, but their active portions are in the dermis.

Table 5–1   EPIDERMIS

Part	Function
Stratum corneum (keratin)	• Prevents loss or entry of water • If unbroken, prevents entry of pathogens and most chemicals
Stratum germinativum	• Continuous mitosis produces new cells to replace worn off surface cells
Melanocytes	• Produce melanin on exposure to ultraviolet (UV) rays
Melanin	• Protects living skin layers from further exposure to UV rays

Hair Follicles

Hair follicles are made of epidermal tissue, and the growth process of hair is very similar to growth

Box 5–2 PREVENTING SKIN CANCER : COMMON SENSE AND SUNSCREENS

Anyone can get skin cancer, and the most important factor is exposure to sunlight. Light-skinned people are, of course, more susceptible to the effects of ultraviolet (UV) rays, which may trigger **mutations** in living epidermal cells.

Squamous cell **carcinoma** and basal cell carcinoma are the most common forms of skin cancer. The lesions are visible as changes in the normal appearance of the skin, and a **biopsy** (microscopic examination of a tissue specimen) is used to confirm the diagnosis. These lesions usually do not metastasize rapidly, and can be completely removed using simple procedures.

Malignant melanoma is a more serious form of skin cancer, which begins in melanocytes. Any change in a pigmented spot or mole **(nevus)** should prompt a person to see a doctor. Melanoma is serious not because of its growth in the skin, but because it may **metastasize** very rapidly to the lungs, liver, or other vital organ.

Although the most common forms of skin cancer are readily curable, prevention is a much better strategy. We cannot, and we would not want to, stay out of the sun altogether, but we may be able to do so when sunlight is most damaging. During the summer months, UV rays are especially intense between 10:00 AM and 2:00 PM. If we are or must be outdoors during this time, use of a sunscreen is recommended by dermatologists.

Sunscreens contain chemicals such as PABA (para-amino benzoic acid) that block UV rays and prevent them from damaging the epidermis. An SPF (sun protection factor) of 15 or higher is considered good protection. Use of a sunscreen on exposed skin not only helps prevent skin cancer but prevents sunburn and its painful effects. It is especially important to prevent children from getting severely sunburned, since such burns have been linked to the development of skin cancer years later.

of the epidermis. At the base of a follicle is the **hair root,** where mitosis takes place (Fig. 5–2). The new cells produce keratin, get their color from melanin, then die and become incorporated into the **hair shaft.** The hair that we comb and brush every day consists of dead, keratinized cells.

Compared to some other mammals, humans do not have very much hair. The actual functions of human hair are quite few. Eyelashes and eyebrows help to keep dust and perspiration out of the eyes, and the hairs just inside the nostrils help to keep dust out of the nasal cavities. Hair of the scalp does provide insulation from cold for the head. The hair on our bodies, however, no longer serves this function, but we have the evolutionary remnants of it. Attached to each hair follicle is a small, smooth muscle called the **pilomotor** or arrector pili muscle.

When stimulated by cold or emotions such as fear, these muscles pull the hair follicles upright. For an animal with fur, this would provide greater insulation. Since people do not have thick fur, all this does for us is give us "goosebumps."

Nail Follicles

Found on the ends of fingers and toes, **nail follicles** produce nails just as hair follicles produce hair. Mitosis takes place in the **nail root** (Fig. 5–3), and the new cells produce keratin (a stronger form of this protein than is found in hair) and then die. Although the nail itself consists of keratinized dead cells, the flat nail bed is living tissue. This is why cutting a nail too short can be quite painful. Nails function to protect the ends of the fingers and toes

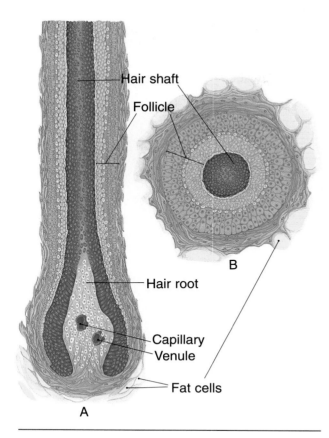

Figure 5–2   Structure of a hair follicle. **(A)**, Longitudinal section. **(B)**, Cross section.

skin. This information may stimulate responses; a simple example would be responding to a feeling of cold by putting on a sweater.

The sensitivity of an area of skin is determined by how many receptors are present. The skin of the fingertips, for example, is very sensitive to touch because there are many receptors per square inch. The skin of the upper arm, with few touch receptors per square inch, is less sensitive.

When receptors detect changes, they generate nerve impulses that are carried to the brain, which interprets the impulses as a particular sensation. Sensation, therefore, is actually a function of the brain (we will return to this in Chapters 8 and 9).

Glands

Glands are made of epithelial tissue. The exocrine glands of the skin have their secretory portions in the dermis. Some of these are shown in Fig. 5–1.

Sebaceous Glands   The ducts of **sebaceous glands** open into hair follicles or directly to the skin surface. Their secretion is **sebum,** a lipid substance that we commonly refer to as oil. The function of sebum is to prevent drying of skin and hair. The importance of this may not be readily apparent, but skin that is dry tends to crack more easily. Even very small breaks in the skin are potential entryways for

from mechanical injury and to give the fingers greater ability to pick up small objects.

Receptors

The sensory **receptors** in the dermis are for the cutaneous senses: touch, pressure, heat, cold, and pain. For each sensation there is a specific type of receptor, which is a structure that will detect a particular change. For pain, the receptors are **free nerve endings.** For the other cutaneous senses, the receptors are called **encapsulated nerve endings,** which means there is a cellular structure around the sensory nerve ending (see Fig. 5–1). The purpose of these receptors and sensations is to provide the central nervous system with information about the external environment and its effect on the

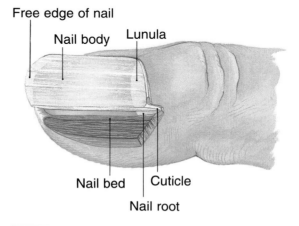

Figure 5–3   Structure of a fingernail shown in longitudinal section.

Box 5–3 COMMON SKIN DISORDERS

Impetigo—a bacterial infection often caused by *streptococci* or *staphylococci*. The characteristic pustules (pus-containing lesions) crust as they heal; the infection is contagious to other children.

Eczema—an allergic reaction more common in children than adults; the rash is itchy (pruritus), and may blister or ooze. Eczema is often related to foods such as fish, eggs, or milk products, or to inhaled allergens such as dust, pollens, or animal dander. Prevention depends upon determining what the child is allergic to and eliminating or at least limiting exposure.

Warts—caused by a virus that makes epidermal cells divide abnormally, producing a growth on the skin that is often raised and has a rough or pitted surface. Warts are probably most common on the hands, but they may be anywhere on the skin. Plantar warts on the sole of the foot may become quite painful because of the constant pressure of standing and walking.

Fever blisters (cold sores)—caused by the Herpes simplex virus, to which most people are exposed as children. An active lesion, usually at the edge of the lip, is painful and oozes. If not destroyed by the immune system, the virus "hides out" and becomes dormant in nerves of the face. Another lesion, weeks or months later, may be triggered by stress or another illness.

---

bacteria. Decreased sebum production is another consequence of getting older, and elderly people often have dry and more fragile skin.

Adolescents may have the problem of overactive sebaceous glands. Too much sebum may trap bacteria within hair follicles and create small infections. Since sebaceous glands are more numerous around the nose and mouth, these are common sites of pimples in young people (see also Box 5–3: Common Skin Disorders).

Ceruminous Glands   These are modified sebaceous glands located in the dermis of the ear canals. Their secretion is called **cerumen** or ear wax. Cerumen keeps the outer surface of the eardrum pliable and prevents drying. However, if excess cerumen accumulates in the ear canal, it may become impacted against the eardrum. This might diminish the acuity of hearing by preventing the ear drum from vibrating properly.

Sweat Glands   There are two types of sweat glands, apocrine and eccrine. **Apocrine glands** are most numerous in the axillae (underarm) and genital areas and are most active in stress and emotional situations. Although their secretion does have an odor, it is barely perceptible to other people. However, animals, such as dogs, can tell people apart by their individual scents. If the apocrine secretions are allowed to accumulate on the skin, bacteria metabolize the chemicals in the sweat and produce waste products which have distinct odors that many people find unpleasant.

Eccrine glands are found all over the body but are especially numerous on the forehead, upper lip, palms, and soles. The secretory portion of these glands is simply a coiled tube in the dermis. The duct of this tube extends to the skin's surface, where it opens into a **pore.**

The sweat produced by eccrine glands is important in the maintenance of normal body temperature. In a warm environment, or during exercise, more sweat is secreted onto the skin surface, where it is then evaporated by excess body heat. Although this is a very effective mechanism of heat loss, it has a potentially serious disadvantage. Loss of too much body water in sweat may lead to **dehydration,** as in heat exhaustion. Increased sweating during exercise or on a hot day should always be accompanied by increased fluid intake.

Blood Vessels

Besides the capillaries in the dermis, the other blood vessels of great importance are the arterioles. **Arterioles** are small arteries, and the smooth muscle in their walls permits them to constrict (close) or dilate (open). This is important in the maintenance of body temperature, because blood carries heat, which is a form of energy.

In a warm environment the arterioles dilate **(vasodilation),** which increases blood flow through the dermis, and brings excess heat close to the body surface to be radiated to the environment. In a cold environment, however, body heat must be conserved if possible, so the arterioles constrict. The **vasoconstriction** decreases the flow of blood through the dermis and keeps heat within the core of the body. This adjusting mechanism is essential for maintaining homeostasis. Regulation of the diameter of the arterioles in response to external temperaure changes is controlled by the nervous system. These changes can often be seen in light-skinned people. Flushing, especially in the face, may be observed in hot weather. In cold, the skin of the extremities may become even paler as blood flow through the dermis decreases. In people with dark skin, such changes are not as readily apparent since they are masked by melanin in the epidermis.

Vasoconstriction in the dermis may also occur during stressful situations. For our ancestors, stress usually demanded a physical response: either stand and fight or run away to safety. This is called the "fight or flight response." Our nervous systems are still programmed to respond as if physical activity were necessary to cope with the stress situation. Vasoconstriction in the dermis will shunt, or redirect, blood to more vital organs such as the muscles, heart, and brain. In times of stress, the skin is a relatively unimportant organ and can function temporarily with a minimal blood flow.

OTHER FUNCTIONS OF THE SKIN

Excretion—small amounts of **urea** (a waste product of protein metabolism) and sodium chloride are excreted in sweat. This is a very minor function of the skin; the kidneys are primarily responsible for removing waste products from the blood.

Formation of **vitamin D**—there is a form of cholesterol in the skin that, on exposure to ultraviolet light, is changed to vitamin D. This is why vitamin D is sometimes referred to as the "sunshine vitamin." People who do not get much sunlight depend more on nutritional sources of vitamin D, such as fortified milk. Vitamin D is important for the absorption of calcium and phosphorus from food in the small intestine. The functions of dermal structures are summarized in Table 5–2.

Table 5–2 DERMIS

Part	Function
Papillary Layer	• Contains capillaries that nourish the stratum germinativum
Hair (Follicles)	• Eyelashes and nasal hair keep dust out of eyes and nasal cavities • Scalp hair provides insulation from cold for the head
Nails (Follicles)	• Protect ends of fingers and toes from mechanical injury
Receptors	• Detect changes that are felt as the cutaneous senses: touch, pressure, heat, cold, and pain
Sebaceous Glands	• Produce sebum, which prevents drying of skin and hair
Ceruminous Glands	• Produce cerumen, which prevents drying of the eardrum
Eccrine Sweat Glands	• Produce watery sweat that is evaporated by excess body heat to cool the body
Arterioles	• Dilate in response to warmth to increase heat loss • Constrict in response to cold to conserve body heat • Constrict in stressful situations to shunt blood to more vital organs
Cholesterol	• Converted to vitamin D on exposure to UV rays

SUBCUTANEOUS TISSUE

The **subcutaneous tissue** may also be called the **superficial fascia,** one of the connective tissue membranes. Made of areolar connective tissue and adipose tissue, the superficial fascia connects the dermis to the underlying muscles. Its other functions are those of its tissues, as you may recall from Chapter 4.

Areolar connective tissue contains collagen and elastin fibers and many white blood cells that have left capillaries to wander around here. These migrating white blood cells destroy pathogens that enter the body through breaks in the skin.

The cells (adipocytes) of adipose tissue are specialized to store fat, and our subcutaneous layer of fat stores excess nutrients as a potential energy source. This layer also cushions bony prominences, such as when sitting, and provides some insulation from cold. For people, this last function is relatively minor, since we do not have a thick layer of fat, as do animals such as whales and seals. The functions of subcutaneous tissue are summarized in Table 5–3.

Table 5–3 SUBCUTANEOUS TISSUE

Part	Function
Areolar Connective Tissue	• Connects skin to muscles • Contains many WBCs to destroy pathogens that enter breaks in the skin
Adipose Tissue	• Contains stored energy in the form of true fats • Cushions bony prominences • Provides some insulation from cold

SUMMARY

The integumentary system is the outermost organ system of the body. You have probably noticed that many of its functions are related to this location. The skin protects the body against pathogens and chemicals, minimizes loss or entry of water, and blocks the harmful effects of sunlight. Sensory receptors in the skin provide information about the external environment, and the skin helps regulate body temperature in response to environmental changes.

STUDY OUTLINE

The integumentary system consists of the skin and its accessory structures and the subcutaneous tissue. The two major layers of the skin are the outer epidermis and the inner dermis.

Epidermis—made of stratified squamous epithelium (see Fig. 5–1, Table 5–1)

1. Stratum germinativum—the innermost layer where mitosis takes place; new cells produce keratin and die as they are pushed toward the surface.
2. Stratum corneum—the outermost layers of dead cells; keratin prevents loss and entry of water and resists entry of pathogens and chemicals.
3. Melanocytes—in the lower epidermis, produce melanin. UV rays stimulate melanin production; melanin prevents further exposure of the stratum germinativum to UV rays by darkening the skin.

Dermis—made of irregular fibrous connective tissue; collagen provides strength, and elastin provides elasticity; capillaries in the papillary layer nourish the stratum germinativum (see Table 5–2 and Fig. 5–1)

1. Hair follicles—mitosis takes place in the hair root; new cells produce keratin, die, and become the hair shaft. Hair of the scalp provides insulation from cold for the head; eyelashes keep dust out of eyes; nostril hairs keep dust out of nasal cavities (see Figs. 5–1 and 5–2).
2. Nail follicles—at the ends of fingers and toes; mitosis takes place in the nail root; the nail itself is dead, keratinized cells. Nails protect the ends

of the fingers and toes and enable the fingers to pick up small objects (see Fig. 5–3).

3. Receptors—detect changes in the skin: touch, pressure, heat, cold, and pain; provide information about the external environment which initiates appropriate responses; sensitivity of the skin depends on the number of receptors present.

4. Sebaceous glands—secrete sebum into hair follicles or to the skin surface; sebum prevents drying of skin and hair.

5. Ceruminous glands—secrete cerumen in the ear canals; cerumen prevents drying of the ear drum.

6. Apocrine sweat glands—modified scent glands in axillae and genital area; activated by stress and emotions.

7. Eccrine sweat glands—most numerous on face, palms, soles. Activated by high external temperature or exercise; sweat on skin surface is evaporated by excess body heat; potential disadvantage is dehydration.

8. Arterioles—smooth muscle permits constriction or dilation. Vasoconstriction in cold temperatures decreases dermal blood flow which conserves heat in the body core. Vasodilation in warm temperatures increases dermal blood flow which brings heat to the surface to be lost. Vasoconstriction during stress shunts blood away from the skin to more vital organs, such as muscles, to permit a physical response, if necessary.

Other Functions of the Skin

1. Excretion of small amounts of urea and NaCl (minor function).
2. Formation of vitamin D from cholesterol on exposure to UV rays of sunlight.

Subcutaneous Tissue—also called the superficial fascia; connects skin to muscles (see Fig. 5–1 and Table 5–3)

1. Areolar tissue—contains WBCs that destroy pathogens that get through breaks in the skin.
2. Adipose tissue—stores fat as potential energy; cushions bony prominences; provides some insulation from cold.

REVIEW QUESTIONS

1. Name the parts of the integumentary system. (p. 88)

2. Name the two major layers of skin, the location of each, and the tissue of which each is made. (pp. 88, 89)

3. In the epidermis: (p. 88)
   a. Where does mitosis take place?
   b. What protein do the new cells produce?
   c. What happens to these cells?

4. Describe the functions of the stratum corneum. (p. 88)

5. Name the cells that produce melanin. What is the stimulus? Describe the function of melanin. (p. 88)

6. Where, on the body, does human hair have important functions? Describe these functions. (p. 92)

7. Describe the functions of nails. (pp. 92–93)

8. Name the cutaneous senses. Describe the importance of these senses. (p. 93)

9. Explain the functions of sebum and cerumen. (pp. 93–94)

10. Explain how sweating helps maintain normal body temperature. (p. 94)

11. Explain how the arterioles in the dermis respond to cold or warm external temperatures and to stress situations. (p. 95)

12. What vitamin is produced in the skin? What is the stimulus for the production of this vitamin? (p. 95)

13. Name the tissues of which the superficial fascia is made. Describe the functions of these tissues. (p. 96)

Chapter 6
The Skeletal System

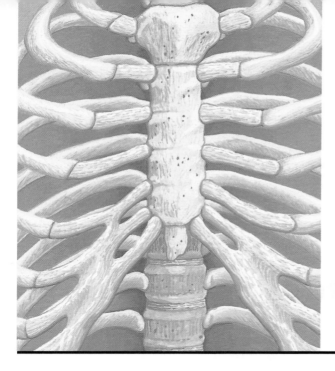

Chapter 6

Student Objectives

- Describe the functions of the skeleton.
- Explain how bones are classified, and give an example of each type.
- Describe how the embryonic skeleton model is replaced by bone.
- Name the nutrients necessary for bone growth, and explain their functions.
- Name the hormones involved in bone growth and maintenance, and explain their functions.
- Explain what is meant by "exercise" for bones, and explain its importance.
- Name all the bones of the human skeleton (be able to point to each on diagrams, skeleton models, or yourself).
- Describe the functions of the skull, vertebral column, rib cage, scapula, and pelvic bone.
- Explain how joints are classified. For each type, give an example, and describe the movement possible.
- Describe the parts of a synovial joint, and explain their functions.

The Skeletal System

New Terminology

Appendicular (AP–en–**DIK**–yoo–lar)
Articulation (ar–TIK–yoo–**LAY**-shun)
Axial (**ACK**–see–uhl)
Bursa (**BURR**–sah)
Diaphysis (dye–**AFF**–i–sis)
Epiphysis (e–**PIFF**–i–sis)
Epiphyseal disc (e–**PIFF**–i–SEE–al DISK)
Fontanel (FON–tah–**NELL**)
Haversian system (ha–**VER**–zhun **SIS**–tem)
Ligament (**LIG**–uh–ment)
Ossification (AHS–i–fi–**KAY**–shun)
Osteoblast (**AHS**–tee–oh–BLAST)
Osteoclast (**AHS**–tee–oh–KLAST)
Paranasal sinus (PAR–uh–**NAY**–zuhl **SIGH**–nus)
Periosteum (**PER**–ee–**AHS**–tee–um)
Suture (**SOO**–cher)
Symphysis (**SIM**–fi–sis)
Synovial fluid (sin–**OH**–vee–al **FLOO**–id)

Related Clinical Terminology

Autoimmune disease (AW–toh–im–**YOON** di–**ZEEZ**)
Bursitis (burr–**SIGH**–tiss)
Cleft palate (KLEFT **PAL**–uht)
Fracture (**FRAK**–chur)
Herniated disc (**HER**–nee–ay–ted DISK)
Kyphosis (kye–**FOH**–sis)
Lordosis (lor–**DOH**–sis)
Osteoarthritis (AHS–tee–oh–ar–**THRY**–tiss)
Osteomyelitis (AHS–tee–oh–my–uh–**LYE**–tiss)
Osteoporosis (AHS–tee–oh–por–**OH**–sis)
Rheumatoid arthritis (**ROO**–muh–toyd ar–**THRY**–tiss)
Rickets (**RICK**–ets)
Scoliosis (SKOH–lee–**OH**–sis)

Terms that appear in **bold type** in the chapter text are defined in the glossary, which begins on p. 549.

Imagine for a moment that people did not have skeletons. What comes to mind? Probably that each of us would be a little heap on the floor, much like a jellyfish out of water. Such an image is accurate and reflects the most obvious function of the skeleton: to support the body. Although it is a framework for the body, the skeleton is not at all like the wooden beams that support a house. Bones are living organs that actively contribute to the maintenance of the internal environment of the body.

The **skeletal system** consists of bones and other structures that make up the joints of the skeleton. The types of tissue present are bone tissue, cartilage, and fibrous connective tissue, which forms the ligaments that connect bone to bone.

FUNCTIONS OF THE SKELETON

1. Provides a framework that supports the body; the muscles that are attached to bones move the skeleton.
2. Protects some internal organs from mechanical injury; the rib cage protects the heart and lungs, for example.
3. Contains and protects the red bone marrow, one of the hemopoietic (blood-forming) tissues.
4. Provides a storage site for excess calcium. Calcium may be removed from bone to maintain a normal blood calcium level, which is essential for blood clotting and proper functioning of muscles and nerves.

TYPES OF BONE TISSUE

Bone was described as a tissue in Chapter 4. Recall that bone cells are called **osteocytes,** and the **matrix** of bone is made of calcium salts and collagen. The **calcium salts** are calcium carbonate ($CaCO_3$) and calcium phosphate ($Ca_3(PO_4)_2$), which give bone the strength for its supportive and protective functions. The function of osteocytes is to regulate the amount of calcium that is deposited in, or removed from, the bone matrix.

In bone as an organ, two types of bone tissue are present (Fig. 6–1). **Compact bone** is made of **haversian systems:** cylinders of bone matrix with osteocytes in concentric rings around central **haversian canals.** In the haversian canals are blood vessels; the osteocytes are in contact with these blood vessels and with one another through microscopic channels **(canaliculi)** in the matrix.

The second type of bone tissue is **spongy bone,** which does look rather like a sponge. Osteocytes, matrix, and blood vessels are present but are not arranged in haversian systems. The cavities in spongy bone often contain **red bone marrow,** which produces red blood cells, platelets, and the five types of white blood cells.

CLASSIFICATION OF BONES

1. **Long bones**—the bones of the arms, legs, hands, and feet (but not the wrists and ankles). The shaft of a long bone is the **diaphysis,** and the ends are called **epiphyses** (see Fig. 6–1). The diaphysis is made of compact bone and is hollow, forming a canal within the shaft. This **marrow canal** (or medullary cavity) contains **yellow bone marrow,** which is mostly adipose tissue. The epiphyses are made of spongy bone covered with a thin layer of compact bone. Although red bone marrow is present in the epiphyses of children's bones, it is largely replaced by yellow bone marrow in adult bones.
2. **Short bones**—the bones of the wrists and ankles.
3. **Flat bones**—the ribs, shoulder blades, hipbones, and cranial bones.
4. **Irregular bones**—the vertebrae and facial bones.

Short, flat, and irregular bones are all made of spongy bone covered with a thin layer of compact bone. Red bone marrow is found within the spongy bone.

The joint surfaces of bones are covered with **articular cartilage,** which provides a smooth sur-

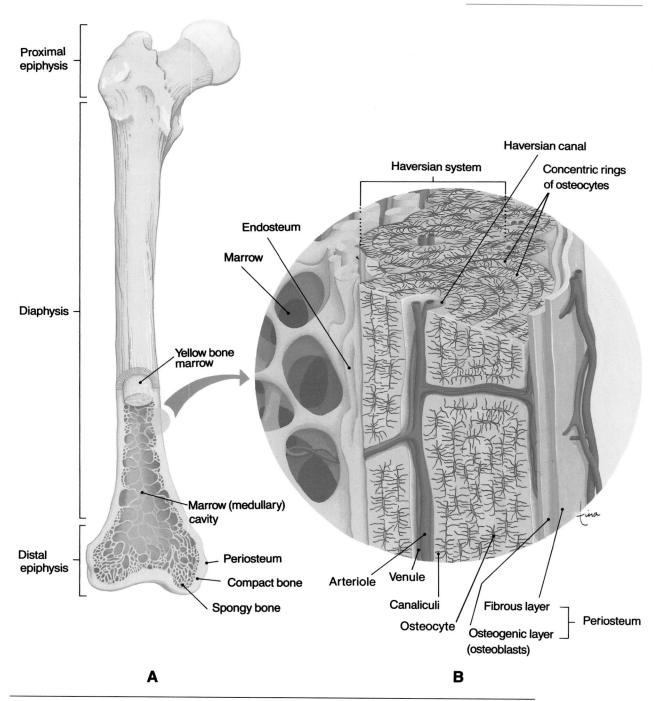

Proximal epiphysis

Diaphysis

Distal epiphysis

Endosteum

Marrow

Yellow bone marrow

Marrow (medullary) cavity

Periosteum

Compact bone

Spongy bone

Haversian canal

Haversian system

Concentric rings of osteocytes

Arteriole

Venule

Canaliculi

Osteocyte

Fibrous layer

Osteogenic layer (osteoblasts)

Periosteum

A

B

Figure 6–1 Bone tissue. (**A**), Femur with distal end cut in longitudinal section. (**B**), Compact bone showing haversian systems.

face. Covering the rest of the bone is the **periosteum,** a fibrous connective tissue membrane whose collagen fibers merge with those of the tendons and ligaments that are attached to the bone. The periosteum anchors these structures and also contains the blood vessels that enter the bone itself.

EMBRYONIC GROWTH OF BONE

During embryonic development, the skeleton is first made of cartilage and fibrous connective tissue, which are gradually replaced by bone. Bone matrix is produced by cells called **osteoblasts** (a blast cell is a "producing" cell, and "osteo" means bone). In the embryonic model of the skeleton, osteoblasts differentiate from the fibroblasts that are present. The production of bone matrix, called **ossification,** begins in a **center of ossification** in each bone.

The cranial and facial bones are first made of fibrous connective tissue. In the third month of fetal development, fibroblasts (spindle-shaped connective tissue cells) become more specialized and differentiate into osteoblasts, which produce bone matrix. From each center of ossification, bone growth radiates outward as calcium salts are deposited in the collagen of the model of the bone. This process is not complete at birth; a baby has areas of fibrous connective tissue remaining between the bones of the skull. These are called **fontanels** (Fig. 6–2), which permit compression of the baby's head during birth without breaking the still thin cranial bones. You may have heard fontanels referred to as

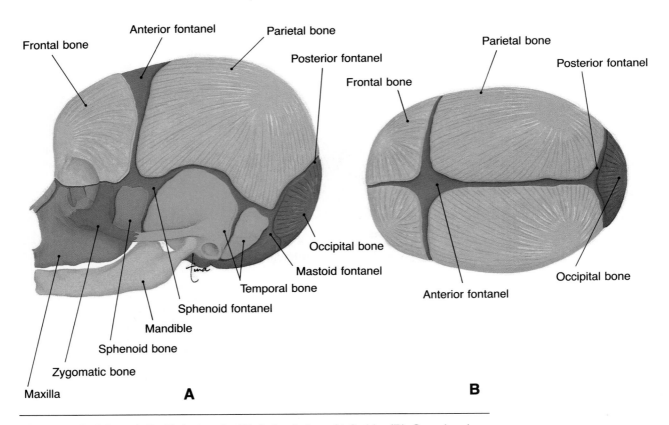

Figure 6–2   Infant skull with fontanels. (**A**), Lateral view of left side. (**B**), Superior view.

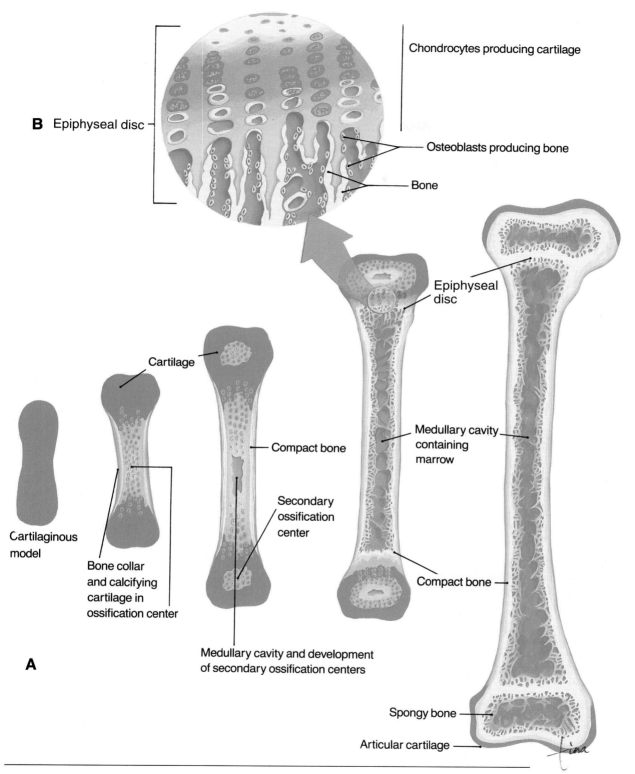

B Epiphyseal disc

Chondrocytes producing cartilage

Osteoblasts producing bone

Bone

Cartilage

Compact bone

Secondary ossification center

Epiphyseal disc

Medullary cavity containing marrow

Compact bone

Cartilaginous model

Bone collar and calcifying cartilage in ossification center

Medullary cavity and development of secondary ossification centers

Spongy bone

Articular cartilage

A

Figure 6–3 The ossification process in a long bone. (**A**), Progression of ossification from the cartilage model of the embryo to the bone of a young adult. (**B**), Microscopic view of an epiphyseal disc showing cartilage production and bone replacement.

"soft spots," and indeed they are. A baby's skull is still quite fragile and must be protected from trauma. By the age of 2 years, all the fontanels have become ossified, and the skull becomes a more effective protective covering for the brain.

The rest of the embryonic skeleton is first made of cartilage, and ossification begins in the third month of gestation in the long bones. Osteoblasts produce bone matrix in the center of the diaphyses of the long bones and in the center of short, flat, and irregular bones. Bone matrix gradually replaces the original cartilage (Fig. 6–3).

The long bones also develop centers of ossification in their epiphyses. At birth, ossification is not yet complete and continues throughout childhood. In long bones, growth occurs in the **epiphyseal discs** at the junction of the diaphysis with each epiphysis. An epiphyseal disc is still cartilage, and the bone grows in length as more cartilage is produced on the epiphysis side (see Fig. 6–3). On the diaphysis side, osteoblasts produce bone matrix to replace the cartilage. Between the ages of 16 and 25 years, all of the cartilage of the ephiphyseal discs is replaced by bone. This is called closure of the epiphyseal discs, and the bone lengthening process stops.

Also in long bones are specialized cells called **osteoclasts.** These calls reabsorb bone matrix in the center of the diaphysis to form the **marrow canal.** Blood vessels grow into the marrow canals of embryonic long bones, and red bone marrow is established. After birth, the red bone marrow is replaced by yellow bone marrow. Red bone marrow remains in the spongy bone of short, flat, and irregular bones. For other functions of osteoclasts and osteoblasts, see Box 6–1: Fractures and Their Repair.

FACTORS THAT AFFECT BONE GROWTH AND MAINTENANCE

1. Heredity—each person has a genetic potential for height, with genes inherited from both parents. There are many genes involved, and their interactions are not well understood. Some of these genes are probably those for the en-zymes involved in cartilage and bone production, for this is how bones grow.

2. Nutrition—nutrients are the raw materials of which bones are made. Calcium, phosphorus, and protein become part of the bone matrix itself. Vitamin D is needed for the efficient absorption of calcium and phosphorus by the small intestine. Vitamins A and C do not become part of bone but are necessary for the process of bone matrix formation (ossification).

   Without these and other nutrients, bones cannot grow properly. Children who are malnourished grow very slowly and may not reach their genetic potential for height.

3. Hormones—endocrine glands produce hormones that stimulate specific effects in certain cells. Several hormones have important roles in bone growth and maintenance. These include growth hormone, thyroxine, parathyroid hormone, and insulin, which help regulate cell division, protein synthesis, calcium metabolism, and energy production. The hormones and their specific functions are listed in Table 6–1.

4. Exercise or "Stress"—for bones, exercise means bearing weight, which is just what bones are specialized to do. Without this stress (which is normal), bones will lose calcium faster than it is replaced. Exercise need not be strenuous, it can be as simple as the walking involved in everday activities. Bones that do not get this exercise, such as those of bed-ridden patients, will become thinner and more fragile. This condition is discussed further in Box 6–2: Osteoporosis.

THE SKELETON

The human skeleton has two divisions: the **axial skeleton,** which forms the axis, and the **appendicular skeleton,** which supports the appendages or limbs. The axial skeleton consists of the skull, vertebral column, and rib cage. The bones of the arms and legs and the shoulder and pelvic girdles make up the appendicular skeleton. There are 206

Box 6–1 FRACTURES AND THEIR REPAIR

A fracture means that a bone has been broken. There are different types of fractures classified as to extent of damage.

Simple (closed)—the broken parts are still in normal anatomical position; surrounding tissue damage is minimal (skin is not pierced).

Compound (open)—the broken end of a bone has been moved, and it pierces the skin; there may be extensive damage to surrounding blood vessels, nerves, and muscles.

Greenstick—the bone splits longitudinally. The bones of children contain more collagen than do adult bones and tend to splinter rather than break completely.

Comminuted—two or more intersecting breaks create several bone fragments.

Impacted—the broken ends of a bone are forced into one another; many bone fragments may be created.

Spontaneous (pathologic)—a bone breaks without apparent trauma; may accompany bone disorders such as osteoporosis.

The Repair Process

Even a simple fracture involves significant bone damage that must be repaired if the bone is to resume its normal function. Fragments of dead or damaged bone must first be removed. This is accomplished by osteoclasts, which dissolve and reabsorb the calcium salts of bone matrix. Imagine a building that has just collapsed; the rubble must be removed before reconstruction can take place. This is what the osteoclasts do. Then, new bone must be produced. The inner layer of the periosteum contains osteoblasts that are activated when bone is damaged. The osteoblasts produce bone matrix to knit the broken ends of the bone together.

Since most bone has a good blood supply, the repair process is usually relatively rapid, and a simple fracture often heals within 6 weeks. Some parts of bones, however, have a poor blood supply, and repair of fractures takes longer. These areas are the neck of the femur (the site of a "fractured hip") and the lower third of the tibia.

Other factors that influence repair include the age of the person, general state of health, and nutrition. The elderly and those in poor health often have slow healing of fractures. A diet with sufficient calcium, phosphorus, vitamin D, and protein is also important. If any of these nutrients is lacking, bone repair will be a slower process.

Box Figure 6–A Types of fractures. Several types of fractures are depicted in the right arm.

Table 6–1   HORMONES INVOLVED IN BONE GROWTH AND MAINTENANCE

Hormone (Gland)	Functions
Growth Hormone (anterior pituitary gland)	• Increases the rate of mitosis of chondrocytes and osteoblasts • Increases the rate of protein synthesis (collagen, cartilage matrix, and enzymes for cartilage and bone formation)
Thyroxine (thyroid gland)	• Increases the rate of protein synthesis • Increases energy production from all food types
Insulin (pancreas)	• Increases energy production from glucose
Parathyroid Hormone (parathyroid glands)	• Increases the reabsorption of calcium from bones to the blood (raises blood calcium level) • Increases the absorption of calcium by the small intestine and kidneys (to the blood)
Calcitonin (thyroid gland)	• Decreases the reabsorption of calcium from bones (lowers blood calcium level)
Estrogen (ovaries) or Testosterone (testes)	• Promotes closure of the epiphyses of long bones (growth stops) • Helps retain calcium in bones to maintain a strong bone matrix

bones in total, and the complete skeleton is shown in Fig. 6–4.

SKULL

The **skull** consists of eight cranial bones and 14 facial bones. Also in the head are three small bones in each middle ear cavity and the hyoid bone that supports the base of the tongue. The **cranial bones** form the braincase that encloses and protects the brain, eyes, and ears. The names of some of these bones will be familiar to you; they are the same as the terminology used (see Chapter 1) to describe areas of the head. These are the **frontal**

Box 6–2   OSTEOPOROSIS

Bone is an active tissue; calcium is constantly being removed to maintain normal blood calcium levels. Usually, however, calcium is replaced in bones at a rate equal to its removal, and the bone matrix remains strong.

Osteoporosis is characterized by excessive loss of calcium from bones without sufficient replacement. Possible causes include insufficient dietary intake of calcium, inactivity, and lack of the sex hormones. Osteoporosis is most common among elderly women, since estrogen secretion decreases sharply at menopause (in older men, testosterone is still secreted in significant amounts). Factors such as bed rest or inability to get even minimal exercise will make calcium loss even more rapid.

As bones lose calcium and become thin and brittle, fractures are much more likely to occur. Among elderly women, a fractured hip (the neck of the femur) is an all-too-common consequence of this degenerative bone disorder.

Osteoporosis may be minimized by adequate intake of calcium and protein, moderate exercise, and estrogen replacement therapy after menopause. However, some osteoporosis is probably inevitable as women reach their 60s and 70s. Young women, even teenagers, should make sure they get adequate dietary calcium to form strong bone matrix, since this will delay the serious effects of osteoporosis later in life.

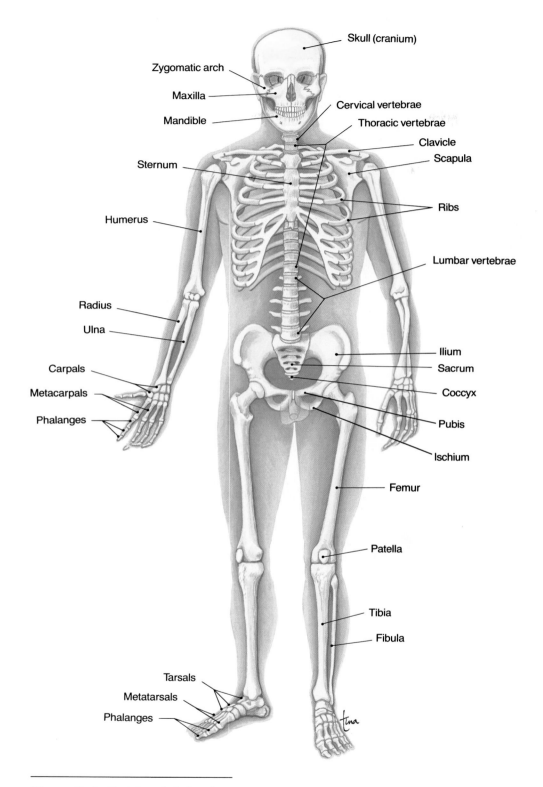

Figure 6–4 Skeleton. Anterior view.

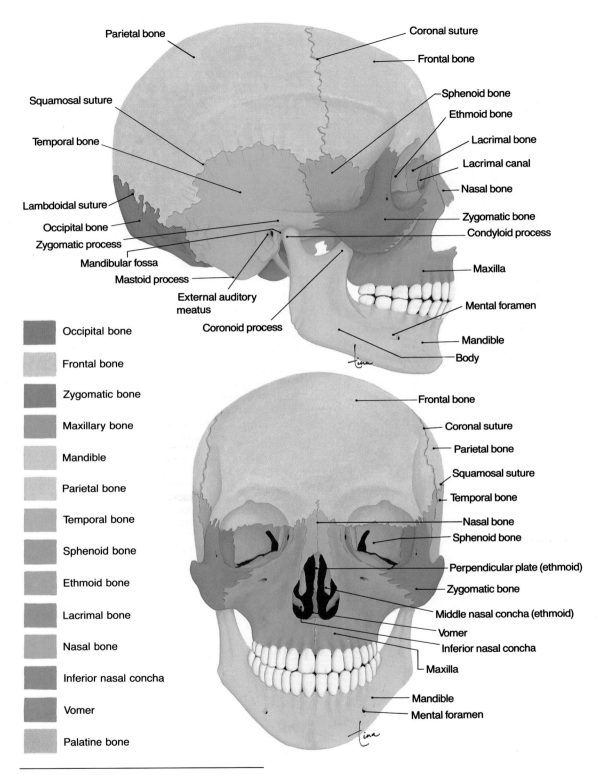

Parietal bone

Coronal suture

Frontal bone

Squamosal suture

Sphenoid bone

Ethmoid bone

Temporal bone

Lacrimal bone

Lacrimal canal

Nasal bone

Lambdoidal suture

Zygomatic bone

Occipital bone

Condyloid process

Zygomatic process

Mandibular fossa

Maxilla

Mastoid process

External auditory meatus

Mental foramen

Coronoid process

Mandible

Body

Occipital bone

Frontal bone

Zygomatic bone

Maxillary bone

Mandible

Parietal bone

Temporal bone

Sphenoid bone

Ethmoid bone

Lacrimal bone

Nasal bone

Inferior nasal concha

Vomer

Palatine bone

Frontal bone

Coronal suture

Parietal bone

Squamosal suture

Temporal bone

Nasal bone

Sphenoid bone

Perpendicular plate (ethmoid)

Zygomatic bone

Middle nasal concha (ethmoid)

Vomer

Inferior nasal concha

Maxilla

Mandible

Mental foramen

Figure 6–5 Skull. Lateral view of right side.

Figure 6–6 Skull. Anterior view.

bone, **parietal bones** (two), **temporal bones** (two), and **occipital bone.** The **sphenoid bone** and **ethmoid bone** are part of the floor of the braincase and the orbits (sockets) for the eyes. All the joints between cranial bones are immovable joints called **sutures.** It may seem strange to refer to a joint without movement, but the term joint is used for any "joining together" or junction of two bones. The classification of joints will be covered later in this chapter. All the bones of the skull, as well as the large sutures, are shown in Figs. 6–5 through 6–8. Their anatomically important parts are described in Table 6–2.

Of the 14 **facial bones,** only the **mandible** (lower jaw) is movable; it forms a **condyloid joint** with each temporal bone. The other joints between facial bones are all sutures. The **maxillae** are the upper jaw bones, which also form the anterior portion of the hard palate (roof of the mouth). Sockets for the roots of the teeth are found in the maxillae and the mandible. The other facial bones are described in Table 6–2.

Paranasal sinuses are air cavities located in the maxillae, and frontal, sphenoid, and ethmoid bones (Fig. 6–9). As the name "paranasal" suggests, they open into the nasal cavities and are lined with **ciliated epithelium** continuous with the mucosa of the nasal cavities. We are aware of our sinuses only when they become "stuffed up," which means that the mucus they produce cannot drain into the nasal

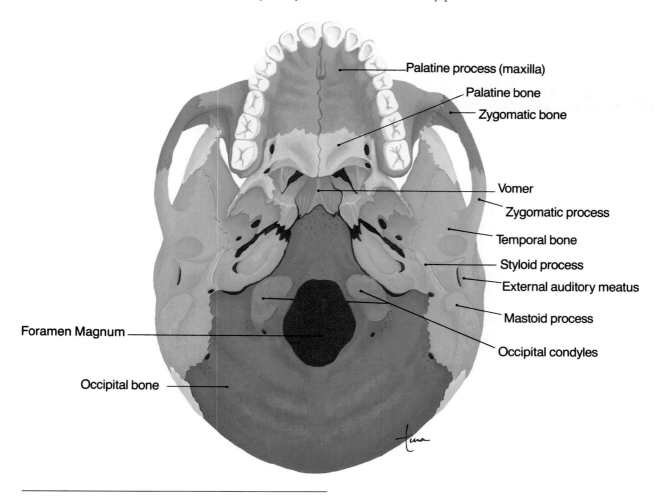

Figure 6–7   Skull. Inferior view with mandible removed.

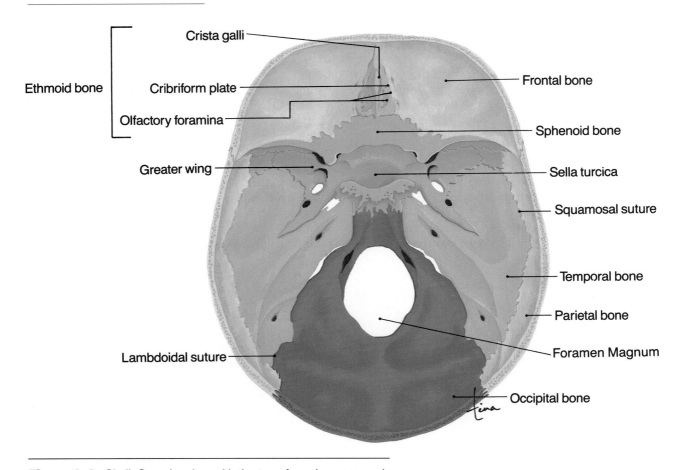

Crista galli

Cribriform plate

Olfactory foramina

Ethmoid bone

Greater wing

Lambdoidal suture

Frontal bone

Sphenoid bone

Sella turcica

Squamosal suture

Temporal bone

Parietal bone

Foramen Magnum

Occipital bone

Figure 6–8 Skull. Superior view with the top of cranium removed.

cavities. This may happen during upper respiratory infections such as colds, or with allergies such as hay fever. These sinuses, however, do have functions: they make the skull lighter in weight, since air is lighter than bone, and they provide resonance for the voice.

The **mastoid sinuses** are air cavities in the mastoid process of each temporal bone; they open into the middle ear. Before the availability of antibiotics, middle ear infections often caused mastoiditis, infection of these sinuses.

Within each middle ear cavity are three **auditory bones:** the malleus, incus, and stapes. As part of the hearing process, these bones transmit vibrations from the ear drum to the receptors in the inner ear.

VERTEBRAL COLUMN

The **vertebral column** (spinal column or backbone) is made of individual bones called **vertebrae.** The names of vertebrae indicate their location along the length of the spinal column. There are seven cervical vertebrae, 12 thoracic, five lumbar, five sacral fused into one sacrum, and four to five small coccygeal vertebrae fused into one coccyx (Fig. 6–10).

The seven **cervical vertebrae** are those within the neck. The first vertebra is called the **atlas,** which supports the skull and forms a **pivot joint** with the **axis,** the second cervical vertebra. This pivot joint allows us to turn our heads from side to side. The

Table 6–2   BONES OF THE SKULL—IMPORTANT PARTS

Terminology of Bone Markings

Foramen—a hole or opening	Meatus—a tunnel-like cavity	Condyle—a rounded projection
Fossa—a depression	Process—a projection	Plate—a flat projection

Bone	Part	Description
Frontal	• Frontal sinus • Coronal suture	• Air cavity that opens into nasal cavity • Joint between frontal and parietal bones
Parietal (2)	• Sagittal suture	• Joint between the 2 parietal bones
Temporal (2)	• Squamosal suture • External auditory meatus • Mastoid process • Mastoid sinus • Mandibular fossa • Zygomatic process	• Joint between temporal and parietal bone • The tunnel-like ear canal • Oval projection behind the ear canal • Air cavity that opens into middle ear • Oval depression anterior to the ear canal; articulates with mandible • Anterior projection that articulates with the zygomatic bone
Occipital	• Foramen magnum • Condyles • Lambdoidal suture	• Large opening for the spinal cord • Oval projections on either side of the foramen magnum; articulate with the atlas • Joint between occipital and parietal bones
Sphenoid	• Greater wing • Sella turcica • Sphenoid sinus	• Flat, lateral portion between the frontal and temporal bones • Central depression that encloses the pituitary gland • Air cavity that opens into nasal cavity
Ethmoid	• Ethmoid sinus • Crista galli • Cribriform plate and Olfactory foramina • Perpendicular plate • Conchae (4 are part of ethmoid; 2 inferior are separate bones)	• Air cavity that opens into nasal cavity • Superior projection for attachment of meninges • On either side of base of crista galli; olfactory nerves pass through foramina • Upper part of nasal septum • Shelf-like projections into nasal cavities which increase surface area of nasal mucosa
Mandible	• Body • Condyles • Sockets	• U-shaped portion with lower teeth • Oval projections that articulate with the temporal bones • Conical depressions that hold roots of lower teeth
Maxilla (2)	• Maxillary sinus • Palatine process • Sockets	• Air cavity that opens into nasal cavity • Projection that forms anterior part of hard palate • Conical depressions that hold roots of upper teeth
Nasal (2)	—	• Forms the bridge of the nose
Lacrimal (2)	Lacrimal canal	• Opening for nasolacrimal duct to take tears to nasal cavity
Zygomatic (2)	—	• Form point of cheek; articulate with frontal, temporal, and maxillae
Palatine (2)	—	• Forms the posterior part of hard palate
Vomer	—	• Lower part of nasal septum

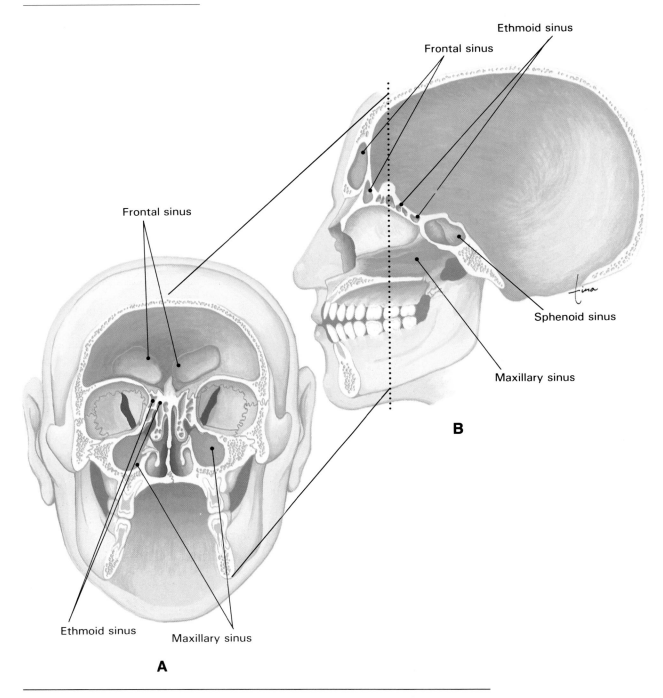

Figure 6–9 Paranasal sinuses. (**A**), Anterior view of skull. (**B**), Left lateral view of skull.

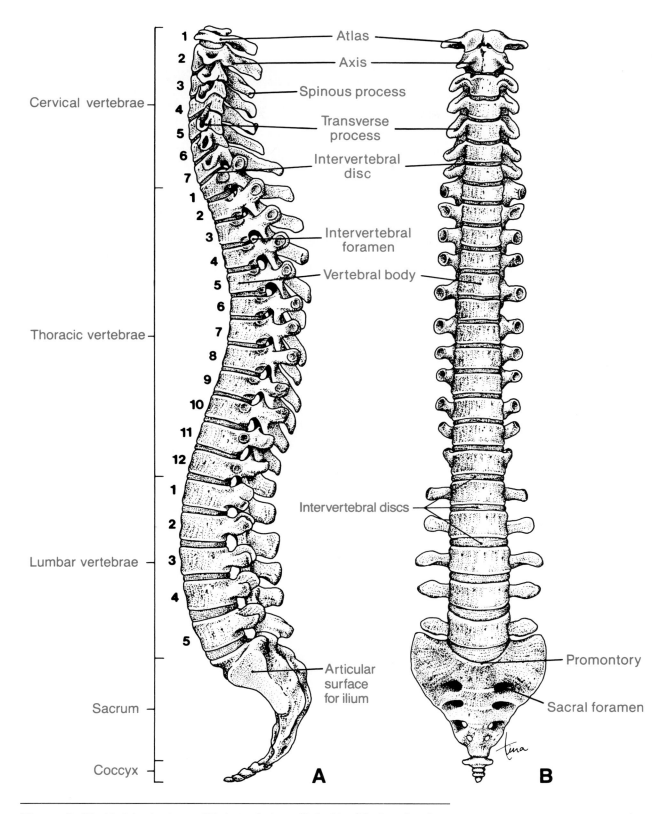

Cervical vertebrae

Thoracic vertebrae

Lumbar vertebrae

Sacrum

Coccyx

Atlas

Axis

Spinous process

Transverse process

Intervertebral disc

Intervertebral foramen

Vertebral body

Intervertebral discs

Articular surface for ilium

Promontory

Sacral foramen

A

B

Figure 6–10   Vertebral column. (**A**), Lateral view of left side. (**B**), Anterior view.

Box 6–3   HERNIATED DISC

The vertebrae are separated by discs of fibrous cartilage that act as cushions to absorb shock. An intervertebral disc has a tough outer covering and a soft center called the nucleus pulposus. Extreme pressure on a disc may rupture the outer layer and force the nucleus pulposus out. This may occur when a person lifts a heavy object improperly, that is, using the back rather than the legs and jerking upward, which puts sudden, intense pressure on the spine. Most often this affects discs in the lumbar region.

Although often called a "slipped disc," the affected disc is usually not moved out of position. The terms **"herniated"** or **"ruptured" disc** more accurately describe what happens. The nucleus pulposus is forced out, usually posteriorly, where it puts pressure on a spinal nerve. For this reason a herniated disc may be very painful or impair function in the muscles supplied by the nerve.

Healing of a herniated disc may occur naturally if the damage is not severe and the person rests and avoids activities that would further compress the disc. Surgery may be required, however, to remove the portion of the nucleus pulposus that is out of place and disrupting nerve functioning.

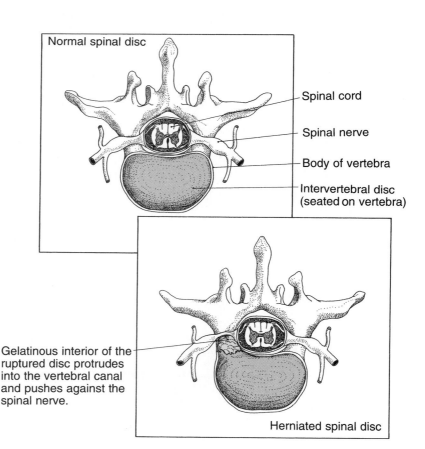

Normal spinal disc

— Spinal cord

— Spinal nerve

— Body of vertebra

— Intervertebral disc (seated on vertebra)

Gelatinous interior of the ruptured disc protrudes into the vertebral canal and pushes against the spinal nerve.

Herniated spinal disc

Box Figure 6–B   Herniated disc. As a result of compression, a ruptured intervertebral disc puts pressure on a spinal nerve.

remaining five cervical vertebrae do not have individual names.

The **thoracic vertebrae** articulate (form joints) with the ribs on the posterior side of the trunk. The **lumbar vertebrae,** the largest and strongest bones of the spine, are found in the small of the back. The **sacrum** permits the articulation of the two hipbones: the **sacroiliac joints.** The **coccyx** is the remnant of tail vertebrae, and some muscles of the perineum (pelvic floor) are anchored to it.

All of the vertebrae articulate with one another in sequence to form a flexible backbone that supports the trunk and head. They also form the **vertebral canal,** a continuous tunnel within the bones that contains the spinal cord and protects it from mechanical injury. The spinous and transverse processes are projections for the attachment of the muscles that bend the vertebral column.

The supporting part of a vertebra is its body; the bodies of adjacent vertebrae are separated by **discs** of fibrous cartilage. These discs cushion and absorb shock and permit some movement between vertebrae **(symphysis joints).** Since there are so many joints, the backbone as a whole is quite flexible (see also Box 6–3: Herniated Disc).

The normal spine in anatomic position has four natural curves, which are named after the vertebrae that form them. Refer to Fig. 6–10, and notice that the cervical curve is forward, the thoracic curve backward, the lumbar curve forward, and the sacral curve backward. These curves center the skull over the rest of the body, which enables a person to more easily walk upright (see Box 6–4: Abnormalities of the Curves of the Spine).

RIB CAGE

The **rib cage** consists of the 12 pairs of ribs and the sternum, or breast bone. The three parts of the **sternum** are the upper **manubrium,** the central **body,** and the lower **xiphoid process** (Fig. 6–11).

All the **ribs** articulate posteriorly with the thoracic vertebrae. The first seven pairs of ribs are called **true ribs;** they articulate directly with the manubrium and body of the sternum by means of costal cartilages. The next three pairs are called **false ribs;** their cartilages join the 7th rib cartilage. The last two pairs are called **floating ribs** because

Box 6–4 ABNORMALITIES OF THE CURVES OF THE SPINE

Scoliosis—an abnormal lateral curvature, which may be congenital, the result of having one leg longer than the other, or the result of chronic poor posture during childhood while the vertebrae are still growing. Usually the thoracic vertebrae are affected, which displaces the rib cage to one side. In severe cases, the abdominal organs may be compressed, and the expansion of the rib cage during inhalation may be impaired.

Kyphosis*—an exaggerated thoracic curve; sometimes referred to as hunchback.

Lordosis*—an exaggerated lumbar curve; sometimes referred to as swayback.

These abnormal curves are usually the result of degenerative bone diseases such as osteoporosis or tuberculosis of the spine. If osteoporosis, for example, causes the bodies of the thoracic vertebrae to collapse, the normal thoracic curve will be increased. Most often the vertebral body "settles" slowly (rather than collapses suddenly) and there is little, if any, damage to the spinal nerves. The damage to the vertebrae, however, cannot be corrected, so these conditions should be thought of in terms of prevention rather than cure.

*Although descriptive of normal anatomy, the terms kyphosis and lordosis, respectively, are commonly used to describe the abnormal condition associated with each.

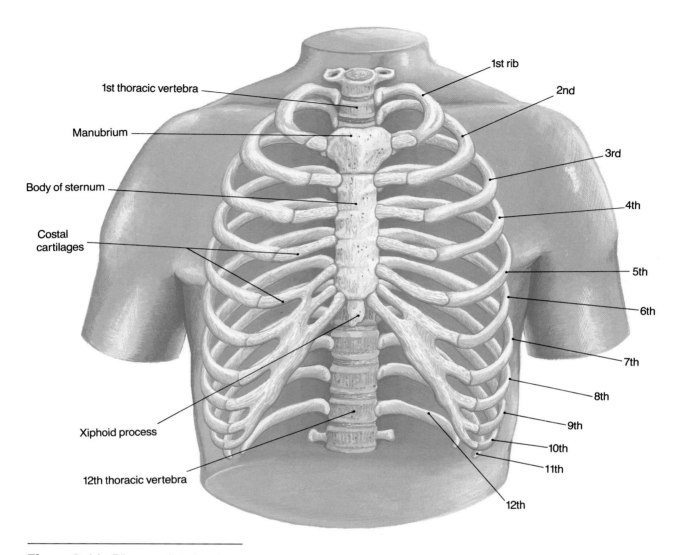

1st thoracic vertebra

Manubrium

Body of sternum

Costal
cartilages

Xiphoid process

12th thoracic vertebra

1st rib

2nd

3rd

4th

5th

6th

7th

8th

9th

10th

11th

12th

Figure 6–11 Rib cage. Anterior view.

they do not articulate with the sternum at all (see Fig. 6–10).

An obvious function of the rib cage is that it encloses and protects the heart and lungs. Keep in mind, though, that the rib cage also protects organs in the upper abdominal cavity, such as the liver and spleen. The other important function of the rib cage depends upon its flexibility: the ribs are pulled upward and outward by the external intercostal mus-

cles. This enlarges the chest cavity, which expands the lungs and contributes to inhalation.

THE SHOULDER AND ARM

The shoulder girdles attach the arms to the axial skeleton. Each consists of a scapula (shoulder blade) and clavicle (collarbone). The **scapula** is a large, flat bone that anchors some of the muscles that move the upper arm. A shallow depression called the glenoid fossa forms a **ball and socket joint** with the humerus, the bone of the upper arm (Fig. 6–12).

Each **clavicle** articulates laterally with a scapula and medially with the manubrium of the sternum. In this position the clavicles act as braces for the scapulae and prevent the shoulders from coming too far forward. Although the shoulder joint is capable of a wide range of movement, the shoulder itself must be relatively stable if these movements are to be effective.

The **humerus** is the long bone of the upper arm. Proximally, the humerus forms a **ball and socket joint** with the scapula. Distally, the humerus forms a **hinge joint** with the ulna of the forearm. This hinge joint, the elbow, permits movement in one plane, that is, back and forth with no lateral movement.

The forearm bones are the **ulna** on the little finger side and the **radius** on the thumb side. The radius and ulna articulate proximally to form a **pivot joint** which permits turning the hand palm up to palm down. You can demonstrate this yourself by holding your arm palm up in front of you, and noting that the radius and ulna are parallel to each other. Then turn your hand palm down, and notice that your upper arm does not move. The radius crosses over the ulna, which permits the hand to perform a great variety of movements without moving the entire arm.

The **carpals** are eight small bones in the wrist; **gliding joints** between them permit a sliding movement. The carpals also articulate with the distal ends of the ulna and radius, and with the proximal ends of the **metacarpals,** the five bones of the hand.

The **phalanges** are the bones of the fingers. There are two phalanges in each thumb and three in each of the fingers. Between phalanges are **hinge joints,** which permit movement in one plane. The thumb, however, is more movable than the fingers because of its carpometacarpal joint. This is a **saddle joint,** which enables the thumb to cross over the palm, and permits gripping. Important parts of these bones are described in Table 6–3.

THE HIP AND LEG

The pelvic girdle consists of the two **hip bones** (coxae or innominate bones), which articulate with the axial skeleton at the sacrum. Each hip bone has three major parts (Fig. 6–13): the ilium, ischium, and pubis. The **ilium** is the flared, upper portion that forms the sacroiliac joint. The **ischium** is the lower, posterior part that we sit on. The **pubis** is the lower, most anterior part. The two **pubic bones** articulate with one another at the **pubic symphysis,** with a disc of fibrous cartilage between them.

The **acetabulum** is the socket in the hip bone that forms a **ball and socket joint** with the femur. Compared to the glenoid fossa of the scapula, the acetabulum is a much deeper socket. This has great functional importance because the hip is a weight-bearing joint, whereas the shoulder is not. Since the acetabulum is deep, the hip joint is not easily dislocated, even by activities such as running and jumping (landing) which put great stress on the joint.

The **femur** is the long bone of the thigh. As mentioned, the femur forms a very movable ball and socket joint with the hipbone. At its distal end, the femur forms a **hinge joint,** the knee, with the tibia of the lower leg. The **patella,** or knee cap, is anterior to the knee joint, enclosed in the tendon of the quadriceps femoris, a large muscle group of the thigh.

The **tibia** is the weight-bearing bone of the lower leg. Notice in Fig. 6–14 that the **fibula** is not part of the knee joint and does not bear weight. The fibula is important, however, in that leg muscles are attached and anchored to it, and it helps stabilize the ankle. The tibia and fibula do not form a pivot joint as do the radius and ulna in the arm. This makes the lower leg and foot more stable, and thus able to support the body.

The **tarsals** are the seven bones in the ankle. The largest is the **calcaneus,** or heel bone; the **talus** transmits weight between the calcaneus and the tibia. **Metatarsals** are the five long bones of each foot, and **phalanges** are the bones of the toes. There are two phalanges in the big toe and three in each of the other toes. The phalanges of the toes form hinge joints with each other. Since there is no

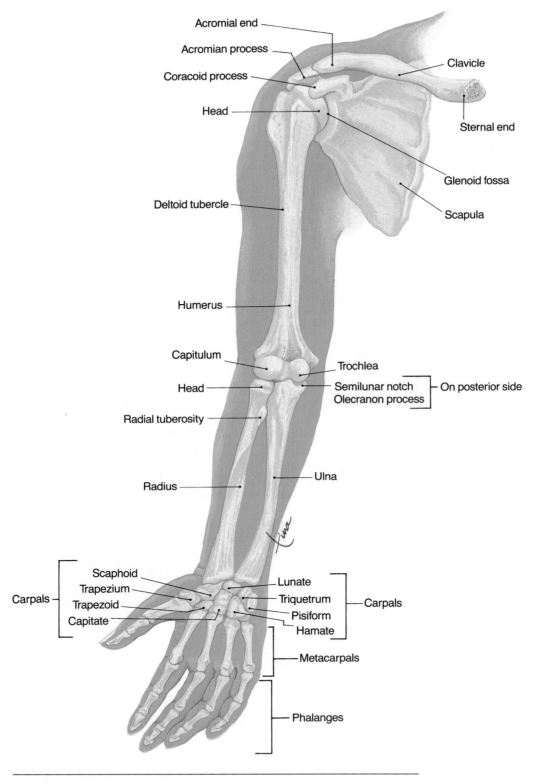

Acromial end

Acromian process

Coracoid process

Head

Deltoid tubercle

Humerus

Capitulum

Head

Radial tuberosity

Radius

Carpals

Scaphoid
Trapezium
Trapezoid
Capitate

Clavicle

Sternal end

Glenoid fossa

Scapula

Trochlea

Semilunar notch
Olecranon process

On posterior side

Ulna

Lunate
Triquetrum
Pisiform
Hamate

Carpals

Metacarpals

Phalanges

Figure 6–12 Bones of arm and shoulder girdle. Anterior view of right arm.

Table 6–3 BONES OF THE SHOULDER AND ARM—IMPORTANT PARTS

Bone	Part	Description
Scapula	• Glenoid fossa • Spine • Acromian process	• Depression that articulates with humerus • Long, posterior process for muscle attachment • Articulates with clavicle
Clavicle	• Acromial end • Sternal end	• Articulates with scapula • Articulates with manubrium of sternum
Humerus	• Head • Olecranon fossa • Capitulum • Trochlea	• Round process that articulates with scapula • Posterior, oval depression for the olecranon process of the ulna • Round process superior to radius • Concave surface that articulates with ulna
Radius	• Head	• Articulates with the ulna
Ulna	• Olecranon process • Semilunar notch	• Fits into olecranon fossa of humerus • "Half-moon" depression that articulates with the trochlea of ulna
Carpals (8)	• Scaphoid • Lunate • Triquetrum • Pisiform • Trapezium • Trapezoid • Capitate • Hamate	• Proximal Row • Distal Row

saddle joint in the foot, the big toe is not as movable as is the thumb. Important parts of these bones are described in Table 6–4.

JOINTS—ARTICULATIONS

A joint is where two bones meet, or **articulate.**

THE CLASSIFICATION OF JOINTS

The classification of joints is based on the amount of movement possible. A **synarthrosis** is an immovable joint, such as a suture between two cranial bones. An **amphiarthrosis** is a slightly movable joint, such as the symphysis joint between adjacent vertebrae. A **diarthrosis** is a freely movable joint. This is the largest category of joints and includes the ball and socket joint, the pivot, hinge, and others. Examples of each type of joint are described in

Table 6–5, and many of these are illustrated in Fig. 6–15.

SYNOVIAL JOINTS

All diarthroses, or freely movable joints, are **synovial joints** because they share similarities of structure. A typical synovial joint is shown in Fig. 6–16. On the joint surface of each bone is the **articular cartilage,** which provides a smooth surface. The **joint capsule,** made of fibrous connective tissue, encloses the joint in a strong sheath, like a sleeve. Lining the joint capsule is the **synovial membrane,** which secretes synovial fluid into the joint cavity. **Synovial fluid** is thick and slippery and prevents friction as the bones move.

Many synovial joints also have **bursae** (or bursas), which are small sacs of synovial fluid between the joint and the tendons that cross over the joint. Bursae permit the tendons to slide easily as the bones are moved. If a joint is used excessively, the

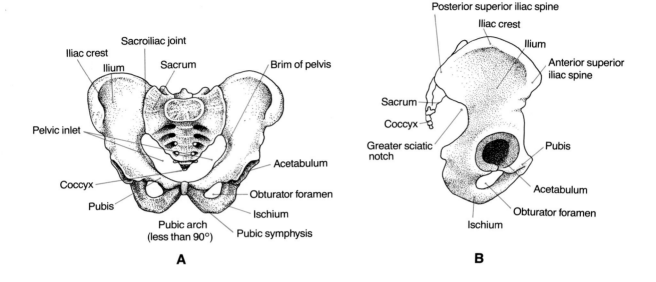

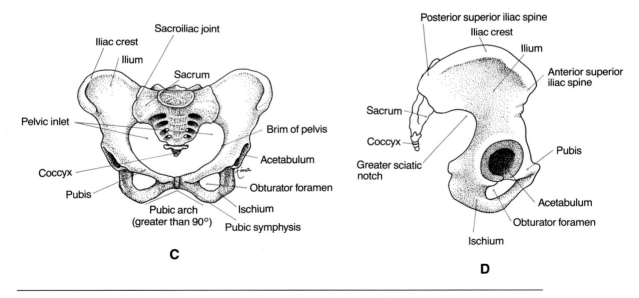

Figure 6–13 Hip bones and sacrum. (**A**), Male pelvis, anterior view. (**B**), Male pelvis, lateral view of right side. (**C**), Female pelvis, anterior view. (**D**), Female pelvis, lateral view of right side.

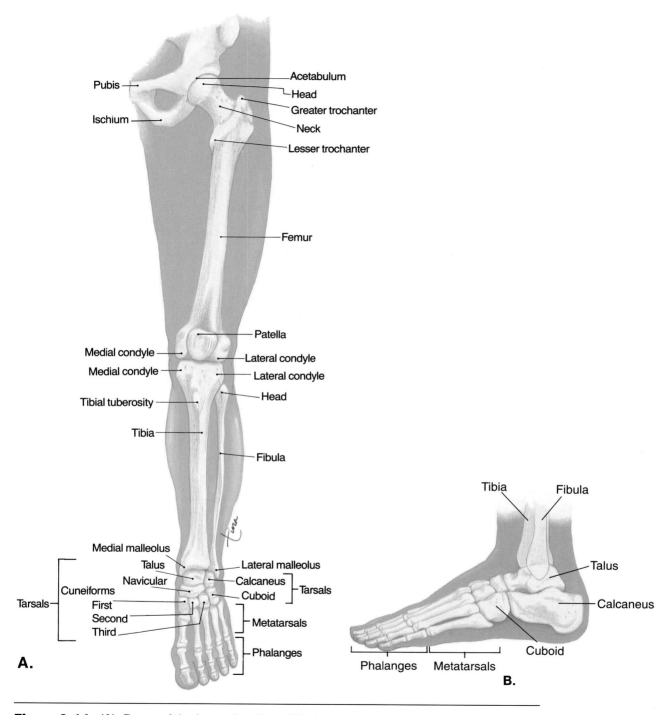

Pubis
Ischium
Acetabulum
Head
Greater trochanter
Neck
Lesser trochanter
Femur
Patella
Medial condyle
Medial condyle
Lateral condyle
Lateral condyle
Tibial tuberosity
Head
Tibia
Fibula
Medial malleolus
Talus
Navicular
Cuneiforms
First
Second
Third
Tarsals
Lateral malleolus
Calcaneus
Cuboid
Tarsals
Metatarsals
Phalanges

A.

Tibia
Fibula
Talus
Calcaneus
Cuboid
Phalanges   Metatarsals

B.

Figure 6–14 (**A**), Bones of the leg and portion of hip bone, anterior view of left leg. (**B**), Lateral view of left foot.

Table 6–4 BONES OF THE HIP AND LEG—IMPORTANT PARTS

Bone	Part	Description
Pelvic (2 hip bones)	• Ilium	Flared, upper portion
	• Iliac crest	Upper edge of ilium
	• Posterior superior iliac spine	Posterior continuation of iliac crest
	• Ischium	Lower, posterior portion
	• Pubis	Anterior, medial portion
	• Pubic symphysis	Joint between the 2 pubic bones
	• Acetabulum	Deep depression that articulates with femur
Femur	• Head	Round process that articulates with hip bone
	• Neck	Constricted portion distal to head
	• Greater trochanter	Large lateral process for muscle attachment
	• Lesser trochanter	Medial process for muscle attachment
	• Condyles	Rounded processes that articulate with tibia
Tibia	• Condyles	Articulate with the femur
	• Medial malleolus	Distal process; medial "ankle bone"
Fibula	• Head	Articulates with tibia
	• Lateral malleolus	Distal process; lateral "ankle bone"
Tarsals (7)	• Calcaneus	Heel bone
	• Talus	Articulates with calcaneus and tibia
	• Cuboid, Navicular	—
	• Cuneiform: 1st, 2nd, 3rd	—

Table 6–5 TYPES OF JOINTS

Category	Type and Description	Examples
Synarthrosis (immovable)	Suture—fibrous connective tissue between bone surfaces	• Between cranial bones; between facial bones
Amphiarthrosis (slightly movable)	Symphysis—disc of fibrous cartilage between bones	• Between vertebrae; between pubic bones
Diarthrosis (freely movable)	Ball and socket—movement in all planes	• Scapula and humerus; pelvic bone and femur
	Hinge—movement in one plane	• Humerus and ulna; femur and tibia; between phalanges
	Condyloid—movement in one plane with some lateral movement	• Temporal bone and mandible
	Pivot—rotation	• Atlas and axis; radius and ulna
	Gliding—side to side movement	• Between carpals
	Saddle—movement in several planes	• Carpometacarpal of thumb

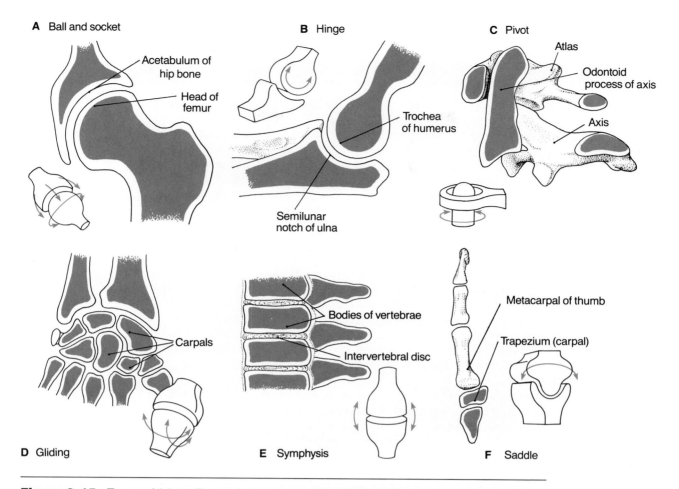

A Ball and socket

Acetabulum of hip bone

Head of femur

B Hinge

Trochea of humerus

Semilunar notch of ulna

C Pivot

Atlas

Odontoid process of axis

Axis

D Gliding

Carpals

E Symphysis

Bodies of vertebrae

Intervertebral disc

F Saddle

Metacarpal of thumb

Trapezium (carpal)

Figure 6–15 Types of joints. For each type, a specific joint is depicted, and a simple diagram shows the position of the joint surfaces. (**A**), Ball and socket. (**B**), Hinge. (**C**), Pivot. (**D**), Gliding. (**E**), Symphysis. (**F**), Saddle.

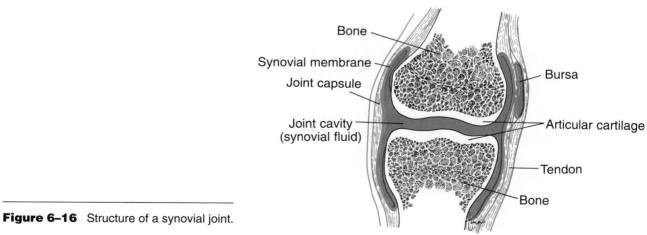

Bone

Synovial membrane

Joint capsule

Joint cavity (synovial fluid)

Bursa

Articular cartilage

Tendon

Bone

Figure 6–16 Structure of a synovial joint.

Box 6–5   ARTHRITIS

The term **arthritis** means inflammation of a joint. Of the many types of arthritis, we will consider two: osteoarthritis and rheumatoid arthritis.

Osteoarthritis is a natural consequence of getting older. In joints that have borne weight for many years, the articular cartilage is gradually worn away. The once smooth joint surface becomes rough, and the affected joint is stiff and painful. As you might guess, the large, weight-bearing joints are most often subjected to this form of arthritis. If we live long enough, most of us can expect some osteoarthritis in knees, hips, or ankles.

Rheumatoid arthritis can be a truly crippling disease that may begin in early middle age or, less commonly, during adolescence. It is believed to be an **autoimmune disease,** which means that the immune system mistakenly directs its destructive capability against part of the body. Exactly what triggers this abnormal response by the immune system is not known with certainty, but certain bacterial and viral infections have been suggested as possibilities.

Rheumatoid arthritis often begins in joints of the extremities, such as those of the fingers. The autoimmune activity seems to affect the synovial membrane, and joints become painful and stiff. Sometimes the disease progresses to total destruction of the synovial membrane and calcification of the joint. Such a joint is then fused and has no mobility at all.

Treatment of rheumatoid arthritis is directed at reducing inflammation as much as possible, for it is the inflammatory process that causes the damage. At present there is no cure for autoimmune diseases.

bursae may become inflamed and painful; this condition is called **bursitis.** Some other disorders of joints are described in Box 6–5: Arthritis.

SUMMARY

Your knowledge of the bones and joints will be useful in the next chapter as you learn the actions of the muscles that move the skeleton. It is important to remember, however, that bones have other functions as well. As a storage site for excess calcium, bones contribute to the maintenance of a normal blood calcium level. The red bone marrow found in flat and irregular bones produces the blood cells: red blood cells, white blood cells, and platelets. Some bones protect vital organs such as the brain, heart, and lungs. As you can see, bones themselves may also be considered vital organs.

STUDY OUTLINE

The skeleton is made of bone and cartilage and has these functions:
1. Is a framework for support, moved by muscles.
2. Protects internal organs from mechanical injury.
3. Contains and protects red bone marrow.
4. Stores excess calcium; important to regulate blood calcium level.

Bone Tissue (see Fig. 6–1)
1. Osteocytes (cells) are found in the matrix of calcium phosphate, calcium carbonate, and collagen.
2. Compact bone—Haversian systems are present.
3. Spongy bone—no Haversian systems; red bone marrow present.

4. Articular cartilage—smooth, on joint surfaces.
5. Periosteum—fibrous connective tissue membrane; anchors tendons and ligaments; has blood vessels that enter the bone.

Classification of Bones

1. Long—arms, legs; shaft is the diaphysis (compact bone) with a marrow cavity containing yellow bone marrow (fat); ends are epiphyses (spongy bone) (see Fig. 6–1).
2. Short—wrists, ankles (spongy bone covered with compact bone).
3. Flat—ribs, pelvic bone, cranial bones (spongy bone covered with compact bone).
4. Irregular—vertebrae, facial bones (spongy bone covered with compact bone).

Embryonic Growth of Bone

1. The embryonic skeleton is first made of other tissues that are gradually replaced by bone. Ossification begins in the third month of gestation; osteoblasts differentiate from fibroblasts and produce bone matrix.
2. Cranial and facial bones are first made of fibrous connective tissue; osteoblasts produce bone matrix in a center of ossification in each bone; bone growth radiates outward; fontanels remain at birth, permit compression of infant skull during birth; fontanels are calcified by age 2 (see Fig. 6–2).
3. All other bones are first made of cartilage; in a long bone the first center of ossification is in the diaphysis, other centers develop in the epiphyses. After birth a long bone grows at the epiphyseal discs: cartilage is produced on the epiphysis side, and bone replaces cartilage on the diaphysis side. Osteoclasts form the marrow cavity by reabsorbing bone matrix in the center of the diaphysis (see Fig. 6–3).

Factors That Affect Bone Growth and Maintenance

1. Heredity—many pairs of genes contribute to genetic potential for height.
2. Nutrition—calcium, phosphorus, and protein become part of the bone matrix; vitamin D is needed for absorption of calcium in the small intestine; vitamins C and A are needed for bone matrix production (calcification).
3. Hormones—produced by endocrine glands; concerned with cell division, protein synthesis, calcium metabolism, and energy production (see Table 6–1).
4. Exercise or stress—weight-bearing bones must bear weight or they will lose calcium and become brittle.

The Skeleton—206 bones in total (see Fig. 6–4)

1. Axial—skull, vertebrae, rib cage.
   - Skull—see Figs. 6–5 through 6–8 and Table 6–2.
     - Eight cranial bones form the braincase, which also protects the eyes and ears; 14 facial bones make up the face; the immovable joints between these bones are called sutures.
     - Paranasal sinuses are air cavities in the maxillae, frontal, sphenoid, and ethmoid bones; lighten the skull and provide resonance for voice (see Fig. 6–9).
     - Three auditory bones in each middle ear cavity transmit vibrations for the hearing process.
   - Vertebral Column—see Fig. 6–10.
     - Individual bones are called vertebrae: seven cervical, 12 thoracic, five lumbar, five sacral (fused into one sacrum), four to five coccygeal (fused into one coccyx). Supports trunk and head, encloses and protects the spinal cord in the vertebral canal. Discs of fibrous cartilage absorb shock between the bodies of adjacent vertebrae, also permit slight movement. Four natural curves center head over body for walking upright (see Table 6–5 for joints).
   - Rib Cage—see Fig. 6–11.
     - Sternum and 12 pairs of ribs; protects thoracic and upper abdominal organs from mechanical injury and is expanded to contribute to inhalation. Sternum consists of manubrium, body, and xiphoid process. All ribs articulate with thoracic vertebrae; true ribs (first seven pairs) articulate directly with sternum by means of costal cartilages; false ribs (next three pairs) articulate with 7th costal cartilage; floating ribs (last two pairs) do not articulate with the sternum.
2. Appendicular—bones of the arms and legs and the shoulder and pelvic girdles.

- Shoulder and Arm—see Fig. 6–12 and Table 6–3.
  - Scapula—shoulder muscles are attached; glenoid fossa articulates with humerus.
  - Clavicle—braces the scapula.
  - Humerus—upper arm; articulates with the scapula and the ulna (elbow).
  - Radius and ulna—forearm—articulate with one another and with carpals.
  - Carpals—eight—wrist; Metacarpals—five—hand; Phalanges—14—fingers (for joints, see Table 6–5).
- Hip and Leg—see Figs. 6–13 and 6–14 and Table 6–4.
  - Pelvic bone—two hip bones; ilium, ischium, pubis; acetabulum articulates with femur.
  - Femur—thigh; articulates with pelvic bone and tibia (knee).
  - Patella—kneecap; in tendon of quadriceps femoris muscle.
  - Tibia and fibula—lower leg; tibia bears weight; fibula does not bear weight, but does anchor muscles and stabilizes ankle.
  - Tarsals—seven—ankle; calcaneus is heel bone.
  - Metatarsals—five—foot; Phalanges—14—toes (see Table 6–5 for joints).

Joints—articulations

1. Classification based on amount of movement:
   - Synarthrosis—immovable.
   - Amphiarthrosis—slightly movable.
   - Diarthrosis—freely movable (see Table 6–5 for examples; see also Fig. 6–15).
2. Synovial joints—all diarthroses have similar structure (see Fig. 6–16):
   - Articular cartilage—smooth on joint surfaces.
   - Joint capsule—strong fibrous connective tissue sheath that encloses the joint.
   - Synovial membrane—lines the joint capsule; secretes synovial fluid that prevents friction.
   - Bursae—sacs of synovial fluid that permit tendons to slide easily across joints.

REVIEW QUESTIONS

1. Explain the differences between compact bone and spongy bone, and state where each type is found. (p. 102)

2. State the locations of red bone marrow, and name the blood cells it produces. (p. 102)

3. Name the tissue of which the embryonic skull is first made. Explain how ossification of cranial bones occurs. (p. 104)

4. State what fontanels are, and explain their function. (p. 104)

5. Name the tissue of which the embryonic femur is first made. Explain how ossification of this bone occurs. Describe what happens in epiphyseal discs to produce growth of long bones. (p. 106)

6. Explain what is meant by "genetic potential" for height, and name the nutrients a child must have in order to attain genetic potential. (p. 106)

7. Explain the functions of calcitonin and parathyroid hormone with respect to bone matrix and to blood calcium level. (p. 108)

8. Explain how estrogen or testosterone affects bone growth, and when. (p. 108)

9. State one way each of the following hormones helps promote bone growth: insulin, thyroxine, growth hormone. (p. 108)

10. Name the bones that make up the braincase. (pp. 108, 111)

11. Name the bones that contain paranasal sinuses and explain the functions of these sinuses. (pp. 111–112)

12. Name the bones that make up the rib cage, and describe two functions of the rib cage. (pp. 117–118)

13. Describe the functions of the vertebral column. State the number of each type of vertebra. (pp. 112, 117)

14. Explain how the shoulder and hip joints are similar and how they differ. (pp. 118–119)

15. Give a specific example (name two bones) for each of the following types of joints: (p. 124)
    a. hinge
    b. symphysis
    c. pivot
    d. saddle
    e. suture
    f. ball and socket

16. Name the part of a synovial joint with each of the following functions: (p. 121)
    a. fluid within the joint cavity that prevents friction
    b. encloses the joint in a strong sheath
    c. provides a smooth surface on bone surfaces
    d. lines the joint capsule and secretes synovial fluid

17. Refer to the diagram of the full skeleton, and point to each bone on yourself.

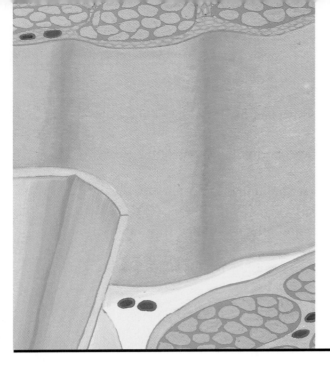

Chapter 7

Student Objectives

- Name the organ systems directly involved in movement, and state how they are involved.
- Describe muscle structure in terms of muscle cells, tendons, and bones.
- Describe the difference between antagonistic and synergistic muscles, and explain why such arrangements are necessary.
- Explain the role of the brain with respect to skeletal muscle.
- Define muscle tone and explain its importance.
- Explain the difference between isotonic and isometric exercise.
- Define muscle sense and explain its importance.
- Name the energy sources for muscle contraction, and state the simple equation for cell respiration.
- Explain the importance of hemoglobin and myoglobin, oxygen debt, lactic acid.
- Describe the neuromuscular junction and state the function of each part.
- Describe the structure of a sarcomere.
- Explain in terms of ions and charges: polarization, depolarization, repolarization.
- Describe the Sliding Filament Theory of muscle contraction.
- Describe some of the body's responses to exercise and explain how each maintains homeostasis.
- Learn the major muscles of the body and their functions.

The Muscular System

New Terminology

Actin (**AK**–tin)
Antagonistic muscles (an–**TAG**–on–ISS–tik **MUSS**–uhls)
Creatine phosphate (**KREE**–ah–tin **FOSS**–fate)
Depolarization (DE–poh–lahr–i–**ZA**–shun)
Fascia (**FASH**–ee–ah)
Insertion (in–**SIR**–shun)
Isometric (EYE–so–**MEH**–trik)
Isotonic (EYE–so–**TAHN**–ik)
Lactic acid (**LAK**–tik **ASS**–id)
Muscle fatigue (**MUSS**–uhl fah–**TEEG**)
Muscle sense (**MUSS**–uhl SENSE)
Muscle tone (**MUSS**–uhl TONE)
Myoglobin (**MYE**–oh–GLOW–bin)
Myosin (**MYE**–oh–sin)
Neuromuscular junction (NYOOR–oh–**MUSS**–kuhl–lar **JUNK**–shun)
Origin (**AHR**–i–jin)
Oxygen debt (**OX**–ah–jen DET)
Polarization (POH–lahr–i–**ZA**–shun)
Prime mover (PRIME **MOO**–ver)
Sarcolemma (SAR–koh–**LEM**–ah)
Sarcomeres (**SAR**–koh–meers)
Synergistic muscles (**SIN**–er–JIS–tik **MUSS**–uhls)
Tendon (**TEN**–dun)

Related Clinical Terminology

Anabolic steroids (an–a–**BOLL**–ik **STEER**–oids)
Atrophy (**AT**–ruh–fee)
Botulism (**BOTT**–yoo–lizm)
Hypertrophy (high–**PER**–truh–fee)
Intramuscular injection (IN–trah–**MUSS**–kuh–ler in–**JEK**–shun)
Muscular dystrophy (**MUSS**–kyoo–ler **DIS**–truh–fee)
Myalgia (my–**AL**–jee–ah)
Myasthenia gravis (MY–ass–**THEE**–nee–yuh **GRAH**–viss)
Myopathy (my–**AH**–puh–thee)
Paralysis (pah–**RAL**–i–sis)
Range-of-motion exercises (RANJE-of-**MOH**–shun **EX**–err–sigh–zez)
Sex-linked trait (SEX LINKED **TRAYT**)
Tetanus (**TET**–uh–nus)

Terms that appear in **bold type** in the chapter text are defined in the glossary, which begins on page 549.

Do you like to dance? Most of us do, or, we may simply enjoy watching good dancers. The grace and coordination involved in dancing result from the interaction of many of the organ systems, but the one you think of first is probably the muscular system.

There are more than 600 muscles in the human body. Most of these muscles are attached to the bones of the skeleton by tendons, although a few muscles are attached to the undersurface of the skin. The primary function of the **muscular system** is to move the skeleton. The other body systems directly involved in movement are the nervous, respiratory, and circulatory systems. The nervous system transmits the electrochemical impulses that cause muscle cells to contract. The respiratory system exchanges oxygen and carbon dioxide between the air and blood. The circulatory system brings oxygen to the muscles and takes carbon dioxide away.

These interactions of body systems will be covered in this chapter, which will focus on the **skeletal muscles.** You may recall from Chapter 4 that there are two other types of muscle tissue: smooth muscle and cardiac muscle. These types of muscle tissue will be discussed in other chapters in relation to the organs of which they are part. Before you continue, you may find it helpful to go back to Chapter 4 and review the structure and characteristics of skeletal muscle tissue. In this chapter we will begin with the gross (large) anatomy and physiology of muscles, then discuss the microscopic structure of muscle cells and the biochemistry of muscle contraction.

MUSCLE STRUCTURE

All muscle cells are specialized for contraction. When these cells contract, they shorten and pull a bone in order to produce movement. Each skeletal muscle is made of thousands of individual muscle cells, which also may be called **muscle fibers** (see Fig. 7–3). Depending on the work a muscle is required to do, variable numbers of muscle fibers contract. When picking up a pencil, for example, only a small portion of the muscle fibers in a muscle

will contract. If the muscle has more work to do, such as picking up a book, more muscle fibers will contract to accomplish the task.

Muscles are anchored firmly to bones by **tendons.** Tendons are made of fibrous connective tissue, which, you may remember, is very strong and merges with the **fascia** that covers the muscle and with the **periosteum,** the fibrous connective tissue membrane that covers bones. A muscle usually has at least two tendons, each attached to a different bone. The more immobile or stationary attachment of the muscle is its **origin;** the more movable attachment is called the **insertion.** The muscle itself crosses the joint of the two bones to which it is attached, and when the muscle contracts it pulls on its insertion and moves the bone in a specific direction.

MUSCLE ARRANGEMENTS

Muscles are arranged so as to bring about a variety of movements. The two general types of arrangements are the opposing **antagonists** and the cooperative **synergists.**

Antagonistic Muscles

Antagonists are opponents, so we use the term **antagonistic muscles** for muscles which have opposing or opposite functions. An example will be helpful here—refer to Fig. 7–1 as you read the following. The biceps brachii is the muscle on the front of the upper arm. The origin of the biceps is on the scapula (there are actually two tendons, hence the name "biceps"), and the insertion is on the radius. When the biceps contracts it **flexes** the forearm, that is, bends the elbow (see Table 7–2). Recall that when a muscle contracts it gets shorter and pulls. Muscles cannot push, for when they relax they exert no force. Therefore, the biceps can bend the elbow but cannot straighten it; another muscle is needed. The triceps brachii is located on the back of the upper arm. Its origins (the prefix "tri" tells you that there are three of them) are on the scapula and humerus, and its insertion is on the ulna. When the triceps contracts and pulls, it **extends** the forearm, that is, straightens the elbow.

Joints that are capable of a variety of movements

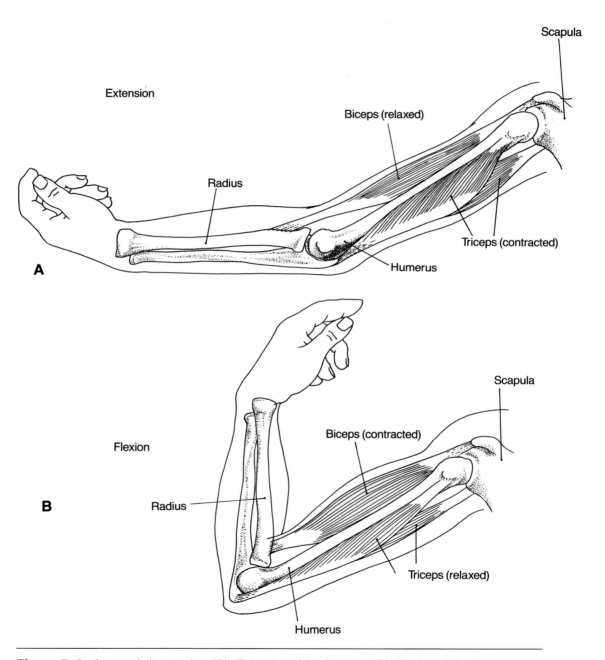

Figure 7–1 Antagonistic muscles. (**A**), Extension of the forearm. (**B**), Flexion of the forearm.

have several sets of antagonists. Notice how many ways you can move your upper arm at the shoulder, for instance. Abducting (laterally raising) the arm is the function of the deltoid. Adducting the arm is brought about by the pectoralis major and latissimus dorsi. Flexion of the arm (across the chest) is also a function of the pectoralis major, and extension of the arm (behind the back) is also a function of the lattisimus dorsi. All of these muscles are described and depicted in the tables and figures later in the chapter. Without antagonistic muscles, this variety of movements would be impossible.

You may be familiar with **range-of-motion** or ROM exercises that are often recommended for bedridden patients. Such exercises are designed to stretch and contract the antagonistic muscles of a joint to preserve as much muscle function and joint mobility as possible.

Synergistic Muscles

Synergistic muscles are those with the same function, or those that work together to perform a particular function. Recall that the biceps brachii flexes the forearm. The brachioradialis, with its origin on the humerus and insertion on the radius, also flexes the forearm. There is even a third flexor of the forearm, the brachialis. You may wonder why we need three muscles to perform the same function, and the explanation lies in the great mobility of the hand. If the hand is palm up, the biceps does most of the work of flexing and may be called the **prime mover.** When the hand is thumb up, the brachioradialis is in position to be the prime mover, and when the hand is palm down, the brachialis becomes the prime mover. If you have ever tried to do chin-ups, you know that it is much easier with your palms toward you than with palms away from you. This is because the biceps is a larger, and usually much stronger, muscle than is the brachialis.

Muscles may also be called synergists if they help to stabilize or steady a joint to make a more precise movement possible. If you drink a glass of water, the biceps brachii may be the prime mover to flex the forearm. At the same time, the muscles of the shoulder keep that joint stable, so that the water gets to your mouth, not over your shoulder or down your chin. The shoulder muscles are considered

synergists for this movement because their contribution makes the movement effective.

THE ROLE OF THE BRAIN

Even our simplest movements require the interaction of many muscles, and the contraction of skeletal muscles depends on the brain. The nerve impulses for movement come from the **frontal lobes** of the **cerebrum.** The cerebrum is the largest part of the brain; the frontal lobes are beneath the frontal bone. The **motor areas** of the frontal lobes generate electrochemical impulses that travel along motor nerves to muscle fibers, causing the muscle fibers to contract.

For a movement to be effective, some muscles must contract while others relax. This is what we call coordination, and it is regulated by the **cerebellum,** which is located below the occipital lobes of the cerebrum.

MUSCLE TONE

Except during certain stages of sleep, most of our muscles are in a state of slight contraction; this is what is known as **muscle tone.** When sitting upright, for example, the tone of your neck muscles keeps your head up, and the tone of your back muscles keeps your back straight. This is an important function of muscle tone for human beings, because it helps us to maintain an upright posture. In order for a muscle to remain slightly contracted, only a few of the muscle fibers in that muscle must contract. Alternate fibers contract so that the muscle as a whole does not become fatigued. This is similar to a pianist continuously rippling her fingers over the keys of the piano—some notes are always sounding at any given moment, but the notes that are sounding are always changing.

Muscle fibers need the energy of ATP in order to contract. When they produce ATP in the process of cell respiration, muscle fibers also produce heat. The heat generated by normal muscle tone is approximately 25% of the total body heat at rest. During exercise, of course, heat production increases significantly.

Box 7–1   ANABOLIC STEROIDS

Anabolic steroids are synthetic drugs very similar in structure and action to the male hormone **testosterone.** Normal secretion of testosterone, beginning in males at puberty, increases muscle size and is the reason men usually have larger muscles than do women.

Some athletes, both male and female, both amateur and professional, take anabolic steroids to build muscle mass and to increase muscle strength. There is no doubt that the use of anabolic steroids will increase muscle size, but there are hazards, some of them very serious. Side effects of such self-medication include liver damage, kidney damage, disruption of reproductive cycles, and mental changes such as irritability and aggressiveness.

Women athletes may develop increased growth of facial and body hair and may become sterile as a result of the effects of a male hormone on their own hormonal cycles.

EXERCISE

Good muscle tone improves coordination. When muscles are slightly contracted, they can react more rapidly if and when greater exertion is necessary. Muscles with poor tone are usually soft and flabby, but exercise will improve muscle tone.

There are two general types of exercise: isotonic and isometric. In **isotonic exercise,** muscles contract and bring about movement. Jogging, swimming and weight-lifting are examples. Isotonic exercise improves muscle tone, muscle strength, and if done repetitively, muscle size. This type of exercise also improves cardiovascular and respiratory efficiency, since movement exerts demands on the heart and respiratory muscles. If done for 30 minutes or longer, such exercise may be called "aerobic," because it strengthens the heart and respiratory muscles as well as the skeletal muscles.

Isometric exercise involves contraction without movement. If you put your palms together and push one hand against the other, you can feel your arm muscles contracting. If both hands push equally, there will be no movement; this is isometric contraction. Such exercises will increase muscle tone and muscle strength but will not increase muscle size very much. Nor is isometric exercise considered aerobic. Without movement, heart rate and breathing do not increase nearly as much as they would during an equally strenuous isotonic exercise. With respect to increasing muscle strength, see Box 7–1: Anabolic Steroids.

MUSCLE SENSE

When you walk up a flight of stairs, do you have to look at your feet to be sure each will get to the next step? Most of us don't (an occasional stumble doesn't count), and for this freedom we can thank our muscle sense. **Muscle sense** is the brain's ability to know where our muscles are and what they are doing, without our having to consciously look at them.

Within muscles are receptors called **stretch receptors** (proprioceptors or muscle spindles). The general function of all sensory receptors is to detect changes. The function of stretch receptors is to detect changes in the length of a muscle as it is stretched. The sensory impulses generated by these receptors are interpreted by the brain as a mental "picture" of where the muscle is.

We can be aware of muscle sense if we choose to be, but usually we can safely take it for granted. In fact, that is what we are meant to do. Imagine what life would be like if we had to watch every move to be sure that a hand or foot performed its intended action. Even simple activities such as walking or eating would require our constant attention.

There are times when we may become aware of our muscle sense. Learning a skill such as typing or playing the guitar involves very precise movements of the fingers, and beginners will often watch their fingers to be sure they are moving properly. With practice, however, muscle sense again becomes unconscious, and the experienced typist or guitarist need not watch every movement.

All sensation is a function of brain activity, and muscle sense is no exception. The impulses for muscle sense are integrated in the **parietal lobes** of the cerebrum (conscious muscle sense) and in the cerebellum (unconscious muscle sense) to be used to promote coordination.

ENERGY SOURCES FOR MUSCLE CONTRACTION

Before discussing the contraction process itself, let us look first at how muscle fibers obtain the energy they need to contract. The direct source of energy for muscle contraction is **ATP.** ATP, however, is not stored in large amounts in muscle fibers and is depleted in a few seconds.

The secondary energy sources are creatine phosphate and glycogen. **Creatine phosphate** is, like ATP, an energy-transfering molecule. When it is broken down (by an enzyme) to creatine, phosphate, and energy, the energy is used to synthesize more ATP. Most of the creatine formed is used to resynthesize creatine phosphate, but some is converted to **creatinine,** a waste product that is excreted by the kidneys.

The most abundant energy source in muscle fibers is **glycogen.** When glycogen is needed to provide energy for sustained contractions (more than a few seconds), it is first broken down into the **glucose** molecules of which it is made. Glucose is then further broken down in the process of cell respira-

tion to produce ATP, and muscle fibers may continue to contract.

Recall from Chapter 2 our simple equation for cell respiration:

$$\text{Glucose} + O_2 \rightarrow CO_2 + H_2O + \text{ATP} + \text{heat}$$

Look first at the products of this reaction. ATP will be used by the muscle fibers for contraction. The heat produced will contribute to body temperature, and if exercise is strenuous, will increase body temperature. The water becomes part of intracellular water, and the carbon dioxide is a waste product that will be exhaled.

Now look at what is needed to release energy from glucose: oxygen. Muscles have two sources of oxygen. The blood delivers a continuous supply of oxygen, which is carried by the **hemoglobin** in red blood cells. Within muscle fibers themselves there is another protein called **myoglobin,** which stores some oxygen within the muscle cells. Both hemoglobin and myoglobin contain the mineral iron, which enables them to bond to oxygen. (Iron also makes both molecules red, and it is myoglobin that gives muscle tissue a red or dark color.)

During strenuous exercise, the oxygen stored in myoglobin is quickly used up, and normal circulation may not deliver oxygen fast enough to permit the completion of cell respiration. Even though the respiratory rate increases, the muscle fibers may literally run out of oxygen. This state is called **oxygen debt,** and in this case, glucose cannot be completely broken down into carbon dioxide and water. If oxygen is not present (or not present in sufficient amounts), glucose is converted to an intermediate molecule called **lactic acid,** which causes **muscle fatigue.**

In a state of fatigue, muscle fibers cannot contract efficiently, and contraction may become painful. To be in oxygen debt means that we owe the body some oxygen. Lactic acid from muscles enters the blood and circulates to the liver, where it is converted back into glucose. This conversion requires ATP, and oxygen is needed to produce the necessary ATP in the liver. This is why, after strenuous exercise, the respiratory rate and heart rate remain high for a time and only gradually return to normal.

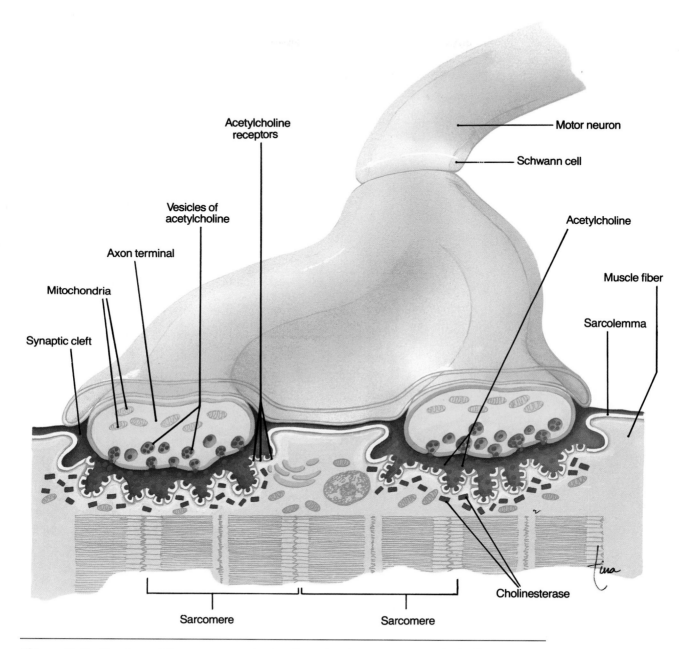

Figure 7–2 Structure of the neuromuscular junction, showing an axon terminal adjacent to the sarcolemma of a muscle fiber.

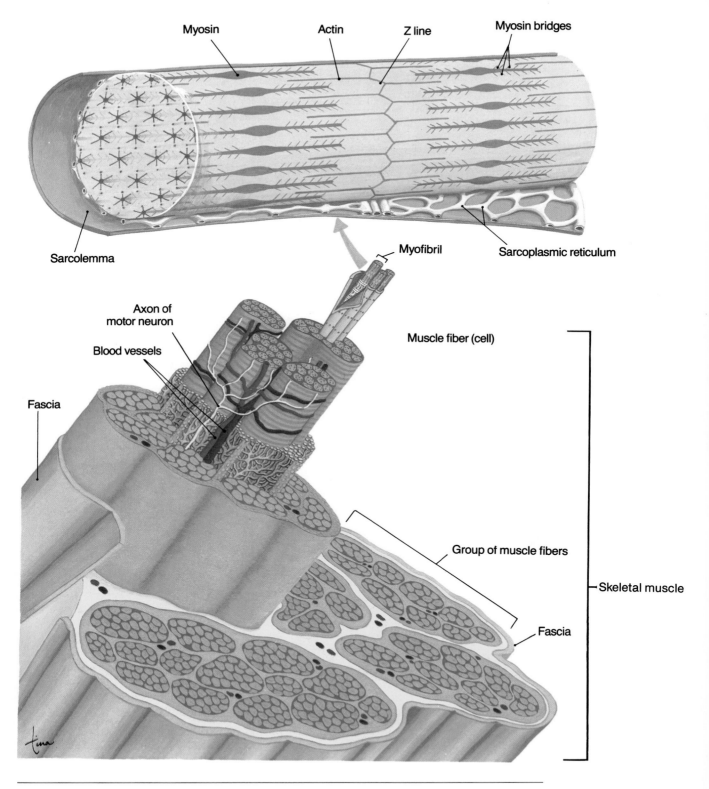

Figure 7–3 Microscopic structure of skeletal muscle. Progressively smaller structure is shown in the expanded portions. The arrow indicates a highly magnified view of the structure of sarcomeres.

MUSCLE FIBER— MICROSCOPIC STRUCTURE

We will now look more closely at a muscle fiber, keeping in mind that there are thousands of these cylindrical cells in one muscle. Each muscle fiber has its own motor nerve ending; the **neuromuscular junction** is where the motor neuron terminates on the muscle fiber (Fig. 7–2). The **axon terminal** is the enlarged tip of the motor neuron; it contains sacs of the neurotransmitter **acetylcholine** (ACh). The membrane of the muscle fiber is the **sarcolemma,** which contains an inactivator called **cholinesterase.** The **synapse** (or synaptic cleft) is the small space between the axon terminal and the sarcolemma.

Within the muscle fiber are thousands of individual contracting units called **sarcomeres,** which are arranged end to end in cylinders called **myofibrils.** The structure of a sarcomere is shown in Fig. 7–3:

the Z lines are the end boundaries of a sarcomere. Filaments of the protein **myosin** are in the center of the sarcomere, and filaments of the protein **actin** are at the ends, attached to the Z lines. Myosin and actin are the contractile proteins of a muscle fiber. Their interactions produce muscle contraction. Also present (not shown) are two inhibitory proteins, **troponin** and **tropomyosin,** which prevent the sliding of myosin and actin when the muscle fiber is relaxed.

Surrounding the sarcomeres is the **sarcoplasmic reticulum,** the endoplasmic reticulum of muscle cells. The sarcoplasmic reticulum is a reservoir for calcium ions (Ca^{+2}), which are essential for the contraction process.

All of these parts of a muscle fiber are involved in the contraction process. Contraction begins when a nerve impulse arrives at the axon terminal and stimulates the release of acetylcholine. Acetylcholine generates electrical changes (the movement of ions) at the sarcolemma of the muscle fiber. These

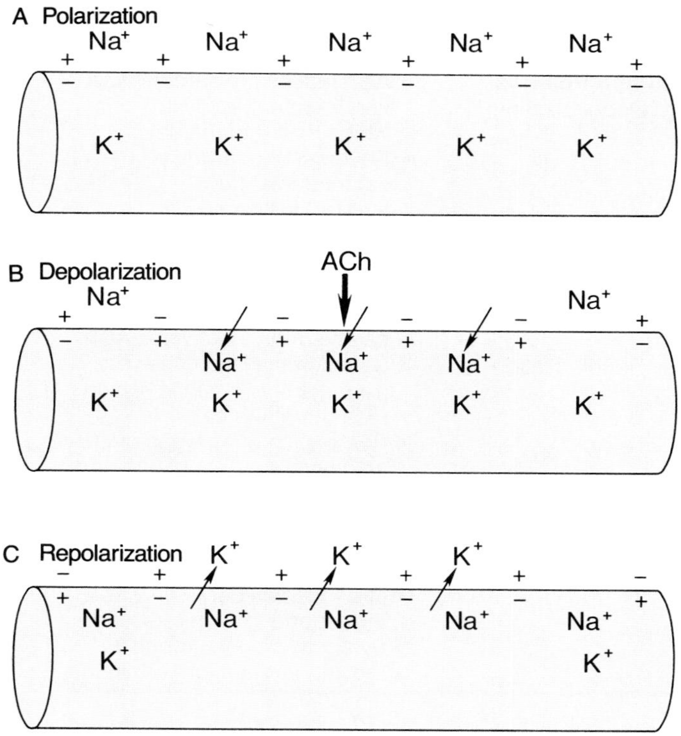

Figure 7–4 Electrical charges and ion concentrations at the sarcolemma. (**A**), Polarization, when the muscle fiber is relaxed. (**B**), Depolarization in response to acetylcholine. (**C**), Repolarization.

electrical changes initiate a sequence of events within the muscle fiber that is called the **Sliding Filament Theory** of muscle contraction. We will begin our discussion with the sarcolemma.

SARCOLEMMA—POLARIZATION

When a muscle fiber is relaxed, the sarcolemma is polarized (has a resting potential), which is a difference in electrical charges between the outside

Table 7–1   SARCOLEMMA— ELECTRICAL CHANGES

State or Event	Description
Resting Potential Polarization	• Sarcolemma has a (+) charge outside and a (−) charge inside. • Na^+ ions are more abundant outside the cell; as they diffuse inward, the sodium pump returns them outside. • K^+ ions are more abundant inside the cell; as they diffuse out the potassium pump returns them inside.
Action Potential Depolarization	• ACh makes the sarcolemma very permeable to Na^+ ions, which rush into the cell. • Reversal of charges on the sarcolemma: now (−) outside and (+) inside. • The reversal of charges spreads along the entire sarcolemma. • Cholinesterase at the sarcolemma inactivates ACh.
Repolarization	• Sarcolemma becomes very permeable to K^+ ions, which rush out of the cell. • Restoration of charges on the sarcolemma: (+) outside and (−) inside. • The sodium and potassium pumps return Na^+ ions outside and K^+ ions inside. • The muscle fiber is now able to respond to ACh released by another nerve impulse arriving at the axon terminal.

and the inside. During **polarization,** the outside of the sarcolemma has a positive charge relative to the inside, which is said to have a negative charge. Sodium ions (Na^+) are more abundant outside the cell, and potassium ions (K^+) and negative ions are more abundant inside (Fig. 7–4).

The Na^+ ions outside tend to diffuse into the cell, and the **sodium pump** transfers them back out. The K^+ ions inside tend to diffuse outside, and the **potassium pump** returns them inside. Both of these pumps are active transport mechanisms which, you may recall, require ATP. Muscle fibers use ATP to maintain a high concentration of Na^+ ions outside the cell and a high concentration of K^+ inside. The pumps, therefore, maintain polarization and relaxation until a nerve impulse stimulates a change.

SARCOLEMMA—DEPOLARIZATION

When a nerve impulse arrives at the axon terminal, it causes the release of acetylcholine, which diffuses across the synapse and bonds to **ACh receptors** on the sarcolemma. By doing so, acetylcholine makes the sarcolemma very permeable to Na^+ ions, which rush into the cell. This makes the inside of the sarcolemma positive relative to the outside, which is now considered negative. This reversal of charges is called **depolarization.** The electrical impulse thus generated (called an action potential) then spreads along the entire sarcolemma of a muscle fiber. Depolarization initiates changes within the cell that bring about contraction. The electrical changes that take place at the sarcolemma are summarized in Table 7–1 and shown in Fig. 7–4.

MECHANISM OF CONTRACTION— SLIDING FILAMENT THEORY

All of the parts of a muscle fiber and the electrical changes described earlier are involved in the contraction process, which is a precise sequence of events.

In summary, a nerve impulse causes depolarization of a muscle fiber, and this electrical change enables the myosin filaments to pull the actin filaments toward the center of the sarcomere, making the sarcomere shorter. All of the sarcomeres shorten and the muscle fiber contracts. A more detailed description of this process is the following:

1. A nerve impulse arrives at the axon terminal; acetylcholine is released and diffuses across the synapse.
2. Acetylcholine makes the sarcolemma more permeable to Na^+ ions, which rush into the cell.
3. The sarcolemma depolarizes, becoming negative outside and positive inside.
4. Depolarization stimulates the release of Ca^{+2} ions from the sarcoplasmic reticulum. Ca^{+2} ions bond to the troponin–tropomyosin complex, which shifts it away from the actin filaments.
5. Myosin splits ATP to release its energy; bridges on the myosin attach to the actin filaments and pull them toward the center of the sarcomere, thus making the sarcomere shorter.

6. All the sarcomeres in a muscle fiber shorten—the entire muscle fiber contracts.
7. The sarcolemma repolarizes: K^+ ions leave the cell, restoring a positive charge outside and a negative charge inside. The pumps then return Na^+ ions outside and K^+ ions inside.
8. Cholinesterase in the sarcolemma inactivates acetylcholine.
9. Subsequent nerve impulses will prolong contraction (more acetycholine is released).
10. When there are no further impulses, the muscle fiber will relax and return to its original length.

The above sequence (1–8) describes a single muscle fiber contraction (called a "twitch") in response to a single nerve impulse. Since all of this takes place in less than a second, useful movements would not be possible if muscle fibers relaxed immediately after contracting. Normally, however, nerve impulses arrive in a continuous stream and produce a sustained contraction called **tetanus,** which is a normal state not to be confused with the disease tetanus (see Box 7–2: Tetanus and Botu-

---

Box 7–2   TETANUS AND BOTULISM

Some bacteria cause disease by producing toxins. A **neurotoxin** is a chemical that in some way disrupts the normal functioning of the nervous system. Since skeletal muscle contraction depends on nerve impulses, the serious consequences for the individual may be seen in the muscular system.

Tetanus is characterized by the inability of muscles to relax. The toxin produced by the tetanus bacteria *(Clostridium tetani)* affects the nervous system in such a way that muscle fibers receive too many impulses, and muscles go into spasms. Lockjaw, the common name for tetanus, indicates one of the first symptoms, which is difficulty opening the mouth because of spasms of the masseter muscles. Treatment requires the antitoxin (an antibody to the toxin) to neutralize the toxin. In untreated tetanus the cause of death is spasm of the respiratory muscles.

Botulism is usually a type of food poisoning, but it is not characterized by typical food poisoning symptoms such as diarrhea or vomiting. The neurotoxin produced by the botulism bacteria *(Clostridium botulinum)* prevents the release of acetylcholine at neuromuscular junctions. Without acetylcholine, muscle fibers cannot contract, and muscles become paralyzed. Early symptoms of botulism include blurred or double vision and difficulty speaking or swallowing. Weakness and paralysis spread to other muscle groups, eventually affecting all voluntary muscles. Without rapid treatment with the antitoxin (the specific antibody to this toxin), botulism is fatal because of paralysis of the respiratory muscles.

lism). When in tetanus, muscle fibers remain contracted and are capable of effective movements. In a muscle such as the biceps brachii that flexes the forearm, an effective movement means that many of its thousands of muscle fibers are in tetanus.

As you might expect with such a complex process, there are many ways muscle contraction may be impaired. Perhaps the most obvious is the loss of nerve impulses to muscle fibers, as occurs when nerves or the spinal cord are severed, or when a **stroke (cerebrovascular accident)** occurs in the frontal lobes of the cerebrum. Without nerve impulses, skeletal muscles become **paralyzed,** unable to contract. Paralyzed muscles eventually **atrophy,** that is, become smaller from lack of use. Other disorders that affect muscle functioning are discussed in Box 7–3: Muscular Dystrophy, and Box 7–4: Myasthenia Gravis.

RESPONSES TO EXERCISE— MAINTAINING HOMEOSTASIS

Although entire textbooks are devoted to exercise physiology, we will discuss it only briefly here as an example of the body's ability to maintain homeostasis. Engaging in moderate or strenuous exercise is a physiological stress situation, a change that the body must cope with and still maintain a normal internal environment, that is, homeostasis.

Some of the body's responses to exercise are diagrammed below; notice how they are related to cell respiration.

As you can see, the respiratory and cardiovascular systems make essential contributions to exercise. The integumentary system also has a role, since it eliminates excess body heat. Although not shown below, the nervous system is also directly involved, as we have seen. The brain generates the impulses for muscle contraction and regulates heart rate, breathing rate, and the diameter of blood vessels. The next time you run up a flight of stairs, hurry to catch a bus, or just go swimming, you might reflect a moment on all of the things that are actually happening to your body . . . after you catch your breath.

MAJOR MUSCLES OF THE BODY

The actions that muscles perform are shown in Fig. 7–5 and are listed in Table 7–2. Most are in pairs as antagonistic functions.

The major muscles are shown in Fig. 7–6. They are listed, according to body area, in Tables 7–3 through 7–6, with associated Figs. 7–7 through 7–10, respectively. Learning the muscles and their functions does involve memorization, but the bones you have already learned will help you. For each muscle, note its origin and insertion. If you know the bones to which a muscle is attached, you can determine the joint the muscle affects when it contracts.

The name of the muscle may also be helpful, and again, many of the terms are ones you have already

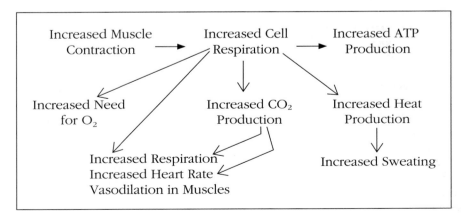

learned. Some examples: "abdominus" refers to an abdominal muscle, "femoris" to a thigh muscle, "brachii" to a muscle of the upper arm, "oculi" to an eye muscle, and so on.

Muscles that are sites for intramuscular injections are shown in Box 7–5.

Box 7–3   MUSCULAR DYSTROPHY

Muscular dystrophy is really a group of genetic diseases in which muscle tissue is replaced by fibrous connective tissue or by fat. Neither of these tissues is capable of contraction, and the result is progressive loss of muscle function. The most common form is Duchenne's muscular dystrophy, in which the loss of muscle function affects not only skeletal muscle but also cardiac muscle. Death usually occurs before the age of 20 due to heart failure, and at present there is no cure.

Duchenne's muscular dystrophy is a **sex-linked** (or x-linked) **trait,** which means that the gene for it is on the X chromosome and is recessive. The female sex chromosomes are XX. If one X chromosome has a gene for muscular dystrophy, and the other X chromosome has a dominant gene for normal muscle function, the woman will not have muscular dystrophy but will be a carrier who may pass the muscular dystrophy gene to her children. The male sex chromosomes are XY, and the Y has no gene at all for muscle function, that is, no gene at all to prevent the expression of the gene on the X chromosome. If the X chromosome has a gene for muscular dystrophy, the male will have the disease. This is why Duchenne's muscular dystrophy is more common in males; the presence of only one gene means the disease will be present.

The muscular dystrophy gene on the X chromosome has recently been located, and the protein the gene codes for has been named dystrophin. Still unknown, however, is the role of this protein in muscular dystrophy. When the precise function of dystrophin is determined, this may provide a basis or starting point for therapy.

Box 7–4   MYASTHENIA GRAVIS

Myasthenia gravis is an **autoimmune** disorder characterized by extreme muscle fatigue even after minimal exertion. Women are affected more often than are men, and symptoms usually begin in middle age. Weakness may first be noticed in the facial or swallowing muscles and may progress to other muscles. Without treatment, the respiratory muscles will eventually be affected, and respiratory failure is the cause of death.

In myasthenia gravis, the autoantibodies (self-antibodies) destroy the **acetylcholine receptors** on the sarcolemma. These receptors are the sites to which acetylcholine bonds and stimulates the entry of Na^+ ions. Without these receptors, the acetylcholine released by the axon terminal cannot cause depolarization of a muscle fiber.

Treatment of myasthenia gravis may involve anticholinesterase medications. Recall that cholinesterase is present in the sarcolemma to inactivate acetylcholine and prevent continuous, unwanted impulses. If this action of cholinesterase is inhibited, acetylcholine remains on the sarcolemma for a longer time and may bond to any remaining receptors to stimulate depolarization and contraction.

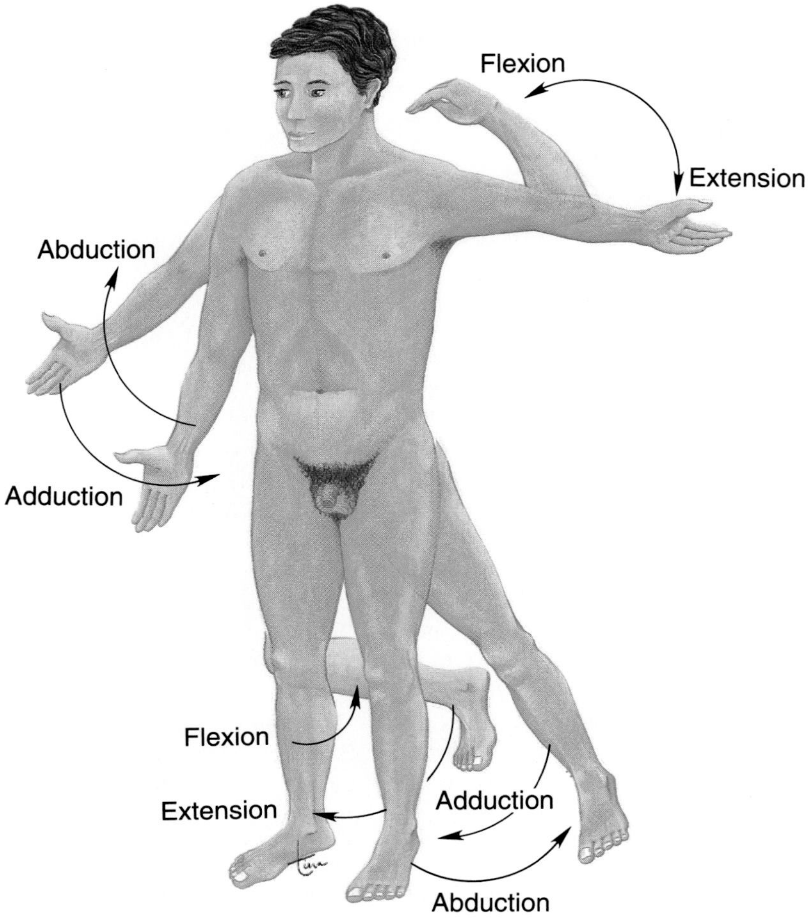

Figure 7–5 Actions of muscles.

Table 7–2 ACTIONS OF MUSCLES

Action	Definition
Flexion	• To decrease the angle of a joint
Extension	• To increase the angle of a joint
Adduction	• To move closer to the midline
Abduction	• To move away from the midline
Pronation	• To turn the palm down
Supination	• To turn the palm up
Dorsiflexion	• To elevate the foot
Plantar flexion	• To lower the foot (point the toes)
Rotation	• To move a bone around its longitudinal axis

Most are grouped in pairs of antagonistic functions.

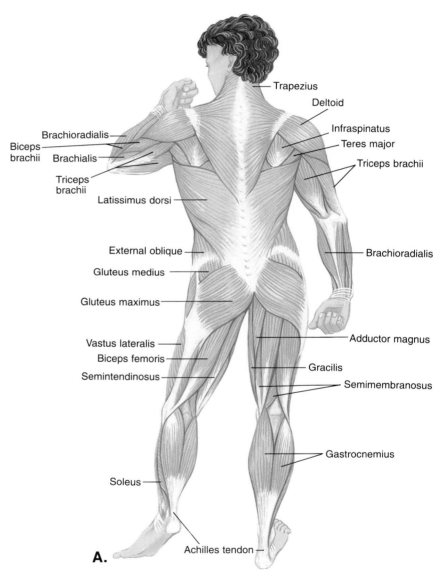

Trapezius

Deltoid

Infraspinatus

Teres major

Triceps brachii

Brachioradialis

Adductor magnus

Gracilis

Semimembranosus

Gastrocnemius

Brachioradialis

Biceps brachii

Brachialis

Triceps brachii

Latissimus dorsi

External oblique

Gluteus medius

Gluteus maximus

Vastus lateralis

Biceps femoris

Semintendinosus

Soleus

Achilles tendon

A.

Figure 7–6 Major muscles of the body. (**A**), Posterior view.

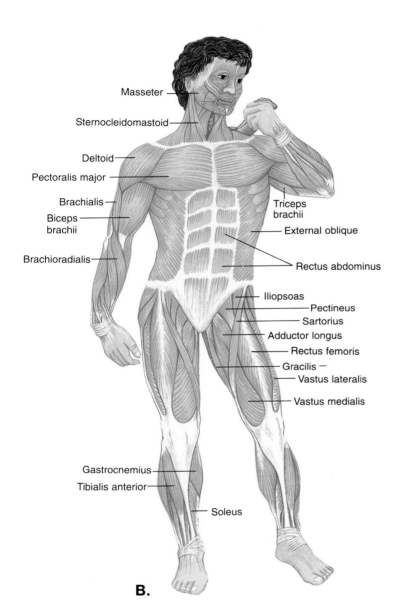

Masseter

Sternocleidomastoid

Deltoid

Pectoralis major

Brachialis

Biceps brachii

Brachioradialis

Triceps brachii

External oblique

Rectus abdominus

Iliopsoas

Pectineus

Sartorius

Adductor longus

Rectus femoris

Gracilis

Vastus lateralis

Vastus medialis

Gastrocnemius

Tibialis anterior

Soleus

B.

Figure 7–6 Continued. (**B**), Anterior view.

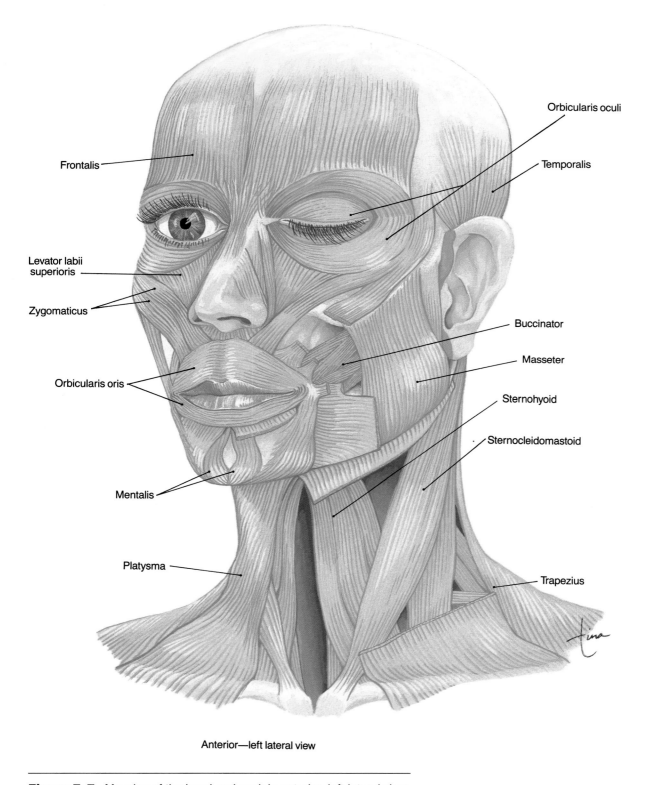

Frontalis

Orbicularis oculi

Temporalis

Levator labii
superioris

Zygomaticus

Buccinator

Masseter

Orbicularis oris

Sternohyoid

Sternocleidomastoid

Mentalis

Platysma

Trapezius

Anterior—left lateral view

Figure 7–7 Muscles of the head and neck in anterior, left-lateral view.

Table 7–3 MUSCLES OF THE HEAD AND NECK

Muscle	Function	Origin	Insertion
Orbicularis oculi	Closes eye	• medial side of orbit	• encircles eye
Orbicularis oris	Puckers lips	• encircles mouth	• skin at corners of mouth
Masseter	Closes jaw	• maxilla and zygomatic	• mandible
Buccinator	Pulls corners of mouth laterally	• maxillae and mandible	• orbicularis oris
Sternocleidomastoid	Turns head to opposite side (both—flex head and neck)	• sternum and clavicle	• temporal bone (mastoid process)
Semispinalis capitis (a deep muscle)	Turns head to same side (both—extend head and neck)	• 7th cervical and first 6 thoracic vertebrae	• occipital bone

Table 7–4 MUSCLES OF THE TRUNK

Muscle	Function	Origin	Insertion
Trapezius	Raises, lowers, and adducts shoulders	• occipital bone and all thoracic vertebrae	• spine of scapula and clavicle
External intercostals	Pull ribs up and out (inhalation)	• superior rib	• inferior rib
Internal intercostals	Pull ribs down and in (forced exhalation)	• inferior rib	• superior rib
Diaphragm	Flattens (down) to enlarge chest cavity for inhalation	• last 6 costal cartilages and lumbar vertebrae	• central tendon
Rectus abdominus	Flexes vertebral column, compresses abdomen	• pubic bones	• 5th–7th costal cartilages and xiphoid process
External oblique	Rotates and flexes vertebral column, compresses abdomen	• lower 8 ribs	• iliac crest and linea alba
Sacrospinalis group (a deep group of muscles)	Extends vertebral column	• ilium, lumbar, and some thoracic vertebrae	• ribs, cervical, and thoracic vertebrae

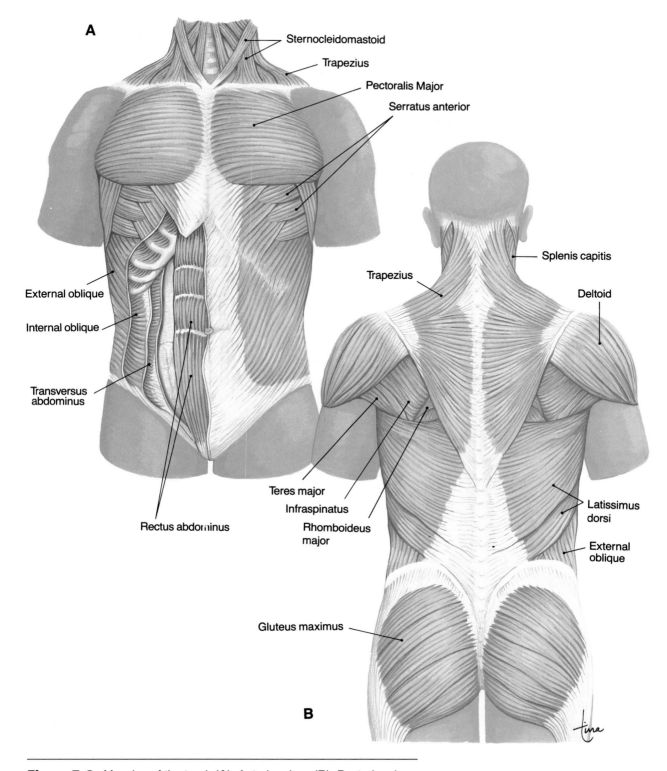

Figure 7–8 Muscles of the trunk (**A**), Anterior view. (**B**), Posterior view.

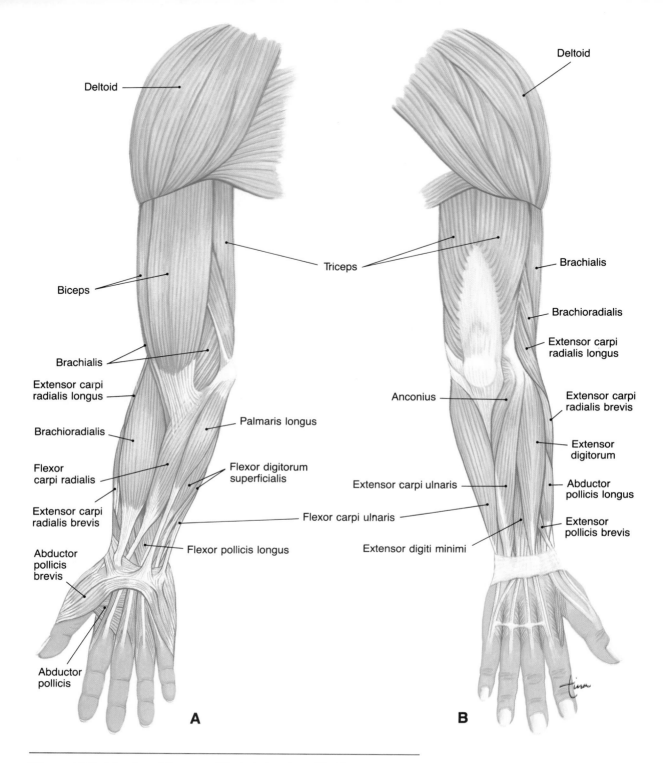

Figure 7–9 Muscles of the arm. (**A**), Anterior view. (**B**), Posterior view.

Table 7–5   MUSCLES OF THE SHOULDER AND ARM

Muscle	Function	Origin	Insertion
Deltoid	Abducts the humerus	• scapula and clavicle	• humerus
Pectoralis major	Flexes and adducts the humerus	• clavicle, sternum, 2nd–6th costal cartilages	• humerus
Latissimus dorsi	Extends and adducts the humerus	• last 6 thoracic vertebrae, all lumbar vertebrae, sacrum, iliac crest	• humerus
Teres major	Extends and adducts the humerus	• scapula	• humerus
Triceps brachii	Extends the forearm	• humerus and scapula	• ulna
Biceps brachii	Flexes the forearm	• scapula	• radius
Brachioradialis	Flexes the forearm	• humerus	• radius

Table 7–6   MUSCLES OF THE HIP AND LEG

Muscle	Function	Origin	Insertion
Iliopsoas	Flexes femur	• ilium, lumbar vertebrae	• femur
Gluteus maximus	Extends femur	• iliac crest, sacrum, coccyx	• femur
Gluteus medius	Abducts femur	• ilium	• femur
Quadriceps femoris group: Rectus femoris Vastus lateralis Vastus medialis Vastus intermedius	Flexes femur and extends lower leg	• ilium and femur	• tibia
Hamstring group: Biceps femoris Semimembranosus Semitendinosus	Extends femur and flexes lower leg	• ischium	• tibia and fibula
Adductor group	Adducts femur	• ischium and pubis	• femur
Sartorius	Flexes femur and lower leg	• ilium	• tibia
Gastrocnemius	Plantar flexes foot	• femur	• calcaneus (Achilles tendon)
Soleus	Plantar flexes foot	• tibia and fibula	• calcaneus (Achilles tendon)
Tibialis anterior	Dorsiflexes foot	• tibia	• metatarsals

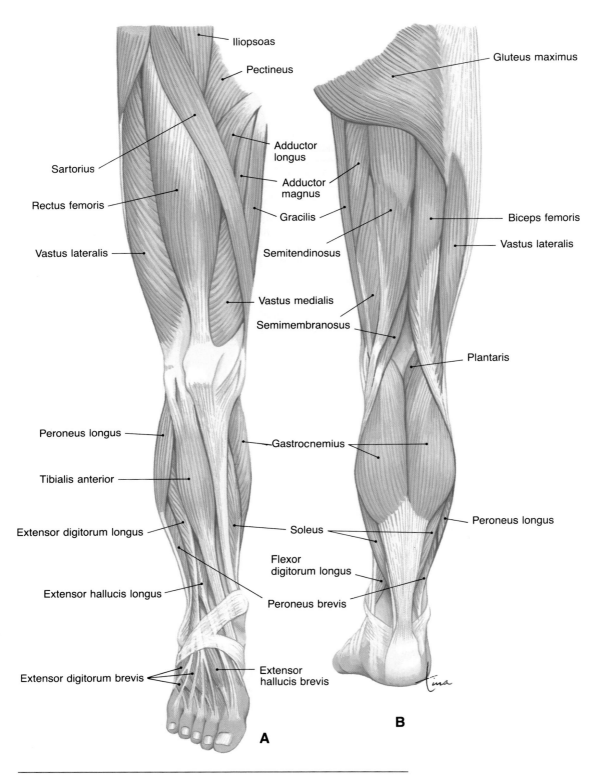

Figure 7-10 Muscles of the leg. (**A**), Anterior view. (**B**), Posterior view.

152

Box 7–5   COMMON INJECTION SITES

Intramuscular injections are used when rapid absorption is needed, because muscle has a good blood supply. Common sites are the buttock (*gluteus medius*), lateral thigh (*vastus lateralis*), and the shoulder (*deltoid*). These sites are shown below; also shown are the large nerves to be avoided when giving such injections.

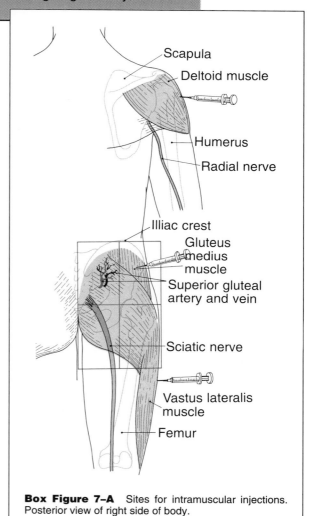

Scapula
Deltoid muscle
Humerus
Radial nerve
Illiac crest
Gluteus medius muscle
Superior gluteal artery and vein
Sciatic nerve
Vastus lateralis muscle
Femur

Box Figure 7–A Sites for intramuscular injections. Posterior view of right side of body.

STUDY OUTLINE

Organ Systems Involved in Movement
1. Muscular—moves the bones.
2. Skeletal—bones are moved, at their joints, by muscles.
3. Nervous—transmits impulses to muscles to cause contraction.
4. Respiratory—exchanges O_2 and CO_2 between the air and blood.
5. Circulatory—transports O_2 to muscles and removes CO_2.

Muscle Structure
1. Muscle fibers (cells) are specialized to contract, shorten, and produce movement.
2. A skeletal muscle is made of thousands of muscle fibers. Varying movements require contraction of variable numbers of muscle fibers in a muscle.
3. Tendons attach muscles to bone; the origin is the more stationary bone, the insertion is the more movable bone. A tendon merges with the fascia of a muscle and the periosteum of a bone; all are made of fibrous connective tissue.

Muscle Arrangements
1. Antagonistic muscles have opposite functions. A muscle pulls when it contracts, but exerts no force when it relaxes and cannot push. When one muscle pulls a bone in one direction, another muscle is needed to pull the bone in the other direction (see also Table 7–2).
2. Synergistic muscles have the same function and alternate as the prime mover depending on the position of the bone to be moved. Synergists also stabilize a joint to make a more precise movement possible.
3. The frontal lobes of the cerebrum generate the impulses necessary for contraction of skeletal muscles. The cerebellum regulates coordination.

Muscle Tone—the state of slight contraction present in muscles
1. Alternate fibers contract to prevent muscle fatigue.

2. Good tone helps maintain posture, produces 25% of body heat (at rest), and improves coordination.
3. Isotonic exercise involves contraction with movement; improves tone and strength and improves cardiovascular and respiratory efficiency (aerobic exercise).
4. Isometric exercise involves contraction without movement; improves tone and strength but is not aerobic.

Muscle Sense—knowing where our muscles are without looking at them
1. Permits us to perform everyday activities without having to concentrate on muscle position.
2. Stretch receptors (proprioceptors) in muscles respond to stretching and generate impulses that the brain interprets as a mental "picture" of where the muscles are. Parietal lobes: conscious muscle sense; cerebellum: unconscious muscle sense used to promote coordination.

Energy Sources for Muscle Contraction
1. ATP is the direct source; the ATP stored in muscles lasts only a few seconds.
2. Creatine phosphate is a secondary energy source; is broken down to creatine + phosphate + energy. The energy is used to synthesize more ATP. Some creatine is converted to creatinine, which must be excreted by the kidneys. Most creatine is used for the resynthesis of creatine phosphate.
3. Glycogen is the most abundant energy source and is first broken down to glucose. Glucose is broken down in cell respiration:

$$Glucose + O_2 \rightarrow CO_2 + H_2O + ATP + heat$$

ATP is used for contraction; heat contributes to body temperature; H_2O becomes part of intracellular fluid; CO_2 is eventually exhaled.
4. Oxygen is essential for the completion of cell respiration. Hemoglobin in RBCs carries oxygen to muscles; myoglobin stores oxygen in muscles;

both these proteins contain iron, which enables them to bond to oxygen.

5. Oxygen debt: muscle fibers run out of oxygen during strenuous exercise, and glucose is converted to lactic acid, which causes fatigue. Breathing rate remains high after exercise to deliver more oxygen to the liver, which converts lactic acid back to glucose (ATP required).

Muscle Fiber—microscopic structure

1. Neuromuscular junction: axon terminal and sarcolemma; the synapse is the space between. The axon terminal contains acetylcholine (a neurotransmitter), and the sarcolemma contains cholinesterase (an inactivator).
2. Sarcomeres are the contracting units of a muscle fiber. Myosin and actin filaments are the contracting proteins of sarcomeres. Troponin and tropomyosin are proteins that inhibit the sliding of myosin and actin when the muscle fiber is relaxed.
3. The sarcoplasmic reticulum surrounds the sarcomeres and is a reservoir for calcium ions.
4. Polarization (resting potential): when the muscle fiber is relaxed, the sarcolemma has a ($+$) charge outside and a ($-$) charge inside. Na^+ ions are more abundant outside the cell and K^+ ions are more abundant inside the cell. The Na^+ and K^+ pumps maintain these relative concentrations on either side of the sarcolemma (see Table 7–1).
5. Depolarization: started by a nerve impulse. Acetylcholine released by the axon terminal makes the sarcolemma very permeable to Na^+ ions, which enter the cell and cause a reversal of charges to ($-$) outside and ($+$) inside. The depolarization spreads along the entire sarcolemma and initiates the contraction process.

Mechanism of Contraction— sliding filament theory

1. Depolarization stimulates a sequence of events that enables myosin filaments to pull the actin filaments to the center of the sarcomere, which shortens.
2. All the sarcomeres in a muscle fiber contract in response to a nerve impulse; the entire cell contracts.
3. Tetanus—a sustained contraction brought about by continuous nerve impulses; all our movements involve tetanus.
4. Paralysis: muscles that do not receive nerve impulses are unable to contract and will atrophy. Paralysis may be the result of nerve damage, spinal cord damage, or brain damage.

Responses to Exercise— maintaining homeostasis

See section in chapter.

Major Muscles

See Tables 7–2 through 7–6 and Figs. 7–5 through 7–10.

REVIEW QUESTIONS

1. Name the organ systems directly involved in movement and for each state how they are involved. (p. 132)

2. State the function of tendons. Name the part of a muscle and a bone to which a tendon is attached. (p. 132)

3. State the term for: (pp. 132, 134)
   a. muscles with the same function
   b. muscles with opposite functions
   c. the muscle that does most of the work in a movement

4. Explain why antagonistic muscle arrangements are necessary. Give two examples. (pp. 132, 134)

5. State three reasons why good muscle tone is important. (p. 134)

6. Explain why muscle sense is important. Name the receptors involved and state what they detect. (pp. 135–136)

7. With respect to muscle contraction, state the role of the cerebellum and the frontal lobes of the cerebrum. (p. 134)

8. Name the direct energy source for muscle contraction. Name the two secondary energy sources. Which of these is more abundant? (p. 136)

9. State the simple equation of cell respiration and what happens to each of the products of this reaction. (p. 136)

10. Name the two sources of oxygen for muscle fibers. State what the two proteins have in common. (p. 136)

11. Explain what is meant by oxygen debt. What is needed to correct oxygen debt, and where does it come from? (p. 136)

12. Name these parts of the neuromuscular junction: (p. 139)
    a. the membrane of the muscle fiber
    b. the end of the motor neuron
    c. the space between neuron and muscle cell.
    State the locations of acetylcholine and cholinesterase.

13. Name the contracting proteins of sarcomeres, and describe their locations in a sarcomere. Where is the sarcoplasmic reticulum and what does it contain? (p. 139)

14. In terms of ions and charges, describe: (p. 140)
    a. polarization
    b. depolarization
    c. repolarization

15. With respect to the Sliding Filament Theory, explain the function of: (p. 140–142)
    a. acetylcholine
    b. calcium ions
    c. myosin and actin
    d. troponin and tropomyosin
    e. cholinesterase

16. State three of the body's physiological responses to exercise, and explain how each helps maintain homeostasis. (p. 142–143)

17. Find the major muscles on yourself, and state a function of each muscle.

Chapter 8
The Nervous System

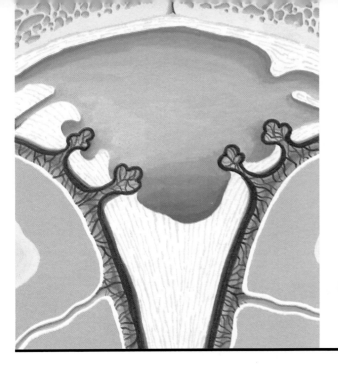

Chapter 8

Box 8–1	MULTIPLE SCLEROSIS
Box 8–2	SHINGLES
Box 8–3	SPINAL CORD INJURIES
Box 8–4	CEREBROVASCULAR ACCIDENTS
Box 8–5	APHASIA
Box 8–6	ALZHEIMER'S DISEASE
Box 8–7	PARKINSON'S DISEASE
Box 8–8	LUMBAR PUNCTURE

Student Objectives

- Name the divisions of the nervous system and the parts of each, and state the general functions of the nervous system.
- Name the parts of a neuron and state the function of each.
- Explain the importance of Schwann cells in the peripheral nervous system and neuroglia in the central nervous system.
- Describe the electrical nerve impulse, and describe impulse transmission at synapses.
- Describe the types of neurons, nerves, and nerve tracts.
- State the names and numbers of the spinal nerves, and their destinations.

The Nervous System

Terms that appear in **bold type** in the chapter text are defined in the glossary, which begins on page 549.

Student Objectives (Continued)

- Explain the importance of stretch reflexes and flexor reflexes.
- State the functions of the parts of the brain; be able to locate each part on a diagram.
- Name the meninges and describe their locations.
- State the locations and functions of cerebrospinal fluid.
- Name the cranial nerves and state their functions.
- Explain how the sympathetic division of the autonomic nervous system enables the body to adapt to a stress situation.
- Explain how the parasympathetic division of the autonomic nervous system promotes normal body functioning in relaxed situations.

New Terminology

Afferent (**AFF**–uh–rent)
Autonomic nervous system (AW–toh–**NOM**–ik)
Cauda equina (**KAW**–dah ee–**KWHY**–nah)
Cerebral cortex (se–**REE**–bruhl **KOR**–tex)
Cerebrospinal fluid (se–**REE**–broh–**SPY**–nuhl)
Choroid plexus (**KOR**–oid **PLEK**–sus)
Corpus callosum (**KOR**–pus kuh–**LOH**–sum)
Cranial nerves (**KRAY**–nee–uhl NERVS)

Efferent (**EFF**–uh–rent)
Gray matter (**GRAY MAH**–TUR)
Neuroglia (new–**ROG**–lee–ah)
Neurolemma (NYOO–ro–**LEM**–ah)
Parasympathetic (PAR–uh–SIM–puh–**THET**–ik)
Reflex (**REE**–flex)
Somatic (sew–**MA**–tik)
Spinal nerves (**SPY**–nuhl NERVS)
Sympathetic (SIM–puh–**THET**–ik)
Ventricles of brain (**VEN**–trick'ls)
Visceral (**VISS**–er–uhl)
White matter (**WIGHT MAH**–TUR)

Related Clinical Terminology

Alzheimer's disease (**ALZ**–high–mer's)
Aphasia (ah–**FAY**–zee–ah)
Blood-brain barrier (BLUHD BRAYNE)
Cerebrovascular accident (CVA) (se–**REE**–broh–**VAS**–kyoo–lur)
Lumbar puncture (**LUM**–bar **PUNK**–chur)
Meningitis (MEN–in–**JIGH**–tis)
Multiple sclerosis (MS) (**MULL**–ti–puhl skle–**ROH**–sis)
Neuralgia (new–**RAL**–jee–ah)
Neuritis (new–**RYE**–tis)
Neuropathy (new–**RAH**–puh–thee)
Parkinson's disease (**PAR**–kin–son's)
Remission (ree–**MISH**–uhn)
Spinal shock (**SPY**–nuhl SHAHK)

Terms that appear in **bold type** in the chapter text are defined in the glossary, which begins on page 549.

Most of us can probably remember being told, when we were children, not to touch the stove or some other source of potential harm. Since children are curious, such warnings often go unheeded. The result? Touching a hot stove brings about an immediate response of pulling away and a vivid memory of painful fingers. This simple and familiar experience illustrates the functions of the **nervous system:**

1. To detect changes and feel sensations
2. To initiate appropriate responses to changes
3. To organize information for immediate use and store it for future use.

The nervous system is one of the regulating systems (the endocrine system is the other and will be discussed in Chapter 10). Electrochemical impulses of the nervous system make it possible to obtain information about the external or internal environment and do whatever is necessary to maintain homeostasis. Some of this activity is conscious, but much of it happens without our awareness.

NERVOUS SYSTEM DIVISIONS

The nervous system has two divisions. The **central nervous system (CNS)** consists of the brain and spinal cord. The **peripheral nervous system (PNS)** consists of cranial nerves and spinal nerves. The PNS includes the autonomic nervous system (ANS).

The peripheral nervous system relays information to and from the central nervous system, and the brain is the center of activity that integrates this information, initiates responses, and makes us the individuals we are.

NERVE TISSUE

Nerve tissue was briefly described in Chapter 4, so we will begin by reviewing what you already know, then adding to it.

Nerve cells are called **neurons,** or **nerve fibers.** Whatever their specific functions, all neurons have the same physical parts. The **cell body** contains the

nucleus (Fig. 8–1) and is essential for the continued life of the neuron. As you will see, neuron cell bodies are found in the central nervous system or close to it in the trunk of the body. In these locations, cell bodies are protected by bone. There are no cell bodies in the arms and legs, which are much more subject to injury.

Dendrites are processes (extensions) that transmit impulses toward the cell body. The one **axon** of a neuron transmits impulses away from the cell body. It is the cell membrane of the dendrites, cell body, and axon that carries the electrical nerve impulse.

In the peripheral nervous system, axons and dendrites are "wrapped" in specialized cells called **Schwann cells** (see Fig. 8–1). During embryonic development, Schwann cells grow to surround the neuron processes, enclosing them in several layers of Schwann cell membrane. These layers are the **myelin sheath;** myelin is a phospholipid that electrically insulates neurons from one another. Without the myelin sheath, neurons would short-circuit, just as electrical wires would if they were not insulated (see Box 8–1: Multiple Sclerosis).

The spaces between adjacent Schwann cells, or segments of the myelin sheath, are called nodes of Ranvier (neurofibral nodes). These nodes are the parts of the neuron cell membrane that depolarize when an electrical impulse is transmitted (see "The Nerve Impulse" section, on pages 165–166).

The nuclei and cytoplasm of the Schwann cells are outside the myelin sheath and are called the **neurolemma,** which becomes very important if nerves are damaged. If a peripheral nerve is severed and reattached precisely by microsurgery, the axons and dendrites may regenerate through the tunnels formed by the neurolemmas. The Schwann cells are also believed to produce a chemical growth factor that stimulates regeneration. Although this regeneration may take months, the nerves may eventually reestablish their proper connections, and the person may regain some sensation and movement in the once-severed limb.

In the central nervous system, the myelin sheaths are formed by **oligodendrocytes,** one of the **neuroglia,** the specialized cells found only in the brain and spinal cord. Since no Schwann cells are present, however, there is no neurolemma, and regenera-

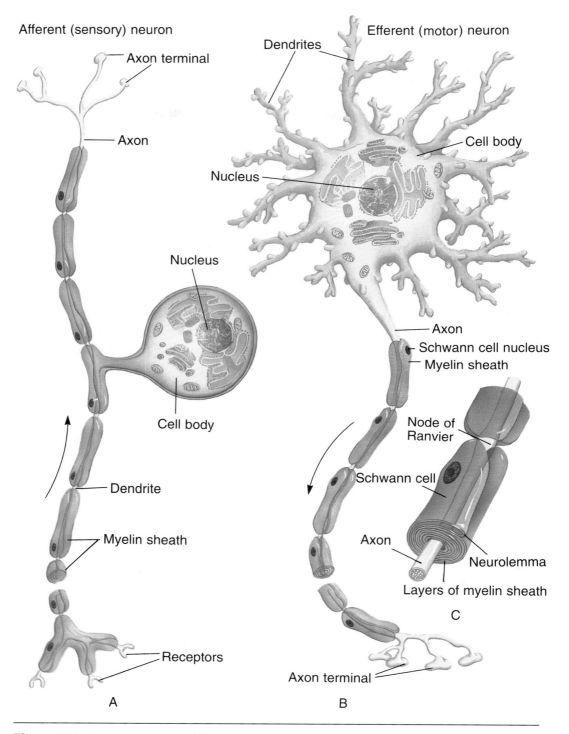

Afferent (sensory) neuron

Axon terminal

Axon

Nucleus

Cell body

Dendrite

Myelin sheath

Receptors

A

Efferent (motor) neuron

Dendrites

Cell body

Nucleus

Axon

Schwann cell nucleus

Myelin sheath

Node of Ranvier

Schwann cell

Axon

Neurolemma

Layers of myelin sheath

C

Axon terminal

B

Figure 8–1 Neuron structure. (**A**), A typical sensory neuron. (**B**), A typical motor neuron. The arrows indicate the direction of impulse transmission. (**C**), Details of the myelin sheath and neurolemma formed by Schwann cells.

Box 8–1 MULTIPLE SCLEROSIS

Multiple sclerosis (MS) is a demyelinating disease, that is, it involves deterioration of the myelin sheath of neurons in the central nervous system. Without the myelin sheath, the impulses of these neurons are short-circuited and do not reach their proper destinations.

There is evidence that multiple sclerosis may be an **autoimmune** disorder that is triggered by a virus. The autoantibodies destroy the oligodendrocytes, the myelin-producing neuroglia of the central nervous system, which results in the formation of scleroses, or plaques of scar tissue, that do not provide electrical insulation. Since loss of myelin may occur in many parts of the central nervous system, the symptoms vary, but they usually include muscle weakness or paralysis, numbness or partial loss of sensation, double vision, and loss of spinal cord reflexes, including those for urination and defecation.

The first symptoms usually appear between the ages of 20 and 40 years, and the disease may progress either slowly or rapidly. Some MS patients have **remissions,** periods of time when their symptoms diminish, but remissions and progression of the disease are not predictable. There is still no cure for MS, but new therapies include suppression of the immune response, and interferon, which seems to prolong remissions in some patients.

---

tion of neurons is not possible. This is why severing of the spinal cord, for example, results in permanent loss of function (see Table 8–1 for other functions of the neuroglia).

SYNAPSES

Neurons that transmit impulses to other neurons do not actually touch one another. The small gap or space between the axon of one neuron and the dendrites or cell body of the next neuron is called the **synapse.** Within the synaptic knob (terminal end) of the axon is a chemical **neurotransmitter** that is released into the synapse by the arrival of an electrical nerve impulse (Fig. 8–2). The neurotransmitter diffuses across the synapse, combines with specific receptor sites on the cell membrane of the next neuron, and there generates an electrical impulse which in turn is carried by this neuron's axon to the next synapse, and so forth. A chemical **inactivator** within the cell body or dendrite of the "receiving" neuron quickly inactivates the neurotransmitter. This prevents unwanted, continuous

Table 8–1 NEUROGLIA

Name	Function
Oligodendrocytes	• Produce the myelin sheath to electrically insulate neurons of the CNS.
Microglia	• Capable of movement and phagocytosis of pathogens and damaged tissue.
Astrocytes	• Contribute to the **blood-brain barrier,** which prevents potentially toxic waste products in the blood from diffusing out into brain tissue. A disadvantage of this barrier, however, is that some useful medications cannot cross it; this becomes important during brain infections, inflammation, or other disease or disorder.
Ependyma	• Line the ventricles of the brain; many of the cells have cilia; involved in circulation of cerebrospinal fluid.

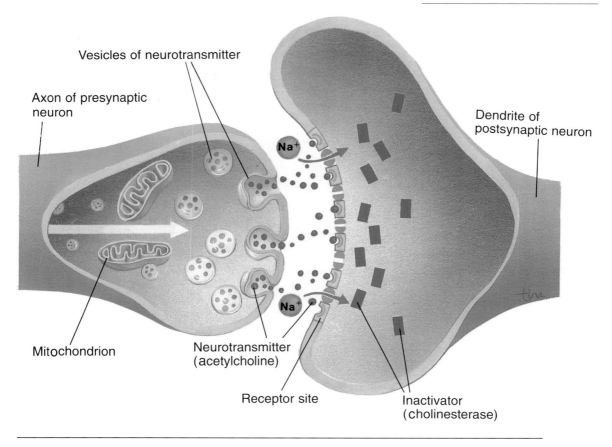

Axon of presynaptic neuron

Vesicles of neurotransmitter

Na⁺

Dendrite of postsynaptic neuron

Mitochondrion

Neurotransmitter (acetylcholine)

Na⁺

Receptor site

Inactivator (cholinesterase)

Figure 8–2   Impulse transmission at a synapse. The arrow indicates the direction of the electrical impulse.

impulses, unless a new impulse from the first neuron releases more neurotransmitter.

One important consequence of the presence of synapses is that they ensure one-way transmission of impulses in a living person. A nerve impulse cannot go backward across a synapse because there is no neurotransmitter released by the dendrites or cell body. Neurotransmitters can only be released by a neuron's axon. Keep this in mind when we discuss the types of neurons, below.

An example of a neurotransmitter is **acetylcholine,** which is found in the CNS, at neuromuscular junctions, and in much of the peripheral nervous system. **Cholinesterase** is the inactivator of acetylcholine. There are many other neurotransmitters, especially in the central nervous system. These include dopamine, norepinephrine, and serotonin. Each of these neurotransmitters has its own chemical inactivator.

TYPES OF NEURONS

Neurons may be classified into three groups: sensory neurons, motor neurons, and interneurons (Fig. 8–3). **Sensory neurons** (or **afferent neurons**) carry impulses from receptors to the central nervous system. **Receptors** detect external or internal changes and send the information to the CNS

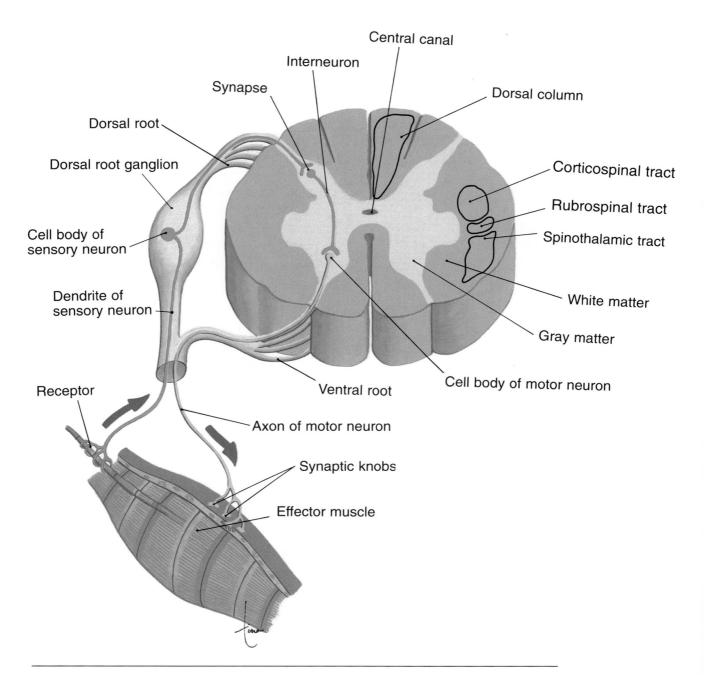

Figure 8–3 Cross section of the spinal cord. Spinal nerve roots and their neurons are shown on the left side. Spinal nerve tracts are shown in the white matter on the right side. All tracts and nerves are bilateral (both sides).

in the form of impulses by way of the afferent neurons. The central nervous system interprets these impulses as a sensation. Sensory neurons from receptors in skin, skeletal muscles, and joints are called **somatic;** those from receptors in internal organs are called **visceral** sensory neurons.

Motor neurons (or **efferent neurons**) carry impulses from the central nervous system to **effectors.** The two types of effectors are muscles and glands. In response to impulses, muscles contract and glands secrete. Motor neurons linked to skeletal muscle are called somatic; those to smooth muscle, cardiac muscle, and glands are called visceral.

Sensory and motor neurons make up the peripheral nervous system. Visceral motor neurons comprise the autonomic nervous system, a specialized subdivision of the PNS that will be discussed later in this chapter.

Interneurons are found entirely within the central nervous system. They are arranged so as to carry only sensory or motor impulses, or to integrate these functions. Some interneurons in the brain are concerned with thinking, learning, and memory.

A neuron carries impulses in only one direction. This is the result of the neuron's structure and location, as well as its physical arrangement with other neurons and the resulting pattern of synapses. The functioning nervous system, therefore, is an enormous network of "one-way streets," and there is no danger of impulses running into and canceling one another out.

NERVES AND NERVE TRACTS

A **nerve** is a group of axons and/or dendrites of many neurons, with blood vessels and connective tissue. **Sensory nerves** are made only of sensory neurons. The optic nerves for vision are examples of nerves with a purely sensory function. **Motor nerves** are made only of motor neurons; autonomic nerves are motor nerves. A **mixed nerve** contains both sensory and motor neurons. Most of our peripheral nerves, such as the sciatic nerves in the legs, are mixed nerves.

The term **nerve tract** refers to groups of neurons within the central nervous system. All the neurons in a nerve tract are concerned with either sensory or motor activity. These tracts are often referred to as white matter; the myelin sheaths of the neurons give them a white color.

THE NERVE IMPULSE

The events of an electrical nerve impulse are the same as those of the electrical impulse generated in muscle fibers which is discussed in Chapter 7. Stated simply, a neuron not carrying an impulse is in a state of **polarization,** with Na^+ ions more abundant outside the cell, and K^+ ions and negative ions more abundant inside the cell. The neuron has a positive charge on the outside of the cell membrane and a relative negative charge inside. A stimulus (such as a neurotransmitter) makes the membrane very permeable to Na^+ ions, which rush into the cell. This brings about **depolarization,** a reversal of charges on the membrane. The outside now has a negative charge, and the inside has a positive charge.

As soon as depolarization takes place, the neuron membrane becomes very permeable to K^+ ions, which rush out of the cell. This restores the positive charge outside and the negative charge inside, and is called **repolarization.** (The term action potential refers to depolarization followed by repolarization.) Then the sodium and potassium pumps return Na^+ ions outside and K^+ ions inside, and the neuron is ready to respond to another stimulus and transmit another impulse. An action potential in response to a stimulus takes place very rapidly and is measured in milliseconds. An individual neuron is capable of transmitting hundreds of action potentials (impulses) each second. A summary of the events of nerve impulse transmission is given in Table 8–2.

Transmission of electrical impulses is very rapid. The presence of an insulating myelin sheath increases the velocity of impulses, since only the nodes of Ranvier depolarize. This is called **saltatory conduction.** Many of our neurons are capable of transmitting impulses at a speed of many meters per second. Imagine a person 6 feet (about 2 meters) tall who stubs his toe; sensory impulses travel from the toe to the brain in less than a second

Table 8–2   THE NERVE IMPULSE

State or Event	Description
Polarization (the neuron is not carrying an electrical impulse)	• Neuron membrane has a (+) charge outside and a (−) charge inside. • Na^+ ions are more abundant outside the cell. • K^+ ions and negative ions are more abundant inside the cell. Sodium and potassium pumps maintain these ion concentrations.
Depolarization (generated by a stimulus)	• Neuron membrane becomes very permeable to Na^+ ions, which rush into the cell. • The neuron membrane then has a (−) charge outside and a (+) charge inside.
Propagation of the impulse from point of stimulus	• Depolarization of part of the membrane makes adjacent membrane very permeable to Na^+ ions, and subsequent depolarization, which similarly affects the next part of the membrane, and so on. • The depolarization continues along the membrane of the neuron to the end of the axon.
Repolarization (immediately follows depolarization)	• Neuron membrane becomes very permeable to K^+ ions, which rush out of the cell. This restores the (+) charge outside and (−) charge inside the membrane. • The Na^+ ions are returned outside and the K^+ ions are returned inside by the sodium and potassium pumps. • The neuron is now able to respond to another stimulus and generate another impulse.

(crossing a few synapses along the way). You can see how the nervous system can communicate so rapidly with all parts of the body, and why it is such an important regulatory system.

At synapses, nerve impulse transmission changes from electrical to chemical and depends on the release of neurotransmitters. Although diffusion across synapses is slow, the synapses are so small that this does not significantly affect the velocity of impulses in a living person.

THE SPINAL CORD

The **spinal cord** transmits impulses to and from the brain and is the integrating center for the spinal cord reflexes. Although this statement of functions is very brief, the spinal cord is of great importance to the nervous system and to the body as a whole.

Enclosed in the vertebral canal, the spinal cord is well protected from mechanical injury. In length, the spinal cord extends from the foramen magnum of the occipital bone to the disc between the first and second lumbar vertebrae.

A cross-section of the spinal cord is shown in Fig. 8–3; refer to it as you read the following. The internal **gray matter** is shaped like the letter H; gray matter consists of the cell bodies of motor neurons and interneurons. The external **white matter** is made of myelinated axons and dendrites of interneurons. These nerve fibers are grouped into nerve tracts based on their functions. **Ascending tracts** (such as the dorsal columns and spinothalamic tracts) carry sensory impulses to the brain. **Descending tracts** (such as the corticospinal and rubrospinal tracts) carry motor impulses away from the brain. Lastly, find the **central canal;** this contains **cerebrospinal fluid** and is continuous with cavities in the brain called ventricles.

SPINAL NERVES

There are 31 pairs of **spinal nerves,** those that emerge from the spinal cord. The nerves are named according to their respective vertebrae: 8 cervical pairs, 12 thoracic pairs, 5 lumbar pairs, 5 sacral pairs, and 1 very small coccygeal pair. These are shown in Fig. 8–4; notice that each nerve is designated by a letter and a number. The 8th cervical nerve is C8, the 1st thoracic nerve is T1, and so on.

In general, the cervical nerves supply the back of

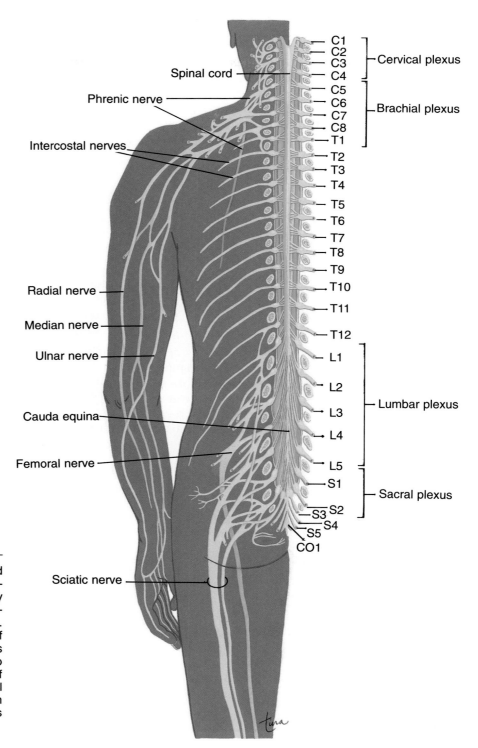

Figure 8–4 The spinal cord and spinal nerves. The distribution of spinal nerves is shown only on the left side. The nerve plexuses are labeled on the right side. A nerve plexus is a network of neurons from several segments of the spinal cord that combine to form nerves to specific parts of the body. For example, the radial and ulnar nerves to the arm emerge from the brachial plexus (see also Table 8–3).

the head, neck, shoulders, arms, and the diaphragm. The first thoracic nerve also contributes to nerves in the arms. The remaining thoracic nerves supply the trunk of the body. The lumbar and sacral nerves supply the hips, pelvic cavity, and legs. Notice that the lumbar and sacral nerves hang below the end of the spinal cord (in order to reach their proper openings to exit from the vertebral canal); this is called the **cauda equina,** literally, the "horse's tail." Some of the important peripheral nerves and their destinations are listed in Table 8–3.

Each spinal nerve has two roots, which are neurons entering or leaving the spinal cord (see Fig. 8–3). The **dorsal root** is made of sensory neurons that carry impulses into the spinal cord. The **dorsal root ganglion** is an enlarged part of the dorsal root that contains the cell bodies of the sensory neurons. The term **ganglion** means a group of cell bodies outside the CNS. These cell bodies are within the vertebral canal and are thereby protected from injury (see Box 8–2: Shingles).

The **ventral root** is the motor root; it is made of motor neurons carrying impulses from the spinal cord to muscles or glands. The cell bodies of these motor neurons, as mentioned above, are in the gray matter of the spinal cord. When the two nerve roots merge, the spinal nerve thus formed is a mixed nerve.

SPINAL CORD REFLEXES

When you hear the term "reflex," you may think of an action that "just happens," and in part this is so. A **reflex** is an involuntary response to a stimulus, that is, an automatic action stimulated by a specific change of some kind. **Spinal cord reflexes** are those that do not depend directly on the brain, although the brain may inhibit or enhance them. We do not have to think about these reflexes, which is very important, as you will see.

Reflex Arc

A **reflex arc** is the pathway nerve impulses travel when a reflex is elicited, and there are five essential parts:

1. **Receptors**—detect a change (the stimulus) and generate impulses.
2. **Sensory neurons**—transmit impulses from receptors to the CNS.
3. **Central nervous system**—contains one or more synapses (interneurons may be part of the pathway).
4. **Motor neurons**—transmit impulses from the CNS to the effector.
5. **Effector**—performs its characteristic action.

Let us now look at the reflex arc of a specific reflex, the **patellar** (or kneejerk) **reflex,** with

Table 8–3 MAJOR PERIPHERAL NERVES

Nerve	Spinal Nerves That Contribute	Distribution
Phrenic	C3–C5	• Diaphragm
Radial	C5–C8, T1	• Skin and muscles of posterior arm, forearm, and hand; thumb and first 2 fingers
Median	C5–C8, T1	• Skin and muscles of anterior arm, forearm, and hand
Ulnar	C8, T1	• Skin and muscles of medial arm, forearm, and hand; little finger and ring finger
Intercostal	T2–T12	• Intercostal muscles, abdominal muscles; skin of trunk
Femoral	L2–L4	• Skin and muscles of anterior thigh, medial leg, and foot
Sciatic	L4–S3	• Skin and muscles of posterior thigh, leg, and foot

Box 8–2   SHINGLES

Shingles is caused by the same virus that causes chickenpox: the Herpes varicella-zoster virus. Varicella is chickenpox, which most of us probably had as children. When a person recovers from chickenpox, the virus may survive in a dormant (inactive) state in the dorsal root ganglia of some spinal nerves. For most people, the immune system is able to prevent reactivation of the virus. With increasing age, however, the immune system is not as effective, and the virus may become active and cause zoster, or shingles.

The virus is present in sensory neurons, often those of the trunk, but the damage caused by the virus is seen in the skin over the affected nerve. The raised, red lesions of shingles are often very painful and follow the course of the nerve on the skin external to it. Occasionally the virus may affect a cranial nerve and cause facial paralysis called Bell's palsy (7th cranial) or extensive facial lesions, or, rarely, blindness. Although it is not a cure, the antiviral medication acyclovir is now being used to lessen the duration of the illness.

which you are probably familiar. In this reflex, a tap on the patellar tendon just below the kneecap causes extension of the lower leg. This is a **stretch reflex,** which means that a muscle that is stretched will automatically contract. Refer now to Fig. 8–5 as you read the following:

In the quadriceps femoris muscle are (1) stretch receptors that detect the stretching produced by striking the patellar tendon. These receptors generate impulses that are carried along (2) sensory neurons in the femoral nerve to (3) the spinal cord. In the spinal cord, the sensory neurons synapse with (4) motor neurons (this is a two–neuron reflex). The motor neurons in the femoral nerve carry impulses back to (5) the quadriceps femoris, the effector, which contracts and extends the lower leg.

The patellar reflex is one of many that are used clinically to determine whether the nervous system is functioning properly. If the patellar reflex were absent in a patient, the problem could be in the thigh muscle, the femoral nerve, or the spinal cord. Further testing would be needed to determine the precise break in the reflex arc. If the reflex is normal, however, that means that all parts of the reflex arc are intact. So the testing of reflexes may be a first step in the clinical assessment of neurological damage.

You may be wondering why we have such reflexes, which are called stretch reflexes. What is their importance in our everyday lives? Imagine a person standing upright—is the body perfectly still? No, it isn't, because gravity exerts a downward pull. However, if the body tilts to the left, the right sides of the leg and trunk are stretched, and these stretched muscles automatically contract and pull the body upright again. This is the purpose of stretch reflexes; they help keep us upright without our having to think about doing so. If the brain had to make a decision every time we swayed a bit, all our concentration would be needed just to remain standing. Since these are spinal cord reflexes, the brain is not directly involved.

Flexor reflexes (or, **withdrawal reflexes**) are another type of spinal cord reflex. The stimulus is something painful and potentially harmful, and the response is to pull away from it. If you inadvertently touch a hot stove, you automatically pull your hand away. Flexor reflexes are three-neuron reflexes, because sensory neurons synapse with interneurons in the spinal cord, which in turn synapse with motor neurons. Again, however, the brain does not have to make a decision to protect the body; the flexor reflex does that automatically (see Box 8–3: Spinal Cord Injuries).

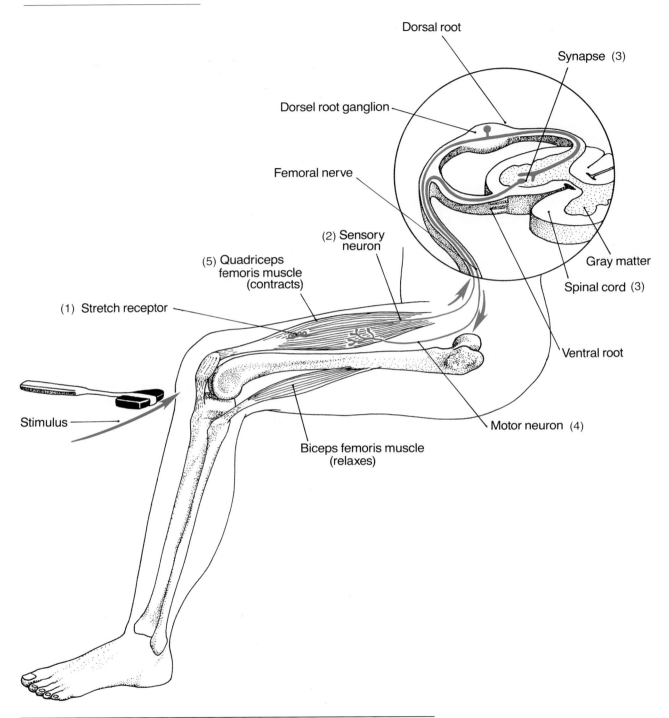

Figure 8–5 Patellar reflex. The reflex arc is shown. See text for description.

Box 8–3 SPINAL CORD INJURIES

Injuries to the spinal cord are most often caused by auto accidents, falls, and gunshot wounds. The most serious injury is transection, or severing, of the spinal cord. If, for example, the spinal cord is severed at the level of the 8th thoracic segment, there will be paralysis and loss of sensation below that level. Another consequence is spinal shock, the at least temporary loss of spinal cord reflexes. In this example, the spinal cord reflexes of the lower trunk and legs will not occur. The stretch reflexes and flexor reflexes of the legs will be at least temporarily abolished, as will the urination and defecation reflexes. Although these reflexes do not depend directly on the brain, spinal cord neurons depend on impulses from the brain to enhance their own ability to generate impulses.

As spinal cord neurons below the injury recover their ability to generate impulses, these reflexes, such as the patellar reflex, often return. Urination and defecation reflexes may also be reestablished, but the person will not have an awareness of the need to urinate or defecate. Nor will voluntary control of these reflexes be possible, since inhibiting impulses from the brain can no longer reach the lower segments of the spinal cord.

Potentially less serious injuries are those in which the spinal cord is crushed rather than severed, and research is providing some promising treatments. Methylprednisolone, a steroid, if given within 8 hours of the injury seems to prevent further damage to the spinal cord by minimizing the inflammation process that occurs in damaged tissue. This helps preserve whatever spinal cord function remains.

A new drug, GM-1 ganglioside, seems to stimulate the production of a nerve growth factor that helps damaged neurons regenerate. Both of these therapies are now undergoing extensive clinical trials in trauma centers and hospitals.

THE BRAIN

The **brain** consists of many parts which function as an integrated whole. The major parts are the medulla, pons, and midbrain (collectively called the **brain stem**); the cerebellum, the hypothalamus and thalamus, and the cerebrum. These parts are shown in Fig. 8–6. We will discuss each part separately, but keep in mind that they are all interconnected and work together.

VENTRICLES

The **ventricles** are four cavities within the brain: two lateral ventricles, the third ventricle, and the fourth ventricle (Fig. 8–7). Each ventricle contains a capillary network called a **choroid plexus,** which forms **cerebrospinal fluid** (CSF) from blood plasma. Cerebrospinal fluid is the tissue fluid of the central nervous system; its circulation and functions will be discussed in the section on meninges.

MEDULLA

The **medulla** extends from the spinal cord to the pons and is anterior to the cerebellum. Its functions are those we think of as vital (as in "vital signs"). The medulla contains cardiac centers that regulate heart rate, vasomotor centers that regulate the diameter of blood vessels and, thereby, blood pressure, and respiratory centers that regulate breathing. You can see why a crushing injury to the occipital bone may be rapidly fatal—we cannot survive without the medulla. Also in the medulla are

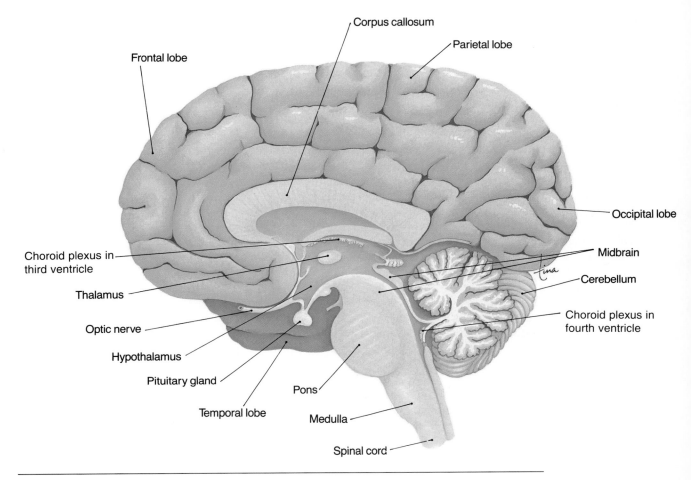

Figure 8–6 Midsagittal section of the brain as seen from the left side. This medial plane shows internal anatomy as well as the lobes of the cerebrum.

reflex centers for coughing, sneezing, swallowing, and vomiting.

PONS

The **pons** bulges anteriorly from the upper part of the medulla. Within the pons are two respiratory centers that work with those in the medulla to produce a normal breathing rhythm. The function of all the respiratory centers will be discussed in Chapter 15.

MIDBRAIN

The **midbrain** extends from the pons to the hypothalamus and encloses the **cerebral aqueduct,** a tunnel that connects the third and fourth ventricles. Several different kinds of reflexes are integrated in the midbrain, including visual and auditory reflexes. If you see a wasp flying toward you, you automatically duck or twist away; this is a visual reflex, as is the coordinated movement of the eyeballs. Turning your head (ear) to a sound is an ex-

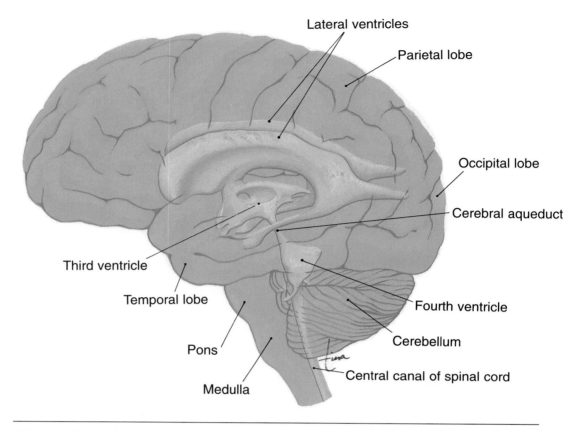

Lateral ventricles

Parietal lobe

Occipital lobe

Cerebral aqueduct

Third ventricle

Temporal lobe

Fourth ventricle

Pons

Cerebellum

Medulla

Central canal of spinal cord

Figure 8–7 Ventricles of the brain as projected into the interior of the brain, which is seen from the left side.

ample of an auditory reflex. The midbrain is also concerned with what are called righting reflexes, those that keep the head upright and maintain balance or equilibrium.

CEREBELLUM

The **cerebellum** is separated from the medulla and pons by the fourth ventricle and is inferior to the occipital lobes of the cerebrum. All the functions of the cerebellum are concerned with movement. These include coordination, regulation of muscle tone, the appropriate trajectory and endpoint of movements, and the maintenance of posture and equilibrium. Notice that these are all involuntary, that is, the cerebellum functions below the level of conscious thought. This is important to permit the conscious brain to work without being overburdened. If you decide to pick up a pencil, for example, the impulses for arm movement come from the cerebrum. The cerebellum then modifies these impulses so that your arm and finger movements are coordinated, and you don't reach past the pencil.

In order to regulate equilibrium, the cerebellum (and midbrain) use information provided by receptors in the inner ears. These receptors will be discussed further in Chapter 9.

HYPOTHALAMUS

Located superior to the pituitary gland and inferior to the thalamus, the **hypothalamus** is a small area of the brain with many diverse functions.

1. Production of **antidiuretic hormone** (ADH) and **oxytocin;** these hormones are then stored in the posterior pituitary gland. ADH enables the kidneys to reabsorb water back to the blood and thus helps maintain blood volume. Oxytocin causes contractions of the uterus to bring about labor and delivery.

2. Production of releasing factors that stimulate the secretion of hormones by the anterior pituitary gland. Since these factors will be covered in Chapter 10, a single example will be given here: the hypothalamus produces **growth hormone releasing hormone** (GHRH), which stimulates the anterior pituitary gland to secrete growth hormone (GH).

3. Regulation of body temperature by promoting responses such as sweating in a warm environment or shivering in a cold environment (see Chapter 17).

4. Regulation of food intake; the hypothalamus is believed to respond to changes in blood nutrient levels. When blood nutrient levels are low, we experience a sensation of hunger, and eat. This raises blood nutrient levels and brings about a sensation of satiety, or fullness, and eating ceases.

5. Integration of the functioning of the autonomic nervous system, which in turn regulates the activity of organs such as the heart, blood vessels, and intestines. This will be discussed in more detail later in this chapter.

6. Stimulation of visceral responses during emotional situations. When we are angry, heart rate usually increases. Most of us, when embarrassed, will blush, which is vasodilation in the skin of the face. These responses are brought about by the autonomic nervous system when the hypothalamus perceives a change in emotional state. The neurological basis of our emotions is not well understood, and the visceral responses to emotions are not something most of us can control.

THALAMUS

The **thalamus** is superior to the hypothalamus and inferior to the cerebrum. The third ventricle is a narrow cavity that passes through both the thalamus and hypothalamus. The functions of the thalamus are concerned with sensation. Sensory impulses to the brain follow neuron pathways that first enter the thalamus, which groups the impulses before relaying them to the cerebrum, where sensations are felt. For example, holding a cup of hot coffee generates impulses for heat, touch and texture, and the shape of the cup (muscle sense), but we do not experience these as separate sensations. The thalamus integrates the impulses, or puts them together, so that the cerebrum feels the whole and is able to interpret the sensation quickly.

The thalamus may also suppress unimportant sensations. If you are reading an enjoyable book, you may not notice someone coming into the room. By temporarily blocking minor sensations, the thalamus permits the cerebrum to concentrate on important tasks.

CEREBRUM

The largest part of the human brain is the **cerebrum,** which consists of two hemispheres separated by the longitudinal fissure. At the base of this deep groove is the **corpus callosum,** a band of 200 million neurons that connects the right and left hemispheres. Within each hemisphere is a lateral ventricle.

The surface of the cerebrum is gray matter called the **cerebral cortex.** Gray matter consists of cell bodies of neurons, which carry out the many functions of the cerebrum. Internal to the gray matter is white matter, made of myelinated axons and dendrites that connect the lobes of the cerebrum to one another and to all other parts of the brain.

In the human brain the cerebral cortex is folded extensively. The folds are called **convolutions** or **gyri,** and the grooves between them are **fissures** or **sulci.** This folding permits the presence of millions more neurons in the cerebral cortex. The cerebral cortex of an animal such as a dog or cat does not have this extensive folding. This difference enables us to read, speak, do long division, and so many other "human" things that dogs and cats cannot.

The cerebral cortex is divided into lobes that have the same names as the cranial bones external to them. Therefore, each hemisphere has a frontal lobe, parietal lobe, temporal lobe, and occipital lobe (Fig. 8–8). These lobes have been mapped, that is, certain areas are known to be associated with specific functions. We will discuss the func-

tions of the cerebrum according to these mapped areas.

Frontal Lobes

Within the **frontal lobes** are the **motor areas** that generate the impulses for voluntary movement. The left motor area controls movement on the right side of the body, and the right motor area controls the left side of the body. This is why a patient who

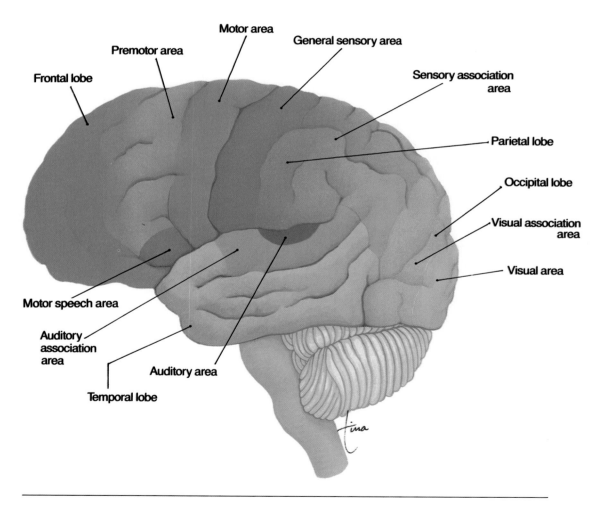

Figure 8–8 Left cerebral hemisphere showing some of the functional areas that have been mapped.

Box 8–4   CEREBROVASCULAR ACCIDENTS

A **cerebrovascular accident (CVA),** or **stroke,** is damage to a blood vessel in the brain, resulting in lack of oxygen to that part of the brain. Possible types of vessel damage are thrombosis or hemorrhage.

A **thrombus** is a blood clot, which most often is a consequence of atherosclerosis, abnormal lipid deposits in cerebral arteries. The rough surface stimulates clot formation, which obstructs the blood flow to the part of the brain supplied by the artery. The symptoms depend on the part of the brain affected and may be gradual in onset if clot formation is slow.

A hemorrhage, the result of arteriosclerosis or **aneurysm** of a cerebral artery, allows blood out into brain tissue, which destroys brain neurons by putting excessive pressure on them as well as depriving them of oxygen. Onset of symptoms in this type of CVA is usually rapid.

If, for example, the CVA is in the left frontal lobe, paralysis of the right side of the body will occur. Speech may also be affected if the speech areas are involved. Some CVAs are fatal because the damage they cause is very widespread or affects vital centers in the medulla or pons.

Recovery from a CVA depends on its location and the extent of damage, as well as other factors. One of these is the redundancy of the brain. Redundancy means repetition; the cerebral cortex has many more neurons than we actually use in daily activities. These neurons are available for use, especially in younger people (less than 50 years of age). When a patient recovers from a disabling stroke, what has often happened is that the brain has established new pathways, with previously little-used neurons now carrying impulses "full time." Such recovery is highly individual and may take months. Yet another important factor seems to be that CVA patients be started on rehabilitation therapy as soon as their condition permits.

has had a cerebrovascular accident, or stroke, in the right frontal lobe will have paralysis of muscles on the left side (see Box 8–4: Cerebrovascular Accidents).

Also in the frontal lobe, usually only the left lobe for most of us, is **Broca's motor speech** area, which controls the movements of the mouth involved in speaking.

Parietal Lobes

The **general sensory areas** in the **parietal lobes** receive impulses from receptors in the skin and feel and interpret the cutaneous sensations. The left area is for the right side of the body and vice versa. These areas also receive impulses from stretch receptors in muscles for conscious muscle

sense. Impulses from taste buds travel to the **taste areas,** which overlap the parietal and temporal lobes.

Temporal Lobes

The **auditory areas,** as their name suggests, receive impulses from receptors in the inner ear for hearing. The **olfactory areas** receive impulses from receptors in the nasal cavities for the sense of smell.

Also in the temporal and parietal lobes in the left hemisphere (for most of us) are other speech areas concerned with the thought that precedes speech. Each of us can probably recall (and regret) times when we have "spoken without thinking," but in actuality that is not possible. The thinking takes

Box 8–5 APHASIA

Our use of language sets us apart from other animals and involves speech, reading, and writing. Language is the use of symbols (words) to designate objects and to express ideas. Damage to the speech areas or interpretation areas of the cerebrum may impair one or more aspects of a person's ability to use language; this is called **aphasia.**

Aphasia may be a consequence of a cerebrovascular accident, or of physical trauma to the skull and brain such as a head injury sustained in an automobile accident. If the motor speech (Broca's) area is damaged, the person is still able to understand written and spoken words and knows what he wants to say, but he cannot say it. Without coordination and impulses from the motor speech area, the muscles used for speech cannot contract to form words properly.

Auditory aphasia is **"word deafness,"** caused by damage to an interpretation area. The person can still hear but cannot comprehend what the words mean. Visual aphasia is **"word blindness";** the person can still see perfectly well, but cannot make sense of written words (the person retains the ability to understand spoken words). Imagine how you would feel if wms qsbbcljw jmqr rfc yzgjgrw rm pcyb. Frustrating isn't it? You know that those symbols are letters, but you cannot "decode" them right away. Those "words" were formed by shifting the alphabet two letters (A = C, B = D, C = E, etc.), and would normally be read as: "you suddenly lost the ability to read." That may give you a small idea of what word blindness is like.

place very rapidly and is essential in order to be able to speak (see Box 8–5: Aphasia).

Occipital Lobes

Impulses from the retinas of the eyes travel along the optic nerves to the **visual areas.** These areas "see" and interpret what is seen. Other parts of the occipital lobes are concerned with spatial relationships; such things as judging distance and seeing in three dimensions.

Association Areas

As you can see in Fig. 8–8, there are many parts of the cerebral cortex not concerned with movement or a particular sensation. These may be called **association areas** and perhaps are what truly make us individuals. It is probably these areas that give each of us a personality, a sense of humor, and the ability to reason and use logic. Learning and

memory are also functions of these areas. The formation of memories is very poorly understood, but it is believed that most, if not all, of what we have experienced or learned is stored somewhere in the brain. Sometimes a trigger may bring back memories; a certain scent or a song are possible triggers. Then we find ourselves recalling something from the past and wondering where it came from.

The loss of personality due to destruction of brain neurons is perhaps most dramatically seen in Alzheimer's disease (see Box 8–6: Alzheimer's Disease).

Basal Ganglia

The **basal ganglia** are paired masses of gray matter within the white matter of the cerebral hemispheres. Their functions are certain subconscious aspects of voluntary movement: regulation of muscle tone and accessory movements such as swinging the arms when walking or gesturing while

Box 8–6 ALZHEIMER'S DISEASE

In the United States, Alzheimer's disease, a progressive, incurable form of mental deterioration, affects approximately 4 million people and is the cause of 100,000 deaths each year. The first symptoms, which usually begin after age 65, are memory lapses and slight personality changes. As the disease progresses, there is total loss of memory, reasoning ability, and personality, and those with advanced disease are unable to perform even the simplest tasks or self-care.

Structural changes in the brains of Alzheimer's patients may be seen at autopsy. Neurofibrillary tangles are abnormal fibrous proteins found in the cerebral cortex in areas important for memory and reasoning. Also present are plaques made of another protein called beta-amyloid.

Recently, a defective gene on chromosome 19 has been found in some patients who have late-onset Alzheimer's disease, the most common type. Yet another gene seems to trigger increased synthesis of beta-amyloid. Current research is focused on the interaction of these genes, much like putting together a puzzle, with the goal of finding drugs to block the overproduction of beta-amyloid.

Box 8–7 PARKINSON'S DISEASE

Parkinson's disease is a disorder of the basal ganglia whose cause is unknown, and which usually begins after the age of 60. Neurons in the basal ganglia that produce the neurotransmitter dopamine begin to degenerate and die, and the deficiency of dopamine causes specific kinds of muscular symptoms. Tremor, or involuntary shaking, of the hands is probably the most common symptom. The accessory movements regulated by the basal ganglia gradually diminish, and the affected person walks slowly without swinging the arms. A mask-like face is characteristic of this disease, as the facial muscles become rigid. Eventually all voluntary movements become slower and much more difficult, and balance is seriously impaired.

Dopamine itself cannot be used to treat Parkinson's disease because it does not cross the blood brain barrier. A substance called L-dopa does cross and can be converted to dopamine by brain neurons. Unfortunately, L-dopa begins to lose its therapeutic effectiveness within 1 or 2 years.

Recently, another medication called deprenyl has been shown to slow the progression of Parkinson's disease in a small number of patients. The researchers believe that deprenyl may actually prevent the death of basal ganglia neurons, but this has yet to be proven. Studies using larger numbers of patients are now being conducted, and deprenyl has been made available to all those with Parkinson's disease.

speaking. The most common disorder of the basal ganglia is Parkinson's disease (see Box 8–7: Parkinson's Disease).

Corpus Callosum

As mentioned previously, the **corpus callosum** is a band of nerve fibers that connects the left and right cerebral hemispheres. This enables each hemisphere to know of the activity of the other. This is especially important for people because for most of us, the left hemisphere contains speech areas and the right hemisphere does not. The corpus callosum, therefore, lets the right hemisphere know

what the left hemisphere is talking about. The "division of labor" of our cerebral hemispheres is beyond the scope of this book, but it is a fascinating subject that you may wish to explore further.

MENINGES AND CEREBROSPINAL FLUID

The connective tissue membranes that cover the brain and spinal cord are called **meninges;** the three layers are illustrated in Fig. 8–9. The thick out-

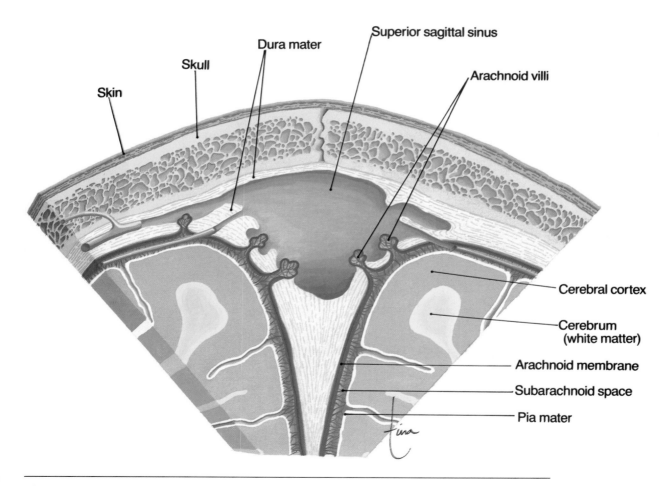

Figure 8–9 Structure of the meninges. Frontal section through the top of the skull showing the double-layered dura matter and one of the cranial venous sinuses.

ermost layer, made of fibrous connective tissue, is the **dura mater,** which lines the skull and vertebral canal. The middle **arachnoid membrane** (arachnids are spiders) is made of web-like strands of connective tissue. The innermost **pia mater** is a very thin membrane on the surface of the spinal cord and brain. Between the arachnoid and the pia mater is the **subarachnoid space,** which contains cerebrospinal fluid (CSF), the tissue fluid of the central nervous system.

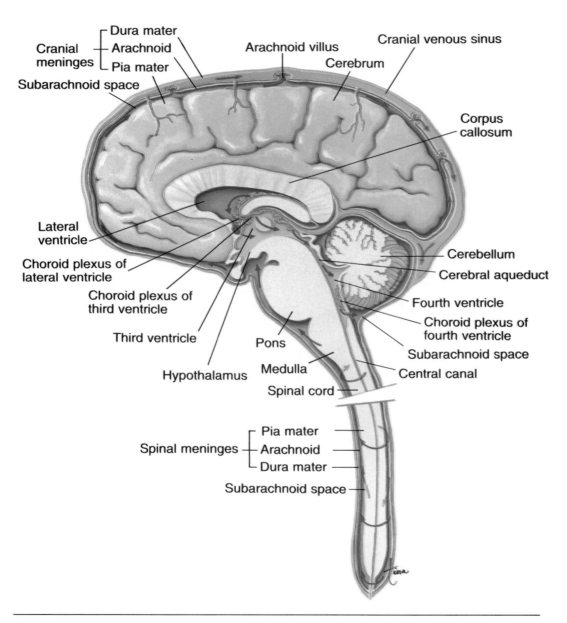

Figure 8–10 Formation, circulation, and reabsorption of cerebrospinal fluid. See text for description.

Box 8–8   LUMBAR PUNCTURE

A **lumbar puncture** (spinal tap) is a diagnostic procedure that involves the removal of cerebrospinal fluid to determine its pressure and constituents. As the name tells us, the removal, using a syringe, is made in the lumbar area. Since the spinal cord ends between the 1st and 2nd lumbar vertebrae, the needle is usually inserted between the 4th and 5th lumbar vertebrae. The meningeal sac containing cerebrospinal fluid extends to the end of the lumbar vertebrae, permitting access to the cerebrospinal fluid with little chance of damaging the spinal cord.

Cerebrospinal fluid is a circulating fluid and has a normal pressure of 70 to 200 mmH₂O. An abnormal pressure usually indicates an obstruction in circulation, which may be caused by infection, a tumor, or mechanical injury. Other diagnostic tests would be needed to determine the precise cause.

Perhaps the most frequent reason for a lumbar puncture is suspected **meningitis,** which may be caused by several kinds of bacteria. If the patient does have meningitis, the cerebrospinal fluid will be cloudy rather than clear and will be examined for the presence of bacteria and many white blood cells. A few WBCs in CSF is normal, since WBCs are found in all tissue fluid.

Another abnormal constituent of cerebrospinal fluid is red blood cells. Their presence indicates bleeding somewhere in the central nervous system. There may be many causes, and again, further testing would be necessary.

Recall the ventricles (cavities) of the brain: two lateral ventricles, the third ventricle, and fourth ventricle. Each contains a choroid plexus, a capillary network that forms cerebrospinal fluid from blood plasma. This is a continuous process, and the cerebrospinal fluid then circulates in and around the central nervous system (Fig. 8–10).

From the lateral and third ventricles, cerebrospinal fluid flows through the fourth ventricle, then to the central canal of the spinal cord, and to the cranial and spinal subarachnoid spaces. As more cerebrospinal fluid is formed, you might expect that some must be reabsorbed, and that is just what happens. From the cranial subarachnoid space, cerebrospinal fluid is reabsorbed through **arachnoid villi** into the blood in **cranial venous sinuses** (large veins within the double-layered cranial dura mater). The cerebrospinal fluid becomes blood plasma again, and the rate of reabsorption normally equals the rate of production.

Since cerebrospinal fluid is tissue fluid, one of its functions is to bring nutrients to CNS neurons and to remove waste products to the blood as the fluid

is reabsorbed. The other function of cerebrospinal fluid is to act as a cushion for the central nervous system. The brain and spinal cord are enclosed in fluid-filled membranes that absorb shock. You can, for example, shake your head vigorously without harming your brain. Naturally, this protection has limits; very sharp or heavy blows to the skull will indeed cause damage to the brain.

Examination of cerebrospinal fluid may be used in the diagnosis of certain diseases (see Box 8–8: Lumbar Puncture).

CRANIAL NERVES

The 12 pairs of **cranial nerves** emerge from the brain stem or other parts of the brain—they are shown in Fig. 8–11. The name "cranial" indicates their origin, and many of them do carry impulses for functions involving the head. Some, however, have more far-reaching destinations.

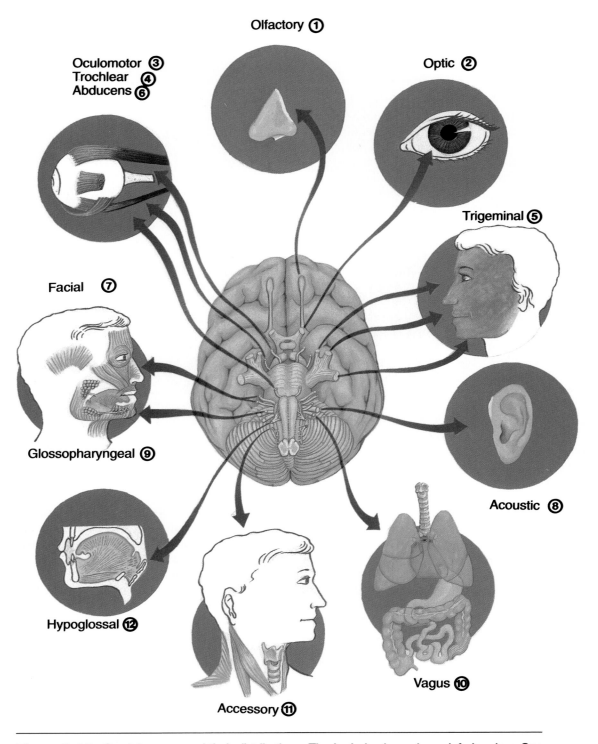

Figure 8–11 Cranial nerves and their distributions. The brain is shown in an inferior view. See Table 8–4 for descriptions.

Table 8–4   CRANIAL NERVES

Number and Name	Function(s)
I Olfactory	• Sense of smell
II Optic	• Sense of sight
III Oculomotor	• Movement of the eyeball; constriction of pupil in bright light or for near vision
IV Trochlear	• Movement of eyeball
V Trigeminal	• Sensation in face, scalp, and teeth; contraction of chewing muscles
VI Abducens	• Movement of the eyeball
VII Facial	• Sense of taste; contraction of facial muscles; secretion of saliva
VIII Acoustic (Vestibulocochlear)	• Sense of hearing; sense of equilibrium
IX Glossopharyngeal	• Sense of taste; sensory for cardiac, respiratory, and blood pressure reflexes; contraction of pharynx; secretion of saliva
X Vagus	• Sensory in cardiac, respiratory, and blood pressure reflexes; sensory and motor to larynx (speaking); decreases heart rate; contraction of alimentary tube (peristalsis); increases digestive secretions
XI Accessory	• Contraction of neck and shoulder muscles; motor to larynx (speaking)
XII Hypoglossal	• Movement of the tongue

The impulses for the senses of smell, taste, sight, hearing, and equilibrium are all carried on cranial nerves to their respective sensory areas in the brain. Some cranial nerves carry motor impulses to muscles of the face and eyes or to the salivary glands. The vagus nerves ("vagus" means "wanderer") branch extensively to the larynx, heart, stomach and intestines, and the bronchial tubes.

The functions of all the cranial nerves are summarized in Table 8–4.

THE AUTONOMIC NERVOUS SYSTEM

The **autonomic nervous system (ANS)** is actually part of the peripheral nervous system in that it consists of motor portions of some cranial and spinal nerves. Since its functioning is so specialized, however, the autonomic nervous system is usually discussed as a separate entity, as we will do here.

Making up the autonomic nervous system are **visceral motor neurons** to smooth muscle, cardiac muscle, and glands. These are the **visceral effectors;** muscle will either contract or relax, and glands will either increase or decrease their secretions.

The ANS has two divisions: **sympathetic** and **parasympathetic.** Often, they function in opposition to one another, as you will see. The activity of both divisions is integrated by the hypothalamus, which ensures that the visceral effectors will respond appropriately to the situation.

AUTONOMIC PATHWAYS

An autonomic nerve pathway from the central nervous system to a visceral effector consists of two motor neurons that synapse in a ganglion outside the CNS (Fig. 8–12). The first neuron is called the **preganglionic neuron,** from the CNS to the ganglion. The second neuron is called the **postganglionic neuron,** from the ganglion to the visceral effector. The ganglia are actually the cell bodies of the postganglionic neurons.

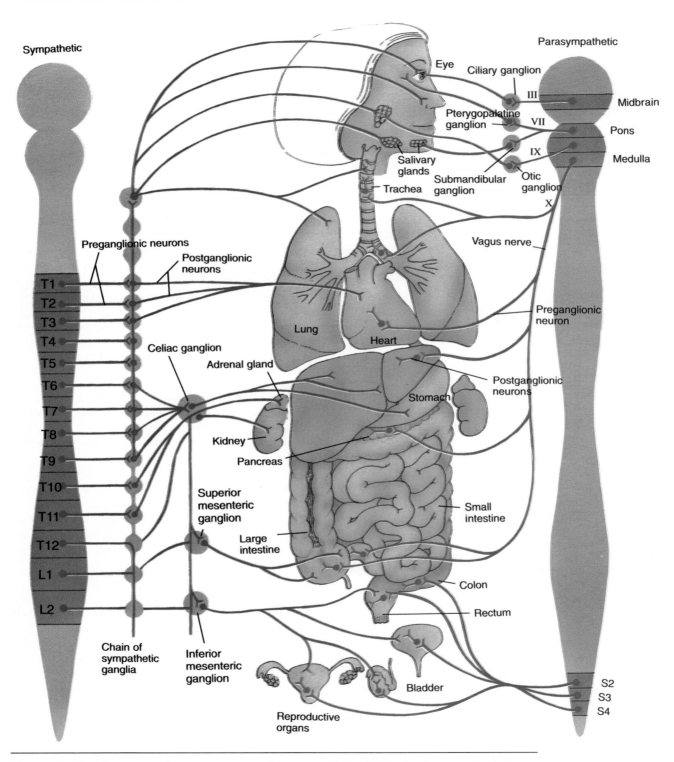

Figure 8–12   The autonomic nervous system. The sympathetic division is shown on the left, and the parasympathetic division is shown on the right (both divisions are bilateral).

SYMPATHETIC DIVISION

Another name for the sympathetic division is Thoracolumbar division, which tells us where the sympathetic preganglionic neurons originate. Their cell bodies are in the thoracic segments and some of the lumbar segments of the spinal cord. Their axons extend to the sympathetic ganglia, most of which are located in two chains just outside the spinal column (see Fig. 8–12). Within the ganglia are the synapses between preganglionic and postganglionic neurons; the postganglionic axons then go to the visceral effectors. One preganglionic neuron often synapses with many postganglionic neurons to many effectors. This anatomic arrangement has physiologic importance: the sympathetic division brings about widespread responses in many organs.

The sympathetic division is dominant in stress situations, which include anger, fear, or anxiety, as well as exercise. For our prehistoric ancestors, stress situations often involved the need for intense physical activity—the "fight or flight response." Our nervous systems haven't changed very much in 50,000 years, and if you look at Table 8–5, you will see the kinds of responses the sympathetic division stimulates. The heart rate increases, vasodilation in skeletal muscles supplies them with more oxygen, the bronchioles dilate to take in more air, and the liver changes glycogen to glucose to supply energy. At the same time, digestive secretions decrease and peristalsis slows; these are not important in a stress situation. Vasoconstriction in the skin and viscera shunts blood to more vital organs such as the heart, muscles, and brain. All these responses enabled our ancestors to stay and fight or to get away from potential danger. Even though we may not always be in life-threatening situations during stress (such as figuring out our income taxes), our bodies are prepared for just that.

PARASYMPATHETIC DIVISION

The other name for the parasympathetic division is the Cranialsacral division. The cell bodies of parasympathetic preganglionic neurons are in the brain stem and the sacral segments of the spinal cord. Their axons are in cranial nerve pairs 3, 7, 9, and

Table 8–5 FUNCTIONS OF THE AUTONOMIC NERVOUS SYSTEM

Organ	Sympathetic Response	Parasympathetic Response
Heart (cardiac muscle)	• Increase rate	• Decrease rate (to normal)
Bronchioles (smooth muscle)	• Dilate	• Constrict (to normal)
Iris (smooth muscle)	• Pupil dilates	• Pupil constricts (to normal)
Salivary glands	• Decrease secretion	• Increase secretion (to normal)
Stomach and intestines (smooth muscle)	• Decrease peristalsis	• Increase peristalsis for normal digestion
Stomach and intestines (glands)	• Decrease secretion	• Increase secretion for normal digestion
Internal anal sphincter	• Contracts to prevent defecation	• Relaxes to permit defecation
Urinary bladder (smooth muscle)	• Relaxes to prevent urination	• Contracts for normal urination
Internal urethral sphincter	• Contracts to prevent urination	• Relaxes to permit urination
Liver	• Changes glycogen to glucose	• None
Sweat glands	• Increase secretion	• None
Blood vessels in skin and viscera (smooth muscle)	• Constrict	• None
Blood vessels in skeletal muscle (smooth muscle)	• Dilate	• None
Adrenal glands	• Increase secretion of epinephrine and norepinephrine	• None

10 and in some sacral nerves and extend to the parasympathetic ganglia. These ganglia are very close to or actually in the visceral effector (see Fig. 8–12), and contain the postganglionic cell bodies, with very short axons to the cells of the effector.

In the parasympathetic division, one preganglionic neuron synapses with just a few postganglionic neurons to only one effector. With this anatomic arrangement, very localized (one organ) responses are possible.

The parasympathetic division dominates in relaxed (non-stress) situations to promote normal functioning of several organ systems. Digestion will be efficient, with increased secretions and peristalsis; defecation and urination may occur; and the heart will beat at a normal resting rate. Other functions of this division are listed in Table 8–5.

Notice that when an organ receives both sympathetic and parasympathetic impulses, the responses are opposites. Notice also that some visceral effectors receive only sympathetic impulses. In such cases, the opposite response is brought about by a decrease in sympathetic impulses.

NEUROTRANSMITTERS

Recall that neurotransmitters enable nerve impulses to cross synapses. In autonomic pathways there are two synapses: one between preganglionic and postganglionic neurons, and the second between postganglionic neurons and visceral effectors.

Acetylcholine is the transmitter released by all preganglionic neurons, both sympathetic and parasympathetic; it is inactivated by **cholinesterase** in postganglionic neurons. Parasympathetic postganglionic neurons all release acetylcholine at the synapses with their visceral effectors. Most sympathetic postganglionic neurons release the transmitter **norepinephrine,** which is inactivated by **COMT** (Catechol-O-methyl transferase).

SUMMARY

The nervous system regulates many of our simplest and our most complex activities. The impulses generated and carried by the nervous system are an example of the chemical level of organization of the body. These nerve impulses then regulate the functioning of tissues, organs, and organ systems, which permits us to perceive and respond to the world around us and the changes within us. The detection of such changes is the function of the sense organs, and they are the subject of our next chapter.

STUDY OUTLINE

Functions of the Nervous System
1. Detect changes and feel sensations.
2. Initiate responses to changes.
3. Organize and store information.

Nervous System Divisions
1. Central nervous system (CNS)—brain and spinal cord.
2. Peripheral nervous system (PNS)—cranial nerves and spinal nerves.

Nerve Tissue—neurons (nerve fibers) and specialized cells (Schwann, neuroglia)
1. Neuron cell body contains the nucleus; cell bodies are in the CNS or in the trunk and are protected by bone.

2. Axon carries impulses away from the cell body; dendrites carry impulses toward the cell body.
3. Schwann cells in PNS: layers of cell membrane form the myelin sheath to electrically insulate neurons; nodes of Ranvier are spaces between adjacent Schwann cells. Nuclei and cytoplasm of Schwann cells form the neurolemma which is essential for regeneration of damaged axons or dendrites.
4. Oligodendrocytes in CNS: form the myelin sheaths (see Table 8–1).
5. Synapse—the space between the axon of one neuron and the dendrites or cell body of the next neuron. A neurotransmitter carries the impulse across a synapse and is then destroyed by a

chemical inactivator. Synapses make impulse transmission one-way in the living person.

Types of Neurons—nerve fibers

1. Sensory—carry impulses from receptors to the CNS; may be somatic (from skin, skeletal muscles, joints) or visceral (from internal organs).
2. Motor—carry impulses from the CNS to effectors; may be somatic (to skeletal muscle) or visceral (to smooth muscle, cardiac muscle, or glands). Visceral motor neurons make up the autonomic nervous system.
3. Interneurons—entirely within the CNS.

Nerves and Nerve Tracts

1. Sensory nerve—made only of sensory neurons.
2. Motor nerve—made only of motor neurons.
3. Mixed nerve—made of both sensory and motor neurons.
4. Nerve tract—a nerve within the CNS; also called white matter.

The Nerve Impulse—see Table 8–2

1. Polarization—neuron membrane has a (+) charge outside and a (−) charge inside.
2. Depolarization—entry of Na^+ ions and reversal of charges on either side of the membrane.
3. Impulse transmission is rapid, often several meters per second.
   - Saltatory conduction—in a myelinated neuron only the nodes of Ranvier depolarize; increases speed of impulses.

The Spinal Cord

1. Functions: transmits impulses to and from the brain, and integrates the spinal cord reflexes.
2. Location: within the vertebral canal; extends from the foramen magnum to the disc between the 1st and 2nd lumbar vertebrae.
3. Cross-section: internal H of gray matter contains cell bodies of motor neurons and interneurons; external white matter is the myelinated axons and dendrites of interneurons.
4. Ascending tracts carry sensory impulses to the brain; descending tracts carry motor impulses away from the brain.
5. Central canal contains cerebrospinal fluid and is continuous with the ventricles of the brain.

Spinal Nerves—see Table 8–3 for major peripheral nerves

1. Eight cervical pairs to head, neck, shoulder, arm, and diaphragm; 12 thoracic pairs to trunk; 5 lumbar pairs and 5 sacral pairs to hip, pelvic cavity and leg; 1 very small coccygeal pair.
2. Cauda equina—the lumbar and sacral nerves that extend below the end of the spinal cord.
3. Each spinal nerve has two roots: dorsal or sensory root; dorsal root ganglion contains cell bodies of sensory neurons; ventral or motor root; the two roots unite to form a mixed spinal nerve.

Spinal Cord Reflexes—do not depend directly on the brain

1. A reflex is an involuntary response to a stimulus.
2. Reflex arc—the pathway of nerve impulses during a reflex (1) receptors, (2) sensory neurons, (3) CNS with one or more synapses, (4) motor neurons, (5) effector which responds.
3. Stretch reflex—a muscle that is stretched will contract; these reflexes help keep us upright against gravity. The patellar reflex is also used clinically to assess neurological functioning, as are many other reflexes (Fig. 8–5).
4. Flexor reflex—a painful stimulus will cause withdrawal of the body part; these reflexes are protective.

The Brain—many parts that function as an integrated whole; see Figs. 8–6 and 8–8 for locations

1. Ventricles—four cavities: two lateral, 3rd, 4th; each contains a choroid plexus that forms cerebrospinal fluid (Fig. 8–7).
2. Medulla—regulates the vital functions of heart rate, breathing, and blood pressure; regulates reflexes of coughing, sneezing, swallowing, and vomiting.
3. Pons—contains respiratory centers that work with those in the medulla.
4. Midbrain—contains centers for visual reflexes, auditory reflexes, and righting (equilibrium) reflexes.
5. Cerebellum—regulates coordination of voluntary movement, muscle tone, stopping movements, and equilibrium.

6. Hypothalamus—produces antidiuretic hormone (ADH) which increases water reabsorption by the kidneys; produces oxytocin which promotes uterine contractions for labor and delivery; produces releasing hormones that regulate the secretions of the anterior pituitary gland; regulates body temperature; regulates food intake; integrates the functioning of the autonomic nervous system (ANS); promotes visceral responses to emotional situations.
7. Thalamus—groups sensory impulses as to body part before relaying them to the cerebrum; suppresses unimportant sensations to permit concentration.
8. Cerebrum—two hemispheres connected by the corpus callosum, which permits communication between the hemispheres. The cerebral cortex is the surface gray matter, which consists of cell bodies of neurons and is folded extensively into convolutions. The internal white matter consists of nerve tracts that connect the lobes of the cerebrum to one another and to other parts of the brain.
   - Frontal lobes—motor areas initiate voluntary movement; Broca's motor speech area (left hemisphere) regulates the movements involved in speech.
   - Parietal lobes—general sensory area feels and interprets the cutaneous senses and conscious muscle sense; taste area extends into temporal lobe, for sense of taste; speech areas (left hemisphere) for thought before speech.
   - Temporal lobes—auditory areas for hearing; olfactory areas for sense of smell; speech areas for thought before speech.
   - Occipital lobes—visual areas for vision; interpretation areas for spatial relationships.
   - Association areas—in all lobes, for abstract thinking, reasoning, learning, memory, and personality.
   - Basal ganglia—gray matter within the cerebral hemispheres; regulate accessory movements and muscle tone.

Meninges and Cerebrospinal Fluid (CSF) (see Figs. 8–9 and 8–10)
1. Three meningeal layers made of connective tissue: outer—dura mater; middle—arachnoid membrane; inner—pia mater; all three enclose the brain and spinal cord.
2. Subarachnoid space contains CSF, the tissue fluid of the CNS.
3. CSF is formed continuously in the ventricles of the brain by choroid plexuses, from blood plasma.
4. CSF circulates from the ventricles to the central canal of the spinal cord and to the cranial and spinal subarachnoid spaces.
5. CSF is reabsorbed from the cranial subarachnoid space through arachnoid villi into the blood in the cranial venous sinuses. The rate of reabsorption equals the rate of production.
6. As tissue fluid, CSF brings nutrients to CNS neurons and removes waste products. CSF also acts as a shock absorber to cushion the CNS.

Cranial Nerves—12 pairs of nerves that emerge from the brain (see Fig. 8–11)
1. Concerned with vision, hearing and equilibrium, taste and smell, and many other functions.
2. See Table 8–4 for the functions of each pair.

The Autonomic Nervous System (ANS) (see Fig. 8–12)
1. Has two divisions: sympathetic and parasympathetic; their functioning is integrated by the hypothalamus.
2. Consists of motor neurons to visceral effectors: smooth muscle, cardiac muscle, and glands.
3. An ANS pathway consists of two neurons that synapse in a ganglion:
   - Preganglionic neurons—from the CNS to the ganglia
   - Postganglionic neurons—from the ganglia to the effectors
   Most sympathetic ganglia are in two chains just outside the vertebral column; parasympathetic ganglia are very near or in the visceral effectors.
4. Neurotransmitters: acetylcholine is released by all preganglionic neurons and by parasympathetic postganglionic neurons; the inactivator is cholinesterase. Norepinephrine is released by most sympathetic postganglionic neurons; the inactivator is COMT.
5. Sympathetic division—dominates the stress situations; responses prepare the body to meet physical demands (see Table 8–5).
6. Parasympathetic division—dominates in relaxed situations to permit normal functioning—(see Table 8–5).

REVIEW QUESTIONS

1. Name the divisions of the nervous system and state the parts of each. (p. 160)

2. State the function of the following parts of nerve tissue: (pp. 160, 162)
   a. axon
   b. dendrites
   c. myelin sheath
   d. neurolemma
   e. microglia
   f. astrocytes

3. Explain the difference between: (pp. 163, 165)
   a. sensory neurons and motor neurons
   b. interneurons and nerve tracts

4. Describe an electrical nerve impulse in terms of charges on either side of the neuron membrane. Describe how a nerve impulse crosses a synapse. (pp. 162, 163, 165)

5. With respect to the spinal cord: (pp. 166, 168)
   a. describe its location
   b. state what gray matter and white matter are made of
   c. state the function of the dorsal root, ventral root, and dorsal root ganglion

6. State the names and number of pairs of spinal nerves. State the part of the body supplied by the: phrenic nerves, radial nerves, sciatic nerves. (pp. 166, 168)

7. Define reflex, and name the five parts of a reflex arc. (p. 168)

8. Define stretch reflexes, and explain their practical importance. Define flexor reflexes, and explain their practical importance. (pp. 168–169)

9. Name the part of the brain concerned with each of the following: (pp. 171–174)
   a. regulates body temperature
   b. regulates heart rate
   c. suppresses unimportant sensations
   d. regulates respiration (two parts)
   e. regulates food intake
   f. regulates coordination of voluntary movement
   g. regulates secretions of the anterior pituitary gland
   h. regulates coughing and sneezing
   i. regulates muscle tone
   j. regulates visual and auditory reflexes
   k. regulates blood pressure

10. Name the part of the cerebrum concerned with each of the following: (pp. 174–179)
    a. feels the cutaneous sensations
    b. contains the auditory areas
    c. contains the visual areas
    d. connects the cerebral hemispheres
    e. regulates accessory movements
    f. contains the olfactory areas

g. initiates voluntary movement

h. contains the speech areas (for most people)

11. Name the three layers of the meninges, beginning with the outermost. (pp. 179–180)

12. State all the locations of cerebrospinal fluid. What is CSF made from? Into what is CSF reabsorbed? State the functions of CSF. (pp. 180–181)

13. State a function of each of the following cranial nerves: (p. 183)
    a. glossopharyngeal
    b. olfactory
    c. trigeminal
    d. facial
    e. vagus (three functions)

14. Explain how the sympathetic division of the ANS helps the body adapt to a stress situation; give three specific examples. (p. 185)

15. Explain how the parasympathetic division of the ANS promotes normal body functioning; give three specific examples. (pp. 185–186)

Chapter 9
The Senses

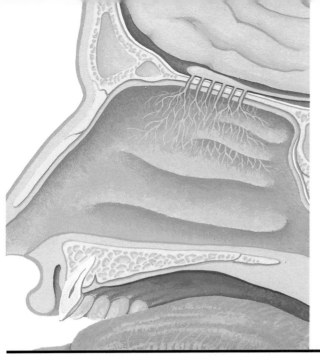

Chapter 9

Student Objectives

- Explain the general purpose of sensations.
- Name the parts of a sensory pathway, and state the function of each.
- Describe the characteristics of sensations.
- Name the cutaneous senses and explain their purpose.
- Explain referred pain and its importance.
- Explain the importance of muscle sense.
- Describe the pathways for the senses of smell and taste, and explain how these senses are interrelated.
- Name the parts of the eye and their functions.
- Describe the physiology of vision.
- Name the parts of the ear and their functions.
- Describe the physiology of hearing.
- Describe the physiology of equilibrium.
- Explain the importance of the arterial pressoreceptors and chemoreceptors.

The Senses

New Terminology

Adaptation (A–dap–**TAY**–shun)
After-image (**AFF**–ter–im–ije)
Aqueous humor (**AY**–kwee–us **HYOO**–mer)
Cochlea (**KOK**–lee–ah)
Cones (**KOHNES**)
Conjunctiva (KON–junk–**TIGH**–vah)
Contrast (**KON**–trast)
Cornea (**KOR**–nee–ah)
Eustachian tube (yoo–**STAY**–shee–un TOOB)
Iris (**EYE**–ris)
Lacrimal glands (**LAK**–ri–muhl)
Organ of Corti (**KOR**–tee)
Olfactory receptors (ohl–**FAK**–toh–ree)
Projection (proh–**JEK**–shun)
Referred pain (ree–**FURRD** PAYNE)
Retina (**RET**–i–nah)
Rhodopsin (roh–**DOP**–sin)
Rods (RAHDS)
Sclera (**SKLER**–ah)
Semicircular canals (SEM–ee–**SIR**–kyoo–lur)
Tympanic membrane (tim–**PAN**–ik)
Vitreous humor (**VIT**–ree–us **HYOO**–mer)

Related Clinical Terminology

Astigmatism (un–**STIG**–mah–TIZM)
Cataract (**KAT**–uh–rackt)
Colorblindness (**KUHL**–or **BLIND**–ness)
Conjunctivitis (kon–JUNK–ti–**VIGH**–tis)
Deafness (**DEFF**–ness)
Detached retina (dee–**TACHD**)
Glaucoma (glaw–**KOH**–mah)
Hyperopia (HIGH–per–**OH**–pee–ah)
Motion sickness (**MOH**–shun)
Myopia (my–**OH**–pee–ah)
Night blindness (NITE **BLIND**–ness)
Otitis media (oh–**TIGH**–tis **MEE**–dee–ah)
Phantom pain (**FAN**–tum)
Presbyopia (PREZ–bee–**OH**–pee–ah)

Terms that appear in **bold type** in the chapter text are defined in the glossary, which begins on page 549.

Our **senses** constantly provide us with information about our surroundings: we see, hear, and touch. The senses of taste and smell enable us to enjoy the flavor of our food or warn us that food has spoiled and may be dangerous to eat. Our sense of equilibrium keeps us upright. We also get information from our senses about what is happening inside the body. The pain of a headache, for example, prompts us to do something about it, such as take aspirin. In general, this is the purpose of sensations: to enable the body to respond appropriately to ever-changing situations and maintain homeostasis.

SENSORY PATHWAY

The impulses involved in sensations follow very precise pathways, which all have the following parts:

1. **Receptors**—detect changes **(stimuli)** and generate impulses. Receptors are usually very specific with respect to the kinds of changes they respond to. Those in the retina detect light rays, those in the nasal cavities detect vapors, and so on. Once a specific stimulus has affected receptors, however, they all respond the same way by generating electrical nerve impulses.
2. **Sensory neurons**—transmit impulses from receptors to the central nervous system. These sensory neurons are found in both spinal nerves and cranial nerves, but each carries impulses from only one type of receptor.
3. **Sensory tracts**—white matter in the spinal cord or brain that transmits the impulses to a specific part of the brain.
4. **Sensory area**—most are in the cerebral cortex. These areas feel and interpret the sensations. Learning to interpret sensations begins in infancy, without our awareness of it, and continues throughout life.

CHARACTERISTICS OF SENSATIONS

1. **Projection**—the sensation seems to come from the area where the receptors were stimulated. If you touch this book, the sensation of touch seems to be in your hand but is actually being felt by your cerebral cortex. That it is indeed the brain that feels sensations is demonstrated by patients who feel **phantom pain** after amputation of a limb. After loss of a hand, for example, the person may still feel that the hand is really there. Why does this happen? The receptors in the hand are no longer present, but the severed nerve endings continue to generate impulses. These impulses arrive in the parietal lobe area for the hand, and the brain does what it has always done and creates the projection, the feeling that the hand is still there. For most amputees, phantom pain diminishes as the severed nerves heal, but the person often experiences a phantom "presence" of the missing part. This may be helpful when learning to use an artificial limb.
2. **Intensity**—some sensations are felt more distinctly and to a greater degree than are others. A weak stimulus such as dim light will affect a small number of receptors, but a stronger stimulus, such as bright sunlight, will stimulate many more receptors. When more receptors are stimulated, more impulses will arrive in the sensory area of the brain. The brain "counts" the impulses and projects a more intense sensation.
3. **Contrast**—the effect of a previous or simultaneous sensation on a current sensation, which may then be exaggerated or diminished. Again, this is a function of the brain, which constantly compares sensations. If, on a very hot day, you jump into a swimming pool, the water may feel quite cold at first. The brain compares the new sensation to the previous one, and since there is a significant difference between the two, the water will seem colder than it actually is.

4. **Adaptation**—becoming unaware of a continuing stimulus. Receptors detect changes, but if the stimulus continues it may not be much of a change, and the receptors will generate fewer impulses. Most of us wear a watch and are probably unaware of its presence on the arm most of the time. The cutaneous receptors for touch or pressure adapt very quickly to a continuing stimulus, and if there is no change, there is nothing for the receptors to detect.

5. **After-image**—the sensation remains in the consciousness even after the stimulus has stopped. A familiar example is the bright after-image seen after watching a flashbulb go off. The very bright light strongly stimulates receptors in the retina, which generate many impulses that are perceived as an intense sensation that lasts longer than the actual stimulus.

CUTANEOUS SENSES

The dermis of the skin contains receptors for the sensations of touch, pressure, heat, cold, and pain. The receptors for pain are **free nerve endings,** which respond to any intense stimulus. Intense cold, for example, may be felt as pain. The receptors for the other cutaneous senses are **encapsulated nerve endings,** meaning that there is a cellular structure around the nerve ending (Fig. 9–1).

The **cutaneous senses** provide us with infor-

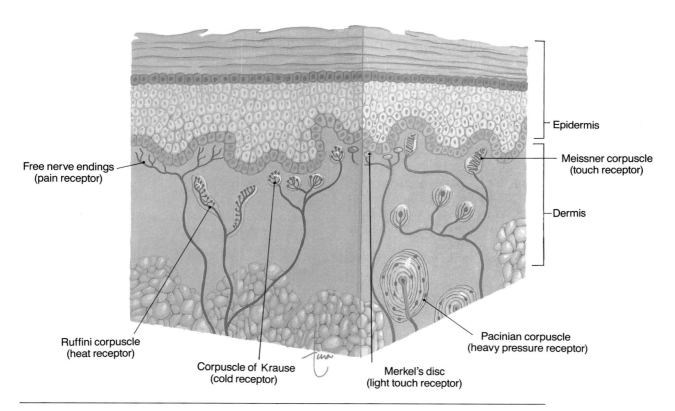

Free nerve endings
(pain receptor)

Ruffini corpuscle
(heat receptor)

Corpuscle of Krause
(cold receptor)

Merkel's disc
(light touch receptor)

Pacinian corpuscle
(heavy pressure receptor)

Meissner corpuscle
(touch receptor)

Epidermis

Dermis

Figure 9–1 Cutaneous receptors in a section of the skin. Free nerve endings and the types of encapsulated nerve endings are shown.

mation about the external environment and also about the skin itself. If you have ever had chickenpox, you may remember the itching sensation of the rash. An itch is actually a mild pain sensation, which may become real pain if not scratched (how scratching relieves the itch has not yet been discovered).

The sensory areas for the skin are in the parietal lobes. The largest parts of this sensory cortex are for the parts of the skin with the most receptors, that is, the hands and face.

REFERRED PAIN

Free nerve endings are also found in internal organs. The smooth muscle of the small intestine, for example, has free nerve endings that are stimulated by excessive stretching or contraction; the resulting pain is called visceral pain. Sometimes pain that originates in an internal organ may be felt in a cutaneous area; this is called **referred pain.** The pain of a heart attack (myocardial infarction) may be felt in the left arm and shoulder, or the pain of gallstones may be felt in the right shoulder.

This referred pain is actually a creation of the brain. Within the spinal cord are sensory tracts that are shared by cutaneous impulses and visceral impulses. Cutaneous impulses are much more frequent, and the brain correctly projects the sensation to the skin. When the impulses have come from an organ such as the heart, however, the brain may still project the sensation to the "usual" cutaneous area. The brain projects sensation based on past experience, and cutaneous pain is far more common than visceral pain. Knowledge of referred pain, as in the examples mentioned earlier, may often be helpful in diagnosis.

MUSCLE SENSE

Muscle sense (also called kinesthetic sense) was discussed in Chapter 7 and will be reviewed only briefly here. Stretch receptors (also called proprioceptors or muscle spindles) detect stretching of muscles and generate impulses, which enable the brain to create a mental picture to know where the muscles are and how they are positioned. Conscious muscle sense is felt by the parietal lobes. Unconscious muscle sense is used by the cerebellum to coordinate voluntary movements. We do not have to see our muscles to be sure that they are performing their intended actions.

SENSE OF TASTE

The receptors for taste are found in **taste buds,** most of which are in papillae on the tongue (Fig. 9–2). These **chemoreceptors** detect chemicals in solution in the mouth. The chemicals are foods and the solvent is saliva (if the mouth is very dry, taste is very indistinct). It is believed that there are four types of taste receptors: sweet, sour, salty, and bitter. We experience many different tastes, however, because foods stimulate different combinations of the four receptors, and the sense of smell also contributes to our perception of food.

The impulses from taste buds are transmitted by the facial and glossopharyngeal (7th and 9th cranial) nerves to the taste areas in the parietal-temporal cortex. The sense of taste is important because it makes eating enjoyable. Some medications may interfere with the sense of taste, and this sense becomes less acute as we get older. These may be contributing factors to poor nutrition in certain patients and in the elderly.

SENSE OF SMELL

The receptors for smell **(olfaction)** are **chemoreceptors** which detect vaporized chemicals that have been sniffed into the upper nasal cavities (see Fig. 9–2). Just as there are basic tastes, there are also believed to be basic scents, but their number is not known and estimates range from 7 to 50. When stimulated by vapor molecules, **olfactory receptors** generate impulses carried by the olfactory nerves (1st cranial) through the ethmoid bone to the olfactory bulbs. The pathway for these impulses ends in the olfactory areas of the temporal lobes.

The human sense of smell is very poorly devel-

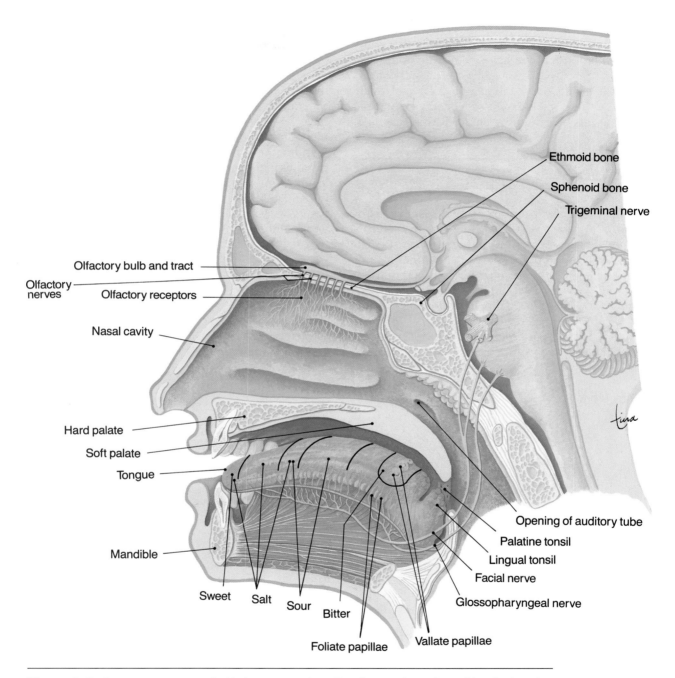

Figure 9–2 Structures concerned with the senses of smell and taste, shown in a midsagittal section of the head.

oped compared to those of other animals. Dogs, for example, have a sense of smell at least 200 times more acute than that of people. As mentioned earlier, however, much of what we call taste is actually the smell of food. If you have a cold and your nasal cavities are stuffed up, food just doesn't taste as good as it usually does. Adaptation occurs relatively quickly with odors. Pleasant scents may be sharply distinct at first but rapidly seem to dissipate or fade.

HUNGER AND THIRST

Hunger and thirst may be called **visceral sensations,** in that they are triggered by internal changes. The receptors are thought to be specialized cells in the hypothalamus. Receptors for hunger are believed to detect changes in blood nutrient levels, and receptors for thirst detect changes in body water content (actually the water-salt proportion).

Naturally we do not feel these sensations in the hypothalamus: they are projected. Hunger is pro-

jected to the stomach, which contracts. Thirst is projected to the mouth and pharynx, and less saliva is produced.

If not satisfied by eating, the sensation of hunger gradually diminishes, that is, adaptation occurs. The reason is that after blood nutrient levels decrease, they become stable as fat in adipose tissue is used for energy. With no sharp fluctuations, the receptors have few changes to detect, and hunger becomes much less intense.

In contrast, the sensation of thirst, if not satisfied by drinking, continues to worsen. As body water is lost, the amount keeps decreasing and does not stabilize. Therefore, there are constant changes for the receptors to detect, and prolonged thirst may be very painful.

THE EYE

The eye contains the receptors for vision and a refracting system that focuses light rays on the receptors in the retina. We will begin our discussion,

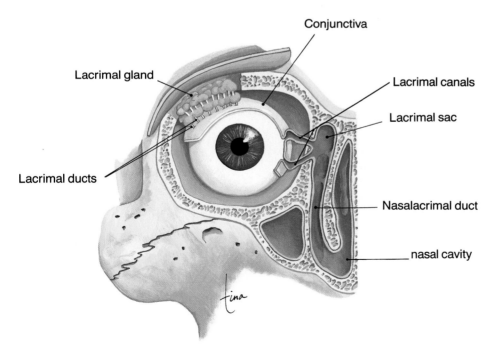

Conjunctiva

Lacrimal gland

Lacrimal canals

Lacrimal sac

Lacrimal ducts

Nasalacrimal duct

nasal cavity

Figure 9–3 Lacrimal apparatus shown in an anterior view of the right eye.

however, with the accessory structures of the eye, then later return to the eye itself and the physiology of vision.

EYELIDS AND THE LACRIMAL APPARATUS

The eyelids contain skeletal muscle that enables the eyelids to close and cover the front of the eyeball. Eyelashes along the border of each eyelid help keep dust out of the eyes. The eyelids are lined with a thin membrane called the **conjunctiva,** which is also folded over the white of the eye. Inflammation of this membrane, called **conjunctivitis,** is often caused by allergies and makes the eyes red, itchy, and watery.

Tears are produced by the **lacrimal glands,** located at the upper, outer corner of the eyeball, within the orbit (Fig. 9–3). Small ducts take tears to the anterior of the eyeball, and blinking spreads the tears and washes the surface of the eye. Tears are mostly water and contain **lysozyme,** an enzyme that inhibits the growth of most bacteria on the wet, warm surface of the eye. At the medial corner of the eyelids are two small openings into the superior and inferior lacrimal canals. These ducts take tears

to the **lacrimal sac** (in the lacrimal bone), which leads to the **nasolacrimal duct** that empties tears into the nasal cavity. This is why crying often makes the nose run.

EYEBALL

Most of the eyeball is within and protected by the **orbit,** formed by the maxilla, zygomatic, frontal, sphenoid, and ethmoid bones. The six **extrinsic muscles** of the eye are attached to this bony socket and to the surface of the eyeball. There are four rectus muscles that move the eyeball up and down or side to side and two oblique muscles that rotate the eye. These are shown in Fig. 9–4. The cranial nerves that innervate these muscles are the oculomotor, trochlear, and abducens (3rd, 4th, and 6th cranial).

Layers of the Eyeball

In its wall, the eyeball has three layers: the outer sclera, middle choroid layer, and inner retina (Fig. 9–5). The **sclera** is the thickest layer and is made of fibrous connective tissue which is visible as the white of the eye. The most anterior portion is the

Figure 9–4 Extrinsic muscles of the eye. Lateral view of left eye (the medial rectus and superior oblique are not shown).

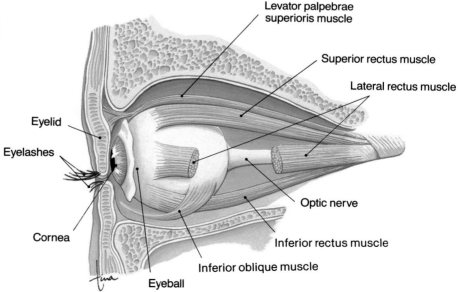

Levator palpebrae superioris muscle

Superior rectus muscle

Lateral rectus muscle

Eyelid

Eyelashes

Optic nerve

Cornea

Inferior rectus muscle

Inferior oblique muscle

Eyeball

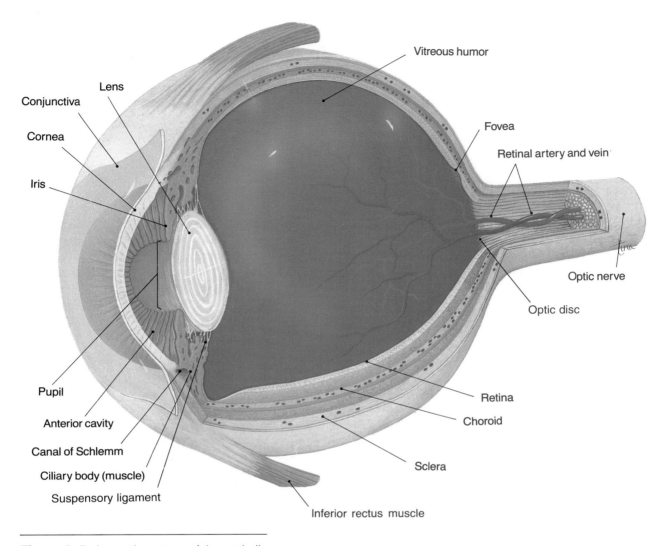

Conjunctiva

Lens

Cornea

Iris

Vitreous humor

Fovea

Retinal artery and vein

Optic nerve

Optic disc

Pupil

Anterior cavity

Canal of Schlemm

Ciliary body (muscle)

Suspensory ligament

Inferior rectus muscle

Retina

Choroid

Sclera

Figure 9–5   Internal anatomy of the eyeball.

Box 9–1   CATARACTS

The lens of the eye is normally transparent but may become opaque; this cloudiness or opacity is called a **cataract.** Cataract formation is most common among elderly people. With age, the proteins of the lens break down and lose their transparency. Long-term exposure to ultraviolet light (sunlight) also seems to be a contributing factor.

The cloudy lens does not refract light properly, and blurred vision is often an early symptom. Treatment of cataracts requires surgical removal of the lens and the use of glasses or contact lenses to replace the refractive function of the lens. The use of very precise laser surgery to destroy small cataracts may permit the lens to retain a portion of its function.

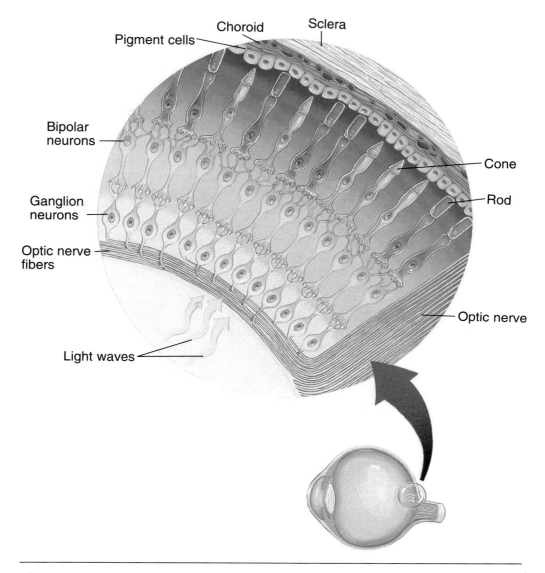

Figure 9–6 Microscopic structure of the retina in the area of the optic disc. See text for description.

cornea, which differs from the rest of the sclera in that it is transparent and has no capillaries. The cornea is the first part of the eye that **refracts,** or bends, light rays.

The **choroid layer** contains blood vessels and a dark blue pigment that absorbs light within the eyeball and thereby prevents glare. The anterior por-

tion of the choroid is modified into more specialized structures: the ciliary body and the iris. The ciliary body (muscle) is a circular muscle that surrounds the edge of the lens and is connected to the lens by **suspensory ligaments.** The **lens** is made of a transparent, elastic protein, and like the cornea, has no capillaries (see Box 9–1: Cataracts). The shape of the lens is changed by the ciliary muscle, which enables the eye to focus light from objects at varying distances from the eye.

Just in front of the lens is the circular **iris,** the colored part of the eye. Two sets of smooth muscle fibers change the diameter of the **pupil,** the central opening. Contraction of the radial fibers dilates the pupil; this is a sympathetic response. Contraction of the circular fibers constricts the pupil; this is a parasympathetic response (oculomotor nerves). Pupillary constriction is a reflex that protects the retina from intense light or that permits more acute near vision, as when reading.

The **retina** lines the posterior two thirds of the eyeball and contains the visual receptors, the rods and cones (Fig. 9–6). **Rods** detect only the presence of light, whereas **cones** detect colors which, as you may know from physics, are the different wavelengths of visible light. Cones are most abundant in the center of the retina. The **fovea,** which contains only cones, is a small depression directly behind the center of the lens and is the area for best color vision. Rods are proportionally more abundant toward the periphery, or edge, of the retina. Our best vision in dim light or at night, for which we depend on the rods, is at the sides of our visual fields.

Neurons called **ganglion neurons** carry the impulses generated by the rods and cones. These neurons all converge at the **optic disc** (see Figs. 9–5 and 9–6) and pass through the wall of the eyeball as the **optic nerve.** There are no rods or cones in the optic disc, so this part of the retina is sometimes called the "blind spot." We are not aware of a blind spot in our field of vision, however, in part because the eyes are constantly moving, and in part because the brain "fills in" the blank spot to create a "complete" picture.

Cavities of the Eyeball

There are two cavities within the eye: the posterior cavity and the anterior cavity. The larger, **posterior cavity** is found between the lens and retina and contains **vitreous humor.** This semisolid substance keeps the retina in place. If the eyeball is punctured and vitreous humor is lost, the retina may fall away from the choroid; this is one possible cause of a **detached retina.**

The **anterior cavity** is found between the front of the lens and the cornea and contains **aqueous humor,** the tissue fluid of the eyeball. Aqueous humor is formed by capillaries in the ciliary body, flows anteriorly through the pupil, and is reabsorbed by the **canal of Schlemm** (small veins also called the scleral venous sinus) at the junction of the iris and cornea. Since aqueous humor is tissue

Box 9–2   GLAUCOMA

The presence of aqueous humor in the anterior cavity of the eye creates a pressure called intraocular pressure. **Glaucoma** is an increase in intraocular pressure due to decreased reabsorption of aqueous humor into the Canal of Schlemm. There are several types of glaucoma, but if untreated, all have the same outcome. Increased pressure in the anterior cavity is transmitted to the lens, vitreous humor, and retina. As pressure on the retina increases, halos may be seen around bright lights. Frequently, however, a person with glaucoma has no symptoms before severe visual impairment or blindness occurs.

Glaucoma may often be controlled with medications that constrict the pupil and flatten the iris, thus opening up access to the Canal of Schlemm. If such medications are not effective, surgery may be required to create a larger Canal of Schlemm.

Anyone over the age of 40 should have a test for glaucoma; anyone with a family history of glaucoma should have this test annually.

Box 9–3   ERRORS OF REFRACTION

Normal visual acuity is referred to as 20/20, that is, the eye should and does clearly see an object 20 feet away. **Nearsightedness (myopia)** means that the eye sees near objects well but not distant ones. If an eye has 20/80 vision, this means that what the normal eye can see at 80 feet, the nearsighted eye can see only if the object is brought to 20 feet away. The nearsighted eye focuses images from distant object in front of the retina, because the eyeball is too long or the lens too thick (see diagram below). These structural characteristics of the eye are hereditary. Correction of nearsightedness requires a concave lens to spread out light rays before they strike the eye.

Farsightedness (hyperopia) means that the eye sees distant objects well. Such an eye may have an acuity of 20/10, that is, it sees at 20 feet what the normal eye can see only at 10 feet. The farsighted eye focuses light from near objects "behind"

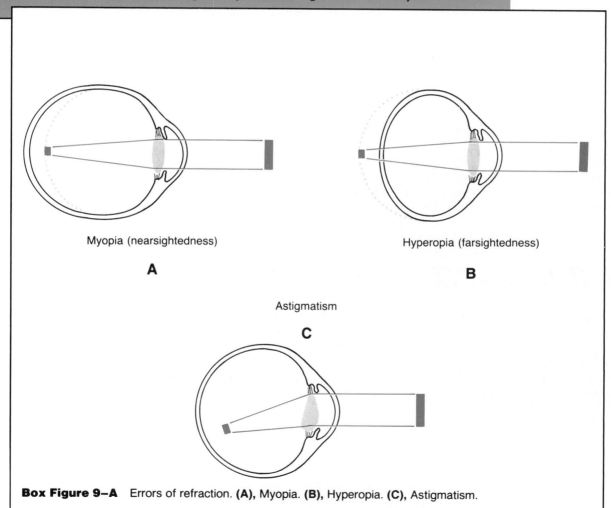

Myopia (nearsightedness)

A

Hyperopia (farsightedness)

B

Astigmatism

C

Box Figure 9–A   Errors of refraction. **(A)**, Myopia. **(B)**, Hyperopia. **(C)**, Astigmatism.

fluid, you would expect it to have a nourishing function, and it does. Recall that the lens and cornea have no capillaries; they are nourished by the continuous flow of aqueous humor (see Box 9–2: Glaucoma).

PHYSIOLOGY OF VISION

In order for us to see, light rays must be focused on the retina, and the resulting nerve impulses must be transmitted to the visual areas of the cerebral cortex in the brain.

Refraction of light rays is the deflection or bending of a ray of light as it passes through one object and into another object of greater or lesser density. The refraction of light within the eye takes place in the following pathway of structures: the cornea, aqueous humor, lens, and vitreous humor. The lens is the only adjustable part of the refraction system. When looking at distant objects, the ciliary muscle is relaxed and the lens is elongated and thin. When looking at near objects, the ciliary muscle contracts to form a smaller circle, the elastic lens recoils and bulges in the middle, and has greater refractive power (see Box 9–3: Errors of Refraction).

When light rays strike the retina, they stimulate chemical reactions in the rods and cones. In rods, the chemical **rhodopsin** breaks down to form scotopsin and retinene (a derivative of vitamin A). This chemical reaction generates an electrical impulse, and rhodopsin is then resynthesized in a slower re-action. Chemical reactions in the cones are brought about by different wavelengths of light. It is believed that there are three types of cones: red-absorbing, blue-absorbing, and green-absorbing. Each type absorbs wavelengths over about a third of the visible light spectrum, so red cones, for example, absorb light of the red, orange, and yellow wavelengths. The chemical reactions in cones also generate electrical impulses (see Box 9–4: Nightblindness and Colorblindness).

The impulses from the rods and cones are transmitted to **ganglion neurons** (see Fig. 9–6); these converge at the optic disc and become the **optic nerve,** which passes posteriorly through the wall of the eyeball.

The optic nerves from both eyes converge at the **optic chiasma,** just in front of the pituitary gland. Here, the medial fibers of each optic nerve cross to the other side. This crossing permits each visual area to receive impulses from both eyes, which is important for binocular vision.

The visual areas are in the **occipital lobes** of the cerebral cortex. Although each eye transmits a slightly different picture, the visual areas put them together or integrate them to make a single image. This is what is called **binocular vision.** The visual areas also right the image, since the image on the retina is upside down. The image on film in a camera is also upside down, but we don't even realize that because we look at the pictures right side up. The brain just as automatically ensures that we see our world right side up.

Box 9–4   NIGHTBLINDNESS AND COLORBLINDNESS

Nightblindness, the inability to see well in dim light or at night is usually caused by a deficiency of vitamin A, although some nightblindness may occur with aging. Vitamin A is necessary for the synthesis of rhodopsin in the rods. Without sufficient vitamin A, there is not enough rhodopsin present to respond to low levels of light.

Colorblindness is a genetic disorder in which one of the three sets of cones is lacking or nonfunctional. Total colorblindness, the inability to see any colors at all, is very rare. The most common form is red-green colorblindness, which is the inability to distinguish between these colors. If either the red cones or green cones are missing, the person will still see most of the colors, but will not have the contrast that the missing set of cones would provide. So red and green shades will look somewhat similar, without the definite difference most of us see. (See the accompanying illustration.) This is a sex-linked trait; the recessive gene is on the X chromosome. A woman with one gene for colorblindness and a gene for normal color vision on her other X chromosome will not be colorblind but may pass the gene for colorblindness to her children. A man with a gene for colorblindness on his X chromosome has no gene at all for color vision on his Y chromosome and will be colorblind.

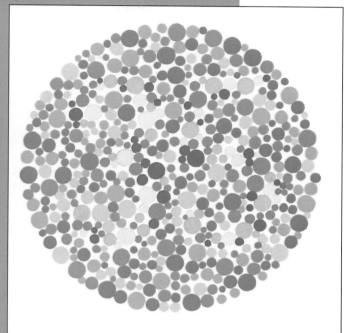

Box Figure 9–B   Example of color patterns used to detect colorblindness.

THE EAR

The ear consists of three areas: the outer ear, the middle ear, and the inner ear (Fig. 9–7). The ear contains the receptors for two senses: hearing and **equilibrium.** These receptors are all found in the inner ear.

OUTER EAR

The **outer ear** consists of the auricle and the ear canal. The **auricle,** or **pinna,** is made of cartilage covered with skin. For animals such as dogs, whose ears are movable, the auricle may act as a funnel for sound waves. For people, however, the stationary auricle is not important. Hearing would not be negatively affected without it, although those of us who wear glasses would have our vision impaired without our auricles. The **ear canal,** also called the **external auditory meatus,** is a tunnel into the temporal bone and curves slightly forward and down.

MIDDLE EAR

The **middle ear** is an air-filled cavity in the temporal bone. The **ear drum,** or **tympanic mem-**

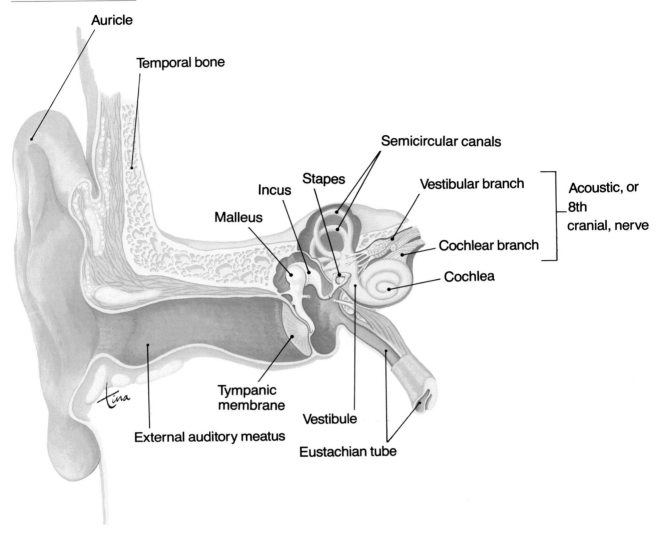

Auricle

Temporal bone

Semicircular canals

Stapes

Incus

Vestibular branch

Acoustic, or
8th
cranial, nerve

Malleus

Cochlear branch

Cochlea

Tympanic
membrane

Vestibule

External auditory meatus

Eustachian tube

Figure 9–7 Outer, middle, and inner ear structures as shown in a frontal section through the right temporal bone.

brane, is stretched across the end of the ear canal and vibrates when sound waves strike it. These vibrations are transmitted to the three auditory bones: the **malleus, incus,** and **stapes.** The stapes then transmits vibrations to the fluid-filled inner ear at the **oval window.**

The **Eustachian tube** (auditory tube) extends from the middle ear to the nasopharynx and permits air to enter or leave the middle ear cavity. The air pressure in the middle ear must be the same as the external atmospheric pressure in order for the ear drum to vibrate properly. You may have noticed your ears "popping" when in an airplane or when driving to a higher or lower altitude. Swallowing or yawning creates the "pop" by opening the Eustachian tubes and equalizing the air pressures.

The Eustachian tubes of children are short and nearly horizontal and may permit bacteria to spread from the pharynx to the middle ear. This is why **otitis media** may be a complication of a strep throat.

INNER EAR

Within the temporal bone, the **inner ear** is a cavity called the **bony labyrinth** (maze), which is lined with membrane called the **membranous labyrinth. Perilymph** is the fluid found between bone and membrane, and **endolymph** is the fluid within the membranous structures of the inner ear. These structures are the cochlea, concerned with hearing, and the utricle, saccule, and semicircular canals, all concerned with equilibrium (Fig. 9–8).

Cochlea

The **cochlea** is shaped like a snail shell with two and a half structural turns. Internally, the cochlea is partitioned into three fluid-filled canals. The medial canal is the cochlear duct, which contains the receptors for hearing in the **Organ of Corti (spiral organ).** The receptors are called hair cells (their

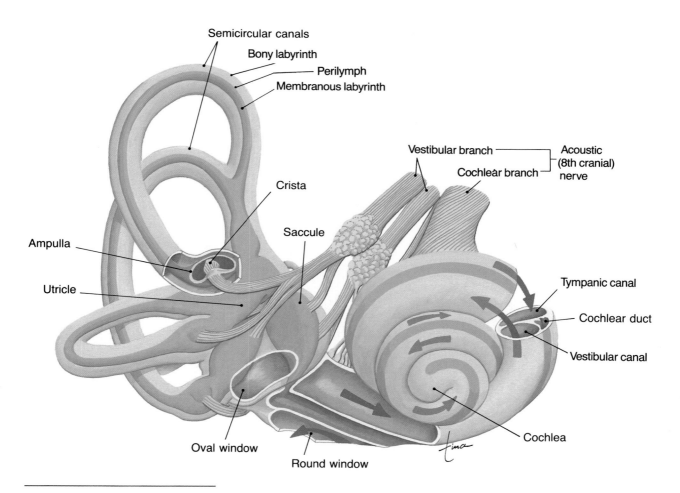

Figure 9–8 Inner ear structures.

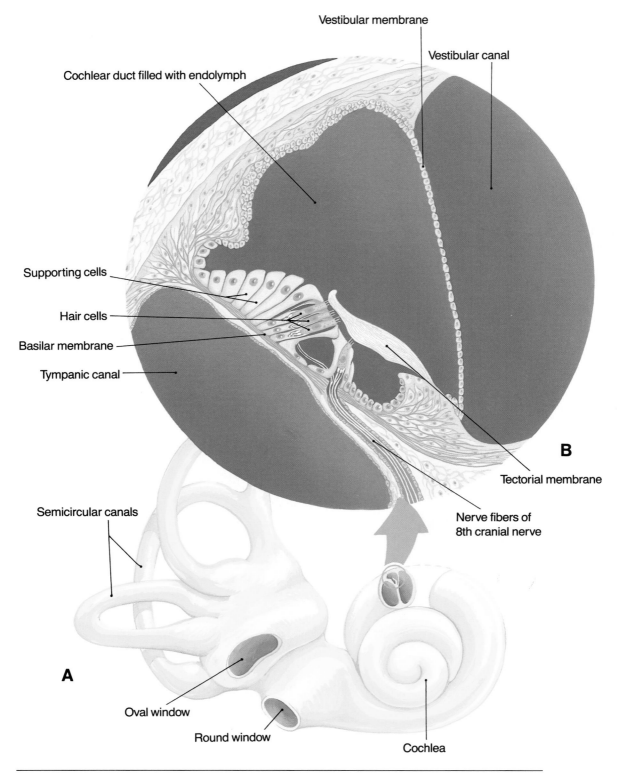

Figure 9–9 Organ of Corti. (**A**), Inner ear structures. (**B**), Magnification of Organ of Corti within the cochlea.

projections are not "hair," of course, but rather are specialized microvilli called stereocilia), which contain endings of the cochlear branch of the 8th cranial nerve. Overhanging the hair cells is the tectorial membrane (Fig. 9–9).

Very simply, the process of hearing involves the transmission of vibrations and the generation of nerve impulses. When sound waves enter the ear canal, vibrations are transmitted by the following sequence of structures: ear drum, malleus, incus, stapes, oval window of the inner ear, perilymph and endolymph within the cochlea, and hair cells of the Organ of Corti. When the hair cells bend, they generate impulses that are carried by the 8th cranial nerve to the brain. As you may recall, the auditory areas are in the **temporal lobes** of the cerebral cortex. It is here that sounds are heard and interpreted (see Box 9–5: Deafness).

Utricle and Saccule

The **utricle** and **saccule** are membranous sacs in an area called the **vestibule,** between the cochlea and semicircular canals. Within the utricle and saccule are hair cells that are moved by gravity as the position of the head changes. The impulses generated by these hair cells are carried by the vestibular portion of the 8th cranial nerve to the cerebellum, midbrain, and the temporal lobes of the cerebrum.

The cerebellum and midbrain use this information to maintain equilibrium at a subconscious level. We can, of course, be aware of the position of the head, and it is the cerebrum that provides awareness.

Semicircular Canals

The three **semicircular canals** are fluid-filled membranous ovals oriented in three different planes. At the base of each is an enlarged portion called the ampulla (see Fig. 9–8), which contains hair cells (the crista) that are affected by movement. As the body moves forward, for example, the hair cells are bent backward at first. The bending of the hair cells generates impulses carried by the vestibular branch of the 8th cranial nerve to the cerebellum, midbrain, and temporal lobes of the cerebrum. These impulses are interpreted as starting or stopping and accelerating or decelerating, and this in-

---

Box 9–5 DEAFNESS

Deafness is the inability to hear properly; the types are classified according to the part of the hearing process that is not functioning normally.

Conduction deafness—impairment of one of the structures that transmits vibrations. Examples of this type are a punctured eardrum, arthritis of the auditory bones, or a middle ear infection in which fluid fills the middle ear cavity.

Nerve deafness—impairment of the 8th cranial nerve or the receptors for hearing in the cochlea. The 8th cranial nerve may be damaged by some antibiotics used to treat bacterial infections. Nerve deafness is a rare complication of some viral infections such as mumps or congenital rubella (German measles). Deterioration of the hair cells in the cochlea is a natural consequence of aging, and the acuity of hearing diminishes as we get older. For example, it may be more difficult for an elderly person to distiguish conversation from background noise. Chronic exposure to loud noise accelerates degeneration of the hair cells and onset of this type of deafness.

Central deafness—damage to the auditory areas in the temporal lobes. This type of deafness is rare but may be caused by a brain tumor, meningitis, or a CVA in the temporal lobe.

Box 9–6 MOTION SICKNESS

Motion sickness is characterized by cold sweats, hyperventilation, nausea, and vomiting when the person is exposed to repetitive motion of some kind. Seasickness is a type of motion sickness, as is carsickness (why children are more often carsick than are adults is not known).

Some people are simply not affected by the rolling of a ship or train; for others, the constant stimulation of the receptors for position first becomes uncomfortable, then nauseating. For those who know they are susceptible to motion sickness, medications are available for use before traveling by plane, train, or car.

---

formation is used to maintain equilibrium while we are moving (see Box 9–6: Motion Sickness).

In summary then, the utricle and saccule provide information about the position of the body at rest, while the semicircular canals provide information about the body in motion.

ARTERIAL RECEPTORS

The aorta and carotid arteries contain receptors that detect changes in the blood. The **aortic arch,** which receives blood pumped by the left ventricle of the heart, curves over the top of the heart. The left and right **carotid arteries** are branches of the aortic arch that take blood through the neck on the way to the brain. In each of these vessels are pressoreceptors and chemoreceptors (see Fig. 12–7).

Pressoreceptors in the carotid sinuses and aortic sinus detect changes in blood pressure. **Chemoreceptors** in the carotid bodies and the aortic body detect changes in the oxygen and carbon dioxide content, and the pH, of blood. The impulses generated by these receptors do not give rise to sensations that we feel but rather are information used to make any necessary changes in respiration or circulation. We will return to this in later chapters, so one example will suffice for now.

If the blood level of oxygen decreases significantly, this change (hypoxia) is detected by carotid and aortic chemoreceptors. The sensory impulses are carried by the glossopharyngeal (9th cranial) and vagus (10th cranial) nerves to the medulla. Cen-

ters in the medulla may then increase the respiratory rate and the heart rate to obtain and circulate more oxygen. These are the respiratory and cardiac reflexes that were mentioned in Chapter 8 as functions of the glossopharyngeal and vagus nerves. The importance of these reflexes is readily apparent: to maintain normal blood levels of oxygen and carbon dioxide and to maintain normal blood pressure.

SUMMARY

Changes take place all around us as well as within us. If the body could not respond appropriately to environmental and internal changes, homeostasis would soon be disrupted, resulting in injury, illness, or even death. In order to respond appropriately to changes, the brain must know what they are. Conveying this information to our brains is the function of our senses. Although we may sometimes take our senses for granted, we could not survive for very long without them.

You have just read about the great variety of internal and external changes that are detected by the sense organs. You are also familiar with the role of the nervous system in the regulation of the body's responses. In the next chapter we will discuss the other regulatory system, the endocrine system. The hormones of the endocrine glands are produced in response to changes, and their regulatory effects all contribute to homeostasis.

STUDY OUTLINE

Purpose of Sensations—to detect changes in the external or internal environment to enable the body to respond appropriately to maintain homeostasis
Sensory Pathway—pathway of impulses for a sensation
1. Receptors—detect a change (usually very specific) and generate impulses.
2. Sensory neurons—transmit impulses from receptors to the CNS.
3. Sensory tracts—white matter in the CNS.
4. Sensory area—most are in the cerebral cortex; feels and interprets the sensation.

Characteristics of Sensations
1. Projection—the sensation seems to come from the area where the receptors were stimulated, even though it is the brain that truly feels the sensation.
2. Intensity—the degree to which a sensation is felt; a strong stimulus affects more receptors, more impulses are sent to the brain and are interpreted as a more intense sensation.
3. Contrast—the effect of a previous or simultaneous sensation on a current sensation as the brain compares them.
4. Adaptation—becoming unaware of a continuing stimulus; if the stimulus remains constant, there is no change for receptors to detect.
5. After-image—the sensation remains in the consciousness after the stimulus has stopped.

Cutaneous Senses—provide information about the external environment and the skin itself
1. In the dermis are free nerve endings for pain and encapsulated nerve endings for touch, pressure, heat, and cold (see Fig. 9–1).
2. Sensory areas are in parietal lobes.
3. Referred pain is visceral pain that is felt as cutaneous pain. Common pathways in the CNS carry both cutaneous and visceral impulses; the brain usually projects sensation to the cutaneous area.

Muscle Sense—knowing where our muscles are without looking at them
1. Stretch receptors in muscles detect stretching.
2. Sensory areas for conscious muscle sense are in parietal lobes.
3. Cerebellum uses unconscious muscle sense to coordinate voluntary movement.

Sense of Taste (see Fig. 9–2)
1. Chemoreceptors are in taste buds on the tongue; detect chemicals (foods) in solution (saliva) in the mouth.
2. Four basic tastes: sweet, sour, salty, bitter; foods stimulate combinations of receptors.
3. Pathway: Facial and glossopharyngeal nerves to taste areas in parietal-temporal lobes.

Sense of Smell (see Fig. 9–2)
1. Chemoreceptors are in upper nasal cavities; detect vaporized chemicals.
2. Pathway: olfactory nerves to olfactory bulbs to olfactory areas in the temporal lobes.
3. Smell contributes greatly to what we call taste.

Hunger and Thirst—visceral (internal) sensations
1. Receptors for hunger: in hypothalamus, detect changes in nutrient levels in the blood; hunger is projected to the stomach; adaptation does occur.
2. Receptors for thirst: in hypothalamus, osmoreceptors detect changes in body water (water-salt proportions); thirst is projected to the mouth and pharynx; adaptation does not occur.

The Eye (see Figs. 9–3 through 9–6)
1. Eyelids and eyelashes keep dust out of eyes; conjunctiva line the eyelids and cover white of eye.
2. Lacrimal glands produce tears which flow across eyeball to two lacrimal ducts, to lacrimal sac to nasolacrimal duct to nasal cavity. Tears wash the anterior eyeball and contain lysozyme to inhibit bacterial growth.

3. The eyeball is protected by the bony orbit (socket).
4. The six extrinsic muscles move the eyeball; innervated by the 3rd, 4th, and 6th cranial nerves.
5. Sclera—outermost layer of the eyeball, made of fibrous connective tissue; anterior portion is the transparent cornea, the first light-refracting structure.
6. Choroid layer—middle layer of eyeball; dark blue pigment absorbs light to prevent glare within the eyeball.
7. Ciliary body (muscle) and suspensory ligaments—change shape of lens, which is made of a transparent, elastic protein and which refracts light.
8. Iris—two sets of smooth muscle fibers regulate diameter of pupil, that is, how much light strikes the retina.
9. Retina—innermost layer of eyeball; contains rods and cones.
   - Rods—detect light; abundant toward periphery of retina.
   - Cones—detect color; abundant in center of retina. Fovea—contains only cones; area of best color vision.
   - Optic disc—no rods or cones; optic nerve passes through eyeball.
10. Posterior cavity contains vitreous humor (semisolid) that keeps the retina in place.
11. Anterior cavity contains aqueous humor that nourishes the lens and cornea; made by capillaries of the ciliary body, flows through pupil, is reabsorbed to blood at the Canal of Schlemm.

Physiology of Vision
1. Refraction (bending and focusing) pathway of light: cornea, aqueous humor, lens, vitreous humor.
2. Lens is adjustable: ciliary muscle relaxes for distant vision, and lens is thin. Ciliary muscle contracts for near vision, and elastic lens thickens and has greater refractive power.
3. Light strikes retina and stimulates chemical reactions in the rods and cones.
4. In rods: rhodopsin breaks down to scotopsin and retinene (from vitamin A), and an electrical impulse is generated. In cones: specific wavelengths of light are absorbed (red, blue, green); chemical reactions generate nerve impulses.
5. Ganglion neurons from the rods and cones form the optic nerve, which passes through the eyeball at the optic disc.
6. Optic chiasma—site of the crossover of medial fibers of both optic nerves, permitting binocular vision.
7. Visual areas in occipital lobes—each area receives impulses from both eyes; both areas create one image from the two slightly different images of each eye; both areas right the upside down retinal image.

The Ear (see Figs. 9–7 through 9–9)
1. Outer ear—auricle or pinna has no real function for people; ear canal curves forward and down into temporal bone.
2. Middle ear—ear drum at end of ear canal vibrates when sound waves strike it. Auditory bones: malleus, incus, stapes; transmit vibrations to inner ear at oval window.
   - Eustachian tube—extends from middle ear to nasopharynx; allows air in and out of middle ear to permit eardrum to vibrate; air pressure in middle ear should equal atmospheric pressure.
3. Inner ear—bony labyrinth in temporal bone, lined with membranous labyrinth. Perilymph is fluid between bone and membrane; endolymph is fluid within membrane. Membranous structures are the cochlea, utricle and saccule, and semicircular canals.
4. Cochlea—snail-shell shaped; three internal canals; cochlear duct contains receptors for hearing: hair cells in the Organ of Corti; these cells contain endings of the cochlear branch of the 8th cranial nerve.
5. Physiology of hearing—sound waves stimulate vibration of ear drum, malleus, incus, stapes, oval window of inner ear, perilymph and endolymph of cochlea, and hair cells of Organ of Corti. When hair cells bend, impulses are generated and carried by the 8th cranial nerve to the auditory areas in the temporal lobes.

6. Utricle and saccule—membranous sacs in the vestibule; each contains hair cells that are affected by gravity. When position of the head changes, hair cells bend and generate impulses along the vestibular branch of the 8th cranial nerve to the cerebellum, midbrain, and cerebrum. Impulses are interpreted as position of the head at rest.

7. Semicircular canals—three membranous ovals in three planes; enlarged base is the ampulla which contains hair cells (crista) that are affected by movement. As body moves, hair cells bend in opposite direction, generate impulses along vestibular branch of 8th cranial nerve to cerebellum, midbrain, and cerebrum. Impulses are interpreted as movement of the body, changing speed, stopping or starting.

Arterial Receptors—in large arteries; detect changes in blood

1. Aortic arch—curves over top of heart. Aortic sinus contains pressoreceptors; aortic body contains chemoreceptors; sensory nerve is vagus (10th cranial).

2. Right and left carotid arteries in the neck; carotid sinus contains pressoreceptors; carotid body contains chemoreceptors; sensory nerve is the glossopharyngeal (9th cranial).

3. Pressoreceptors detect changes in blood pressure; chemoreceptors detect changes in pH or oxygen and CO_2 levels in the blood. This information is used by the vital centers in the medulla to change respiration or circulation to maintain normal blood oxygen and CO_2 and normal blood pressure.

REVIEW QUESTIONS

1. State the two general functions of receptors. Explain the purpose of sensory neurons and sensory tracts. (p. 194)

2. Name the receptors for the cutaneous senses, and explain the importance of this information. (pp. 195–196)

3. Name the receptors for muscle sense and the parts of the brain concerned with muscle sense. (p. 196)

4. State what the chemoreceptors for taste and smell detect. Name the cranial nerve(s) for each of these senses and the lobe of the cerebrum where each is felt. (pp. 196, 198)

5. Name the part of the eye with each of the following functions: (pp. 199–202)
   a. change the shape of the lens
   b. contains the rods and cones
   c. forms the white of the eye
   d. form the optic nerve
   e. keep dust out of eye
   f. changes the size of the pupil
   g. produce tears
   h. absorbs light within the eyeball to prevent glare

6. With respect to vision: (pp. 202, 204)
   a. Name the structures and substances that refract light rays (in order).
   b. State what cones detect and what rods detect. What happens within these receptors when light strikes them?

  c. Name the cranial nerve for vision and the lobe of the cerebrum that contains the visual area.

7. With respect to the ear: (pp. 205–210)
  a. Name the parts of the ear that transmit the vibrations of sound waves (in order).
  b. State the location of the receptors for hearing.
  c. State the location of the receptors that respond to gravity.
  d. State the location of the receptors that respond to motion.
  e. State the two functions of the 8th cranial nerve.
  f. Name the lobe of the cerebrum concerned with hearing.
  g. Name the two parts of the brain concerned with maintaining balance and equilibrium.

8. Name the following (p. 210):
  a. the locations of arterial chemoreceptors, and state what they detect
  b. the locations of arterial pressoreceptors, and state what they detect
  c. the cranial nerves involved in respiratory and cardiac reflexes, and state the part of the brain that regulates these vital functions

9. Explain each of the following: adaptation, after-image, projection, contrast. (pp. 194–195)

Chapter 10
The Endocrine System

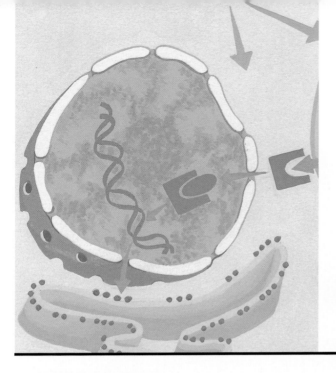

Chapter 10

Student Objectives

- Name the endocrine glands and the hormones secreted by each.
- Explain how a negative feedback mechanism works.
- Explain how the hypothalamus is involved in the secretion of hormones from the posterior pituitary gland and anterior pituitary gland.
- State the functions of oxytocin and antidiuretic hormone, and explain the stimulus for secretion of each.
- State the functions of the hormones of the anterior pituitary gland, and state the stimulus for secretion of each.

The Endocrine System

New Terminology

Alpha cells (**AL**–fah–**SELLS**)
Beta cells (**BAY**–tah–**SELLS**)
Catecholamines (**KAT**–e–kohl–ah–**MEENZ**)
Corpus luteum (**KOR**–pus **LOO**–tee–um)
Gluconeogenesis (GLOO–koh–nee–oh–**JEN**–i–sis)
Glycogenesis (GLIGH–koh–**JEN**–i–sis)
Glycogenolysis (GLIGH–ko–jen–**OL**–i–sis)
Hypercalcemia (HIGH–per–kal–**SEE**–mee–ah)
Hyperglycemia (HIGH–per–gligh–**SEE**–mee–ah)
Hypocalcemia (HIGH–poh–kal–**SEE**–mee–ah)
Hypoglycemia (HIGH–poh–gligh–**SEE**–mee–ah)
Hypophysis (high–**POFF**–e–sis)
Islets of Langerhans (**EYE**–lets of **LAHNG**–er–hanz)
Negative feedback mechanism (**NEG**–ah–tiv **FEED**–bak)
Prostaglandins (PRAHS–tah–**GLAND**–ins)
Releasing hormones (ree–**LEE**–sing **HOR**–mohns)
Renin-angiotensin mechanism (**REE**–nin AN–jee–oh–**TEN**–sin)
Sympathomimetic (SIM–pah–tho–mi–**MET**–ik)
Target organ (**TAR**–get **OR**–gan)

Related Clinical Terminology

Acromegaly (ACK–roh–**MEG**–ah–lee)
Addison's disease (**ADD**–i–sonz)
Cretinism (**KREE**–tin–izm)
Cushing's syndrome (**KOOSH**–ingz **SIN**–drohm)
Diabetes mellitus (DYE–ah–**BEE**–tis mel–**LYE**–tus)
Giantism (**JIGH**–an–tizm)
Goiter (**GOY**–ter)
Grave's disease (GRAYVES)
Ketoacidosis (KEY–toh–ass–i–**DOH**–sis)
Myxedema (MICK–suh–**DEE**–mah)
Pituitary dwarfism (pi–**TOO**–i–TER–ee **DWORF**–izm)

Terms that appear in **bold type** in the chapter text are defined in the glossary, which begins on page 549.

We have already seen how the nervous system regulates body functions by means of nerve impulses and integration of information by the spinal cord and brain. The other regulating system of the body is the **endocrine system,** which consists of endocrine glands that secrete chemicals called **hormones.** These glands are shown in Fig. 10–1.

Endocrine glands are ductless, that is, they do not have ducts to take their secretions to specific sites. Instead, hormones are secreted directly into capillaries and circulate in the blood throughout the body. Each hormone then exerts very specific effects on certain organs, called **target organs** or **target tissues.** Some hormones, such as insulin and thyroxine, have many target organs. Other hormones, such as calcitonin and some pituitary gland hormones, have only one or a few target organs.

In general, the endocrine system and its hormones help regulate growth, the use of foods to produce energy, resistance to stress, the pH of body fluids and fluid balance, and reproduction. In this chapter we will discuss the specific functions of the hormones and how each contributes to homeostasis.

CHEMISTRY OF HORMONES

With respect to their chemical structure, hormones may be classified into three groups: amines, proteins, and steroids.

1. **Amines**—these simple hormones are structural variations of the amino acid tyrosine. This group includes thyroxine from the thyroid gland and epinephrine and norepinephrine from the adrenal medulla.
2. **Proteins**—these hormones are chains of amino acids. Insulin from the pancreas, growth hormone from the anterior pituitary gland, and calcitonin from the thyroid gland are all proteins. Short chains of amino acids may be called **peptides.** Antidiuretic hormone and oxytocin, synthesized by the hypothalamus, are peptide hormones.
3. **Steroids**—cholesterol is the precursor for the steroid hormones, which include cortisol and

aldosterone from the adrenal cortex, estrogen and progesterone from the ovaries, and testosterone from the testes.

REGULATION OF HORMONE SECRETION

Hormones are secreted by endocrine glands when there is a need for them, that is, for their effects on their target organs. The cells of endocrine glands respond to changes in the blood, or perhaps to other hormones in the blood. These stimuli are the information they use to increase or decrease secretion of their own hormones. When a hormone brings about its effects, that reverses the stimulus, and secretion of the hormone decreases until the stimulus reoccurs. A specific example will be helpful here; let us use insulin.

Insulin is secreted by the pancreas when the blood glucose level is high, that is, hyperglycemia is the stimulus for secretion of insulin. Once circulating in the blood, insulin enables cells to remove glucose from the blood to use for energy production and enables the liver to store glucose as glycogen. As a result of these actions of insulin, blood glucose level decreases, reversing the stimulus for secretion of insulin. Insulin secretion then decreases until the blood glucose level increases again.

This is an example of a **negative feedback mechanism,** in which information about the effects of the hormone is "fed back" to the gland, which then decreases its secretion of the hormone. This is why the mechanism is called "negative": the effects of the hormone reverse the stimulus and decrease the secretion of the hormone. The secretion of many other hormones is regulated in a similar way.

The hormones of the anterior pituitary gland are secreted in response to **releasing hormones** secreted by the hypothalamus. You may recall this from Chapter 8. Growth hormone, for example, is secreted in response to growth hormone releasing hormone (GHRH) from the hypothalamus. As growth hormone exerts its effects, the secretion of GHRH decreases, which in turn decreases the se-

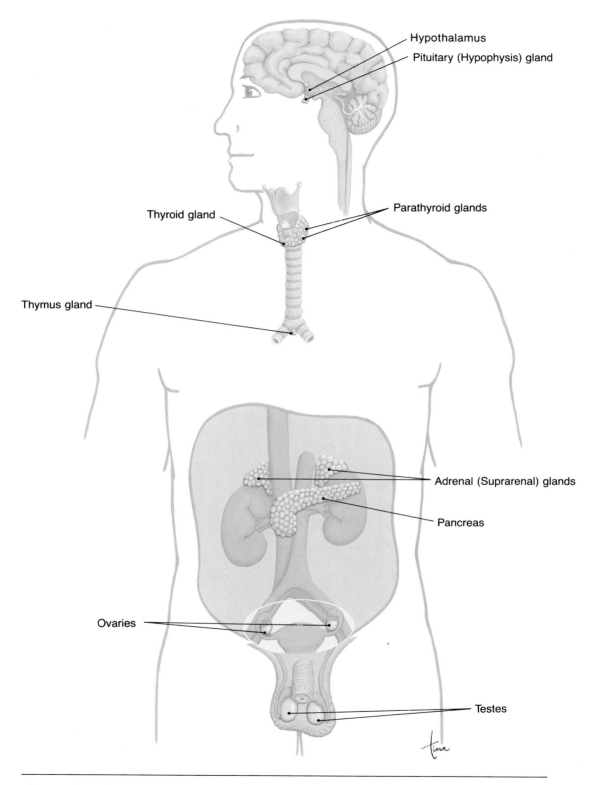

Figure 10–1 The endocrine system. Locations of many endocrine glands. Both male and female gonads (testes and ovaries) are shown.

cretion of growth hormone. This is another type of negative feedback mechanism.

For each of the hormones to be discussed in this chapter, the stimulus for its secretion will also be mentioned. Some hormones function as an **antagonistic pair** to regulate a particular aspect of blood chemistry; these mechanisms will also be covered.

THE PITUITARY GLAND

The **pituitary gland,** (or the **hypophysis**), hangs by a short stalk (infundibulum) from the hypothalamus and is enclosed by the sella turcica of the sphenoid bone. Despite its small size, the pituitary gland regulates many body functions. Its two major portions are the posterior pituitary gland **(neurohypophysis)**, which is actually an exten-

sion of the nerve tissue of the hypothalamus, and the anterior pituitary gland **(adenohypophysis)** which is separate glandular tissue.

POSTERIOR PITUITARY GLAND

The two hormones of the **posterior pituitary gland** are actually produced by the hypothalamus and simply stored in the posterior pituitary until needed. Their release is stimulated by nerve impulses from the hypothalamus (Fig. 10–2).

Antidiuretic Hormone

Antidiuretic hormone (ADH) increases the reabsorption of water by kidney tubules, which decreases the amount of urine formed. The water is reabsorbed into the blood, so as urinary output is

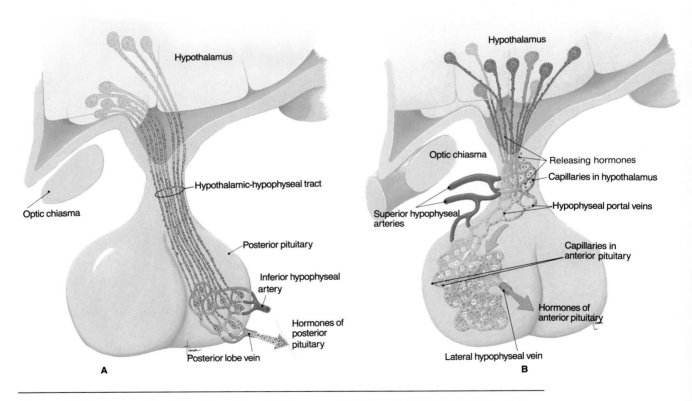

Figure 10–2 Structural relationships of hypothalamus and pituitary gland. (**A**), Posterior pituitary stores hormones produced in the hypothalamus. (**B**), Releasing hormones of the hypothalamus circulate directly to the anterior pituitary and influence its secretions. Notice the two networks of capillaries.

decreased, blood volume is increased, which helps maintain normal blood pressure.

The stimulus for secretion of ADH is decreased water content of the body. If too much water is lost in sweating or diarrhea, for example, **osmoreceptors** in the hypothalamus detect the increased "saltiness" of body fluids. The hypothalamus then transmits impulses to the posterior pituitary to increase the secretion of ADH and decrease the loss of more water in urine.

Any type of dehydration stimulates the secretion of ADH to conserve body water. In the case of severe hemorrhage, ADH is released in large amounts and will also cause vasoconstriction, which will contribute to the maintenance of normal blood pressure.

Ingestion of alcohol inhibits the secretion of ADH and increases urinary output. If alcohol intake is excessive and fluid is not replaced, a person will feel thirsty and dizzy the next morning. The thirst is due to the loss of body water, and the dizziness is the result of low blood pressure.

Oxytocin

Oxytocin stimulates contraction of the uterus at the end of pregnancy and stimulates release of milk from the mammary glands.

As labor begins, the cervix of the uterus is stretched, which generates sensory impulses to the hypothalamus, which in turn stimulates the posterior pituitary to release oxytocin. Oxytocin then causes strong contractions of the smooth muscle (myometrium) of the uterus to bring about delivery of the baby and the placenta.

Recently, it has been discovered that the placenta itself secretes oxytocin at the end of gestation and in an amount far higher than that from the posterior pituitary gland. Research is continuing to determine the exact mechanism and precise role of the placenta in labor.

When a baby is breast-fed, the sucking of the baby stimulates sensory impulses from the mother's nipple to the hypothalamus. Nerve impulses from the hypothalamus to the posterior pituitary cause the release of oxytocin, which stimulates contraction of the smooth muscle cells around the mammary ducts. This release of milk is sometimes called the "milk let-down" reflex. The hormones of the posterior pituitary are summarized in Table 10–1.

ANTERIOR PITUITARY GLAND

The hormones of the **anterior pituitary gland** regulate many body functions. They are in turn regulated by **releasing hormones** from the hypothalamus. These releasing hormones are secreted into capillaries in the hypothalamus and pass through the **hypophyseal portal** veins to another capillary network in the anterior pituitary gland. Here, the releasing hormones are absorbed and stimulate secretion of the anterior pituitary hormones. This small but specialized pathway of circulation is shown in Fig. 10–2. This pathway permits the releasing hormones to rapidly stimulate the anterior pituitary, without having to pass through general circulation.

Growth Hormone

Growth hormone (GH) may also be called **somatotropin,** and does indeed stimulate growth.

Table 10–1 HORMONES OF THE POSTERIOR PITUITARY GLAND

Hormone	Function(s)	Regulation of Secretion
Oxytocin	• Promotes contraction of myometrium of uterus (labor) • Promotes release of milk from mammary glands	Nerve impulses from hypothalamus, the result of stretching of cervix or stimulation of nipple. Secretion from placenta at end of gestation—stimulus unknown
Antidiuretic Hormone (ADH)	• Increases water reabsorption by the kidney tubules (water returns to the blood)	Decreased water content in the body (alcohol inhibits secretion)

GH increases the transport of amino acids into cells, and increases the rate of protein synthesis. It also stimulates cell division in those tissues capable of mitosis. These functions contribute to the growth of the body during childhood, especially growth of bones and muscles.

You may now be wondering if GH is secreted in adults, and the answer is yes. The use of amino acids for the synthesis of proteins is still necessary, even if the body is not growing in height. GH also stimulates the release of fat from adipose tissue and the use of fats for energy production. This is important any time we go for extended periods without eating, no matter what our ages.

The secretion of GH is regulated by two releasing hormones from the hypothalamus. Growth hormone releasing hormone (GHRH), which increases the secretion of GH, is produced during hypoglycemia and during exercise. Another stimulus for GHRH is a high blood level of amino acids; the GH then secreted will ensure the conversion of these amino acids into protein. **Somatostatin** may also be called growth hormone inhibiting hormone (GHIH), and as its name tells us, it decreases the secretion of GH. Somatostatin is produced during states of hyperglycemia. Disorders of GH secretion are discussed in Box 10–1.

Thyroid-Stimulating Hormone

Thyroid-stimulating hormone (TSH) may also be called thyrotropin, and its target organ is the thyroid gland. TSH stimulates the normal growth of the thyroid and the secretion of thyroxine (T_4) and triiodothyronine (T_3). The functions of these thyroid hormones will be covered later in this chapter.

The secretion of TSH is stimulated by thyrotropin releasing hormone (TRH) from the hypothalamus. When metabolic rate (energy production) decreases, TRH is produced.

Adrenocorticotropic Hormone

Adrenocorticotropic hormone (ACTH) stimulates the secretion of cortisol and other hormones by the adrenal cortex. Secretion of ACTH is increased by corticotropin releasing hormone (CRH) from the hypothalamus. CRH is produced in any type of physiological stress situation such as injury, hypoglycemia, or exercise.

Prolactin

Prolactin, as its name suggests, is responsible for lactation. More precisely, prolactin initiates and

Box 10–1 DISORDERS OF GROWTH HORMONE

A deficiency or excess of growth hormone (GH) during childhood will have marked effects on the growth of a child. Hyposecretion of GH results in **pituitary dwarfism,** in which the person may attain a final height of only 3 to 4 feet but will have normal body proportions. GH can now be produced using genetic engineering and may be used to stimulate growth in children with this disorder. Hypersecretion of GH results in **giantism,** in which the long bones grow excessively and the person may attain a height of 8 feet. Most very tall people, such as basketball players, do *not* have this condition; they are tall as a result of their genetic makeup and good nutrition.

In an adult, hypersecretion of GH is caused by a pituitary tumor, and results in **acromegaly.** The long bones cannot grow because the epiphyseal discs are closed, but the growth of other bones is stimulated. The jaw and other facial bones become disproportionally large, as do the bones of the hands and feet. The skin becomes thicker, and the tongue also grows and may protrude. Other consequences include compression of nerves by abnormally growing bones and growth of the articular cartilages, which then erode and bring on arthritis. Treatment of acromegaly requires surgical removal of the tumor or its destruction by radiation.

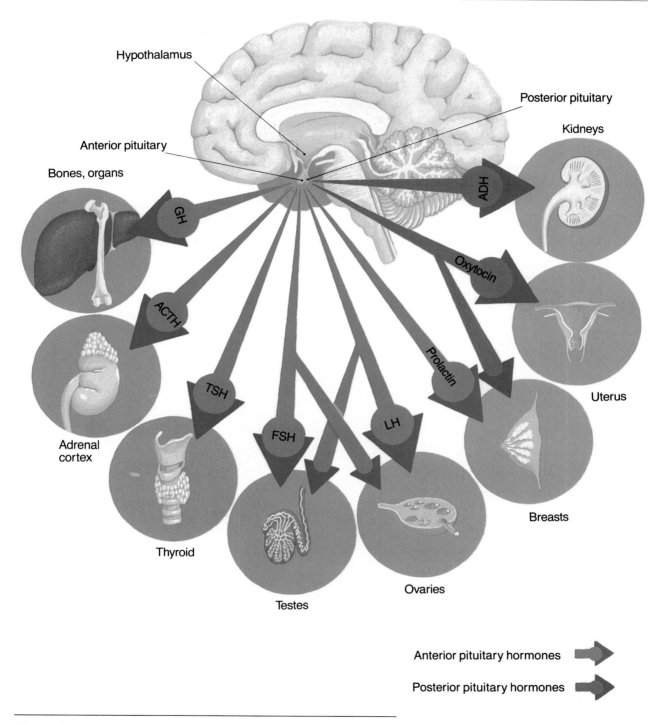

Figure 10–3 Hormones of the pituitary gland and their target organs.

maintains milk production by the mammary glands. The regulation of secretion of prolactin is complex, involving both prolactin releasing hormone (PRH) and prolactin inhibiting hormone (PIH) from the hypothalamus. The mammary glands must first be acted upon by other hormones such as estrogen and progesterone, which are secreted in large amounts by the placenta during pregnancy. Then, after delivery of the baby, prolactin secretion increases and milk is produced. If the mother continues to breast-feed, prolactin levels remain high.

Follicle-Stimulating Hormone

Follicle-stimulating hormone (FSH) is one of the gonadotropic hormones, that is, it has its effects on the gonads: the ovaries or testes. FSH is named for one of its functions in women. Within the ovaries are ovarian follicles that contain potential ova (egg cells). FSH stimulates the growth of ovarian follicles, that is, it initiates egg development in cycles of approximately 28 days. FSH also stimulates secretion of estrogen by the follicle cells. In men, FSH initiates sperm production within the testes.

The secretion of FSH is regulated by the hypothalamus, which produces **gonadotropin releasing hormone (GnRH).**

Luteinizing Hormone

Luteinizing hormone (LH) is another gonadotropic hormone. In women, LH is responsible for

Table 10–2   HORMONES OF THE ANTERIOR PITUITARY GLAND

Hormone	Function(s)	Regulation of Secretion
Growth Hormone (GH)	• Increases rate of mitosis • Increases amino acid transport into cells • Increases rate of protein synthesis • Increases use of fats for energy	• GHRH (hypothalamus) stimulates secretion • GHIH—somatostatin (hypothalamus) inhibits secretion
Thyroid-Stimulating Hormone (TSH)	• Increases secretion of thyroxine and T_3 by thyroid gland	• TRH (hypothalamus)
Adrenocorticotropic Hormone (ACTH)	• Increases secretion of cortisol by the adrenal cortex	• CRH (hypothalamus)
Prolactin	• Stimulates milk production by the mammary glands	• PRH (hypothalamus) stimulates secretion • PIH (hypothalamus) inhibits secretion
Follicle-Stimulating Hormone (FSH)	*In women:* • Initiates growth of ova in ovarian follicles • Increases secretion of estrogen by follicle cells *In men:* • Initiates sperm production in the testes	• GnRH (hypothalamus) • GnRH (hypothalamus)
Luteinizing Hormone (LH) (ICSH)	*In women:* • Causes ovulation • Causes the ruptured ovarian follicle to become the corpus luteum • Increases secretion of progesterone by the corpus luteum *In men:* • Increases secretion of testosterone by the interstitial cells of the testes	• GnRH (hypothalamus) • GnRH (hypothalamus)

ovulation, the release of a mature ovum from an ovarian follicle. LH then stimulates that follicle to develop into the corpus luteum, which secretes progesterone, also under the influence of LH. In men, LH stimulates the interstitial cells of the testes to secrete testosterone (LH is also called ICSH: interstitial cell stimulating hormone).

Secretion of LH is also regulated by GnRH from the hypothalamus. We will return to FSH and LH, as well as to the sex hormones, in Chapter 20.

All the target organs of the pituitary gland are shown in Fig. 10–3. The hormones of the anterior pituitary are summarized in Table 10–2.

THYROID GLAND

The **thyroid gland** is located on the front and sides of the trachea just below the larynx. Its two lobes are connected by a middle piece called the isthmus. The structural units of the thyroid gland are thyroid follicles, which produce **thyroxine (T_4)** and **triiodothyronine (T_3).** Iodine is necessary for the synthesis of these hormones; thyroxine contains four atoms of iodine, and T_3 contains three atoms of iodine.

The third hormone produced by the thyroid gland is **calcitonin,** which is secreted by parafollicular cells. Its function is very different from those of thyroxine and T_3, which you may recall from Chapter 6.

THYROXINE AND T_3

Thyroxine (T_4) and T_3 have the same functions: regulation of energy production and protein synthesis, which contribute to growth of the body and to normal body functioning throughout life. We will use "thyroxine" to designate both hormones. Thyroxine increases cell respiration of all food types (carbohydrates, fats, and excess amino acids) and

Box 10–2   DISORDERS OF THYROXINE

Iodine is an essential component of thyroxine (and T_3), and a dietary deficiency of iodine causes **goiter.** In an attempt to produce more thyroxine, the thyroid cells become enlarged, and the thyroid gland becomes apparent on the front of the neck. The use of iodized salt has made goiter a rare condition.

Hyposecretion of thyroxine in a newborn has devastating effects on the growth of the child. Without thyroxine, physical growth is diminished, as is mental development. This condition is called **cretinism,** characterized by severe physical and mental retardation. If the thyroxine deficiency is detected shortly after birth, the child may be treated with thyroid hormones to promote normal development.

Hyposecretion of thyroxine in an adult is called **myxedema.** Without thyroxine, the metabolic rate (energy production) decreases, resulting in lethargy, muscular weakness, slow heart rate, a feeling of cold, weight gain, and a characteristic puffiness of the face. The administration of thyroid hormones will return the metabolic rate to normal.

Grave's disease is a hypersecretion of thyroxine that is believed to be an immune disorder. The autoantibodies seem to bind to TSH receptors on the thyroid cells and stimulate secretion of excess thyroxine. The symptoms are those that would be expected when the metabolic rate is abnormally elevated: weight loss accompanied by increased appetite, increased sweating, fast heart rate, feeling of warmth, and fatigue. Also present may be goiter and exophthalmos, which is protrusion of the eyes. Treatment is aimed at decreasing the secretion of thyroxine by the thyroid, and medications or radioactive iodine may be used to accomplish this.

thereby increases energy and heat production. Thyroxine also increases the rate of protein synthesis within cells. Normal production of thyroxine is essential for physical growth, normal mental development, and maturation of the reproductive system. Although thyroxine is not a vital hormone, in that it is not crucial to survival, its absence greatly diminishes physical and mental growth and abilities (see Box 10–2: Disorders of Thyroxine).

Secretion of thyroxine and T_3 is stimulated by **thyroid-stimulating hormone (TSH)** from the anterior pituitary gland. When metabolic rate (en-

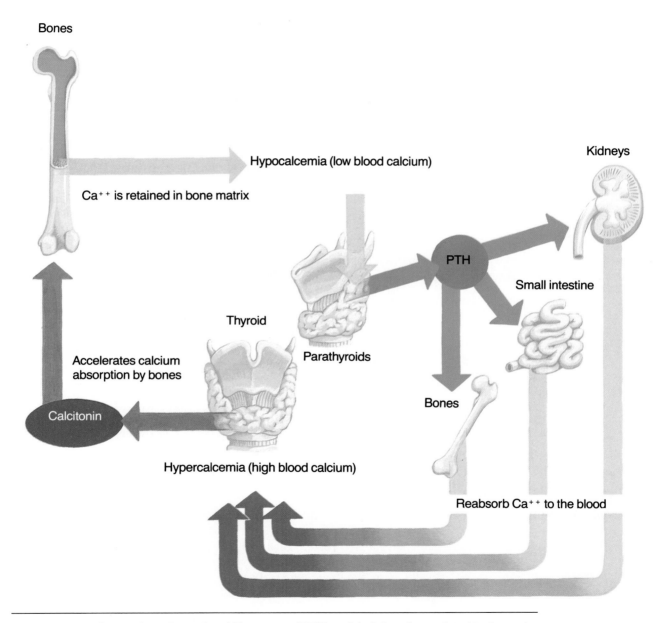

Figure 10–4   Calcitonin and parathyroid hormones (PTH) and their functions related to the maintenance of the blood calcium level.

Table 10–3   HORMONES OF THE THYROID GLAND

Hormone	Function(s)	Regulation of Secretion
Thyroxine (T_4) and Triiodothyronine (T_3)	• Increase energy production from all food types • Increase rate of protein synthesis	TSH (anterior pituitary)
Calcitonin	• Decreases the reabsorption of calcium and phosphate from bones to blood	Hypercalcemia

ergy production) decreases, this change is detected by the hypothalamus, which secretes thyrotropin-releasing hormone (TRH). TRH stimulates the anterior pituitary to secrete TSH, which stimulates the thyroid to release thyroxine and T_3, which raise the metabolic rate by increasing energy production. This negative feedback mechanism then shuts off TRH from the hypothalamus until metabolic rate decreases again.

CALCITONIN

Calcitonin decreases the reabsorption of calcium and phosphate from the bones to the blood, thereby lowering blood levels of these minerals. This function of calcitonin helps maintain normal blood levels of calcium and phosphate and also helps maintain a stable, strong bone matrix.

The stimulus for secretion of calcitonin is **hypercalcemia,** that is, a high blood calcium level. When blood calcium is high, calcitonin ensures that no more will be removed from bones until there is a real need for more calcium in the blood (Fig. 10–4). The hormones of the thyroid gland are summarized in Table 10–3.

PARATHYROID GLANDS

There are four **parathyroid glands:** two on the back of each lobe of the thyroid gland. The hormone they produce is called parathyroid hormone.

PARATHYROID HORMONE

Parathyroid hormone (PTH) is an antagonist to calcitonin and is important for the maintenance of normal blood levels of calcium and phosphate.

The target organs of PTH are the bones, small intestine, and kidneys.

PTH increases the reabsorption of calcium and phosphate from bones to the blood, thereby raising their blood levels. Absorption of calcium and phosphate from food in the small intestine is also increased by PTH. This too raises blood levels of these minerals. In the kidneys, PTH increases the reabsorption of calcium and the excretion of phosphate (more than is obtained from bones). Therefore, the overall effect of PTH is to raise the blood calcium level and lower the blood phosphate level. The functions of PTH are summarized in Table 10–4).

Secretion of PTH is stimulated by **hypocalcemia,** a low blood calcium level. The antagonistic effects of PTH and calcitonin are shown in Fig. 10–4. Together, these hormones maintain blood calcium within a normal range. Calcium in the blood is essential for the process of blood clotting and for normal activity of neurons and muscle cells.

Table 10–4   HORMONE OF THE PARATHYROID GLANDS

Hormone	Functions	Regulation of Secretion
Parathyroid Hormone (PTH)	• Increases the reabsorption of calcium and phosphate from bone to blood • Increases absorption of calcium and phosphate by the small intestine • Increases the reabsorption of calcium and the excretion of phosphate by the kidneys	Hypocalcemia

PANCREAS

The **pancreas** is located in the upper left quadrant of the abdominal cavity, extending from the curve of the duodenum to the spleen. Although the pancreas is both an exocrine (digestive) gland as well as an endocrine gland, only its endocrine function will be discussed here. The hormone-produc-

ing cells of the pancreas are called **islets of Langerhans** (pancreatic islets); they contain **alpha cells** which produce glucagon and **beta cells** which produce insulin.

GLUCAGON

Glucagon stimulates the liver to change glycogen to glucose (this process is called **glycogenol-**

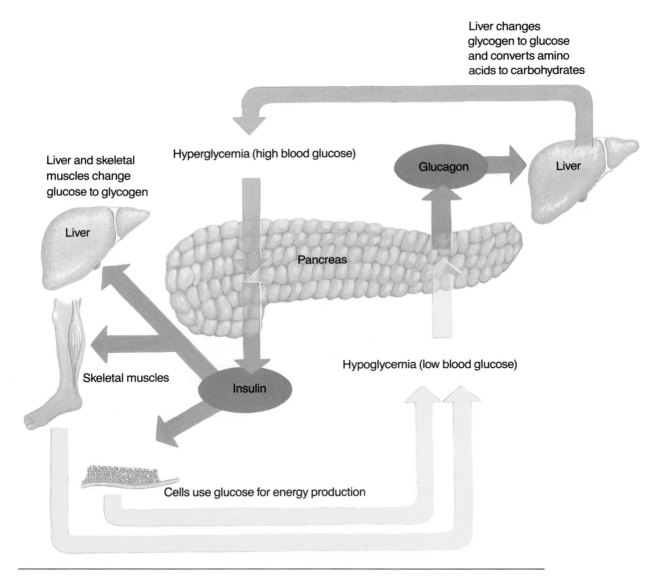

Figure 10–5 Insulin and glucagon and their functions related to the maintenance of the blood glucose level.

Box 10–3 DIABETES MELLITUS

There are two types of **diabetes mellitus:** Type 1 is called insulin-dependent diabetes and its onset is usually in childhood (juvenile onset). Type 2 is called non–insulin-dependent diabetes, and its onset is usually later in life (maturity onset).

Type 1 diabetes is characterized by destruction of the beta cells of the islets of Langerhans and a complete lack of insulin. Destruction of the beta cells is believed to be an autoimmune response, perhaps triggered by a virus; onset of diabetes is usually abrupt. This form of diabetes occurs most often in children, and there may be a genetic predisposition for it (certain HLA types are found more frequently in juvenile-onset diabetics than in other children—See Box 11–5: White Blood Cell Types: HLA). Insulin by injection is essential to control this form of diabetes. Recently, however, an immunosuppressant medication (usually used to prevent rejection of transplanted organs) has been found to slow or stop the progression of type 1 diabetes if given as the disease is developing. By blocking the autoimmune response, this drug seems to permit the survival of some beta cells, which will continue to produce insulin.

In **type 2** diabetes, insulin is produced but cannot exert its effects on cells because of a deficiency of insulin receptors on cell membranes. Onset of type 2 diabetes is usually gradual, and risk factors include a family history of diabetes and being overweight. Control may not require insulin, but rather medications that enable insulin to react with the remaining cell membrane receptors.

Without insulin (or its effects) blood glucose level remains high, and glucose is lost in urine. Since more water is lost as well, symptoms include greater urinary output (polyuria) and thirst (polydipsia).

The long-term effects of hyperglycemia produce distinctive vascular changes. The capillary walls thicken, and exchange of gases and nutrients diminishes. The most damaging effects are seen in the skin (especially of the feet), the retina, and the kidneys. Poorly controlled diabetes may lead to dry gangrene, blindness, and severe kidney damage. In non–insulin-dependent diabetics, atherosclerosis is quite common. It is now possible for diabetics to prevent much of this tissue damage by precise monitoring of blood glucose level and more frequent administration of insulin. The new insulin pumps are able to more closely mimic the natural secretion of insulin.

A very serious potential problem for the insulin-dependent diabetic is **ketoacidosis.** When glucose cannot be used for energy, the body turns to fats and proteins, which are converted by the liver to ketones. Ketones are organic acids (acetone, acetoacetic acid) that can be used in cell respiration, but cells are not able to utilize them rapidly so ketones accumulate in the blood. Since ketones are acids, they lower the pH of the blood as they accumulate. The kidneys excrete excess ketones, but in doing so excrete more water as well, which leads to dehydration and worsens the acidosis. Without administration of insulin to permit the use of glucose for energy, and IV fluids to restore blood volume to normal, ketoacidosis will progress to coma and death.

ysis, which literally means glycogen breakdown) and to increase the use of fats and excess amino acids for energy production. The process of **gluconeogenesis** (literally, making new glucose) is the conversion of excess amino acids into simple carbohydrates that may enter the reactions of cell respiration. The overall effect of glucagon, therefore, is to raise the blood glucose level and to make all types of food available for energy production.

The secretion of glucagon is stimulated by **hypoglycemia,** a low blood glucose level. Such a state may occur between meals or during physiological stress situations such as exercise (Fig. 10–5).

INSULIN

Insulin increases the transport of glucose from the blood into cells by increasing the permeability of cell membranes to glucose (brain cells, however, are not dependent on insulin for glucose intake). Once inside cells, glucose is used in cell respiration to produce energy. The liver and skeletal muscles also change glucose to glycogen (**glycogenesis,**

which literally means glycogen production) to be stored for later use. Insulin also enables cells to take in fatty acids and amino acids to use in the synthesis of lipids and proteins (*not* energy production). With respect to blood glucose, insulin decreases its level by promoting the use of glucose for energy production. The antagonistic functions of insulin and glucagon are shown in Fig. 10–5.

Insulin is a vital hormone; we cannot survive for very long without it. A deficiency of insulin or in its functioning, is called **diabetes mellitus,** which is discussed in Box 10–3: Diabetes Mellitus.

Secretion of insulin is stimulated by **hyperglycemia,** a high blood glucose level. This state occurs after eating, especially of meals high in carbohydrates. As glucose is absorbed from the small intestine into the blood, insulin is secreted to enable cells to use the glucose for immediate energy. At the same time, any excess glucose will be stored in the liver and muscles as glycogen. The hormones of the pancreas are summarized in Table 10–5.

Table 10–5   HORMONES OF THE PANCREAS

Hormone	Functions	Regulation of Secretion
Glucagon	• Increases conversion of glycogen to glucose in the liver • Increases the use of excess amino acids and of fats for energy	Hypoglycemia
Insulin	• Increases glucose transport into cells and the use of glucose for energy production • Increases the conversion of excess glucose to glycogen in the liver and muscles • Increases amino acid and fatty acid transport into cells, and their use in synthesis reactions	Hyperglycemia

ADRENAL GLANDS

The two **adrenal glands** are located one on top of each kidney, which gives them their other name of **suprarenal glands.** Each adrenal gland consists of two parts: an inner adrenal medulla and an outer adrenal cortex. The hormones produced by each part have very different functions.

ADRENAL MEDULLA

The cells of the **adrenal medulla** secrete epinephrine and norepinephrine, which collectively are called **catecholamines,** and are **sympathomimetic.** The secretion of both hormones is stimulated by sympathetic impulses from the hypothalamus, and their functions duplicate those of the sympathetic division of the autonomic nervous system ("mimetic" means "to mimic").

Epinephrine and Norepinephrine

Epinephrine (adrenalin) and norepinephrine (noradrenalin) are both secreted in stress situations and help prepare the body for "fight or flight." **Norepinephrine** is secreted in small amounts, and its most significant function is to cause vasoconstriction in the skin, viscera, and skeletal muscles (that is, throughout the body), which raises blood pressure.

Epinephrine, secreted in larger amounts, increases heart rate and force of contraction and stimulates vasoconstriction in skin and viscera and vasodilation in skeletal muscles. It also dilates the bronchioles, decreases peristalsis, stimulates the liver to change glycogen to glucose, increases the use of fats for energy, and increases the rate of cell respiration. Many of these effects do indeed seem to be an echo of sympathetic responses, don't they? Responding to stress is so important that the body is redundant (it repeats itself) and has both a nervous mechanism and a hormonal mechanism. Epinephrine is actually more effective than sympathetic stimulation, however, because the hormone increases energy production and cardiac output to a greater extent. The hormones of the adrenal medulla are summarized in Table 10–6.

ADRENAL CORTEX

The **adrenal cortex** secretes three types of steroid hormones: mineralocorticoids, glucocorticoids, and sex hormones. The sex hormones, female estrogens and male androgens (similar to testosterone), are produced in very small amounts, and their importance, if any, is not known with certainty. The functions of the other adrenal cortical hormones are well known, however, and these are considered vital hormones.

Aldosterone

Aldosterone is the most abundant of the **mineralocorticoids,** and we will use it as a representative of this group of hormones. The target organs of aldosterone are the kidneys, but there are important secondary effects as well. Aldosterone increases the reabsorption of sodium and the excretion of potassium by the kidney tubules. Sodium ions (Na^+) are returned to the blood, and potassium ions (K^+) are excreted in urine.

As Na^+ ions are reabsorbed, hydrogen ions (H^+) may be excreted in exchange. This is one mechanism to prevent the accumulation of excess H^+ ions which would cause acidosis of body fluids. Also, as Na^+ ions are reabsorbed, negative ions such as chloride (Cl^-) and bicarbonate (HCO_3^-) follow the Na^+ ions back to the blood, and water follows by osmosis. This indirect effect of aldosterone, the reabsorption of water by the kidneys, is very important to maintain normal blood volume and blood pressure. In summary, then, aldosterone maintains normal blood levels of sodium and potassium, and contributes to the maintenance of normal blood pH, blood volume, and blood pressure.

There are a number of factors that stimulate the secretion of aldosterone. These are a deficiency of sodium, loss of blood or dehydration that lowers blood pressure, or an elevated blood level of potassium. Low blood pressure or blood volume ac-

Table 10–6 HORMONES OF THE ADRENAL MEDULLA

Hormone	Function(s)	Regulation of Secretion
Norepinephrine	• Causes vasoconstriction in skin, viscera, and skeletal muscles	
Epinephrine	• Increases heart rate and force of contraction • Dilates bronchioles • Decreases peristalsis • Increases conversion of glycogen to glucose in the liver • Causes vasodilation in skeletal muscles • Causes vasoconstriction in skin and viscera • Increases use of fats for energy • Increases the rate of cell respiration	Sympathetic impulses from the hypothalamus in stress situations

tivates the **renin-angiotensin mechanism** of the kidneys. This mechanism will be discussed in Chapters 13 and 18, so we will say for now that the process culminates in the formation of a chemical called **angiotensin II.** Angiotensin II causes vasoconstriction and stimulates the secretion of aldosterone by the adrenal cortex. Aldosterone then increases sodium and water retention by the kidneys to help restore blood volume and blood pressure to normal.

Cortisol

We will use **cortisol** as a representative of the group of hormones called **glucocorticoids,** since it is responsible for most of the actions of this group. Cortisol increases the use of fats and excess amino acids (gluconeogenesis) for energy, and decreases the use of glucose. This is called the "glucose-sparing effect," and it is important because it conserves glucose for use by the brain. Cortisol is secreted in any type of physiological stress situation: disease, physical injury, hemorrhage, fear or anger, exercise, and hunger. While most body cells easily use fatty acids and amino acids in cell respiration, the brain does not and must have glucose. By enabling other cells to use the alternative energy sources, cortisol ensures that whatever glucose is present will be available to the brain.

Cortisol also has an **anti-inflammatory effect.** During inflammation, **histamine** from damaged tissues makes capillaries more permeable, and the lysosomes of damaged cells release their enzymes, which help break down damaged tissue but may also cause destruction of nearby healthy tissue. Cortisol blocks the effects of histamine and stabilizes

---

Box 10–4 DISORDERS OF THE ADRENAL CORTEX

Addison's disease is the result of hyposecretion of the adrenal cortical hormones. Most cases are idiopathic, that is, of unknown cause; atrophy of the adrenal cortex decreases both cortisol and aldosterone secretion.

Deficiency of cortisol is characterized by hypoglycemia, decreased gluconeogenesis, and depletion of glycogen in the liver. Consequences are muscle weakness and the inability to resist physiological stress. Aldosterone deficiency leads to retention of potassium and excretion of sodium and water in urine. The result is severe dehydration, low blood volume and low blood pressure. Without treatment, circulatory shock and death will follow. Treatment involves administration of hydrocortisone; in high doses this will also compensate for the aldosterone deficiency.

Cushing's syndrome is the result of hypersecretion of the adrenal cortex, primarily cortisol. The cause may be a pituitary tumor that increases ACTH secretion or a tumor of the adrenal cortex itself.

Excessive cortisol promotes fat deposition in the trunk of the body, while the extremities remain thin. The skin becomes thin and fragile, and healing after injury is slow. The bones also become fragile as osteoporosis is accelerated. Also characteristic of this syndrome is the rounded appearance of the face. Treatment is aimed at removal of the cause of the hypersecretion, whether it be a pituitary or adrenal tumor.

Cushing's syndrome may also be seen in people who receive corticosteroids for medical reasons. Transplant recipients or people with rheumatoid arthritis or severe asthma who must take corticosteroids may exhibit any of the above symptoms. In such cases, the disadvantages of this medication must be weighed against the benefits provided.

Table 10–7   HORMONES OF THE ADRENAL CORTEX

Hormone	Functions	Regulation of Secretion
Aldosterone	• Increases reabsorption of Na^+ ions by the kidneys to the blood • Increases excretion of K^+ ions by the kidneys in urine	• Low blood Na^+ level • Low blood volume or blood pressure • High blood K^+ level
Cortisol	• Increases use of fats and excess amino acids for energy • Decreases use of glucose for energy (except for the brain) • Increases conversion of glucose to glycogen in the liver • Anti-inflammatory effect: stabilizes lysosomes and blocks the effects of histamine	• ACTH (anterior pituitary) during physiological stress

lysosomal membranes, preventing excessive tissue destruction. Inflammation is a beneficial process up to a point, and is an essential first step if tissue repair is to take place. It may, however, become a vicious cycle of damage—inflammation, more damage, more inflammation, and so on. Normal cortisol secretion seems to limit the inflammation process to what is useful for tissue repair, and to prevent excessive tissue destruction. Too much cortisol, however, decreases the immune response, leaving the body susceptible to infection and significantly slowing the healing of damaged tissue (see Box 10–4: Disorders of the Adrenal Cortex).

The direct stimulus for cortisol secretion is **ACTH** from the anterior pituitary gland, which in turn is stimulated by corticotropin releasing hormone (CRH) from the hypothalamus. CRH is produced in the physiological stress situations mentioned above. Although we often think of epinephrine as a hormone important in stress, cortisol is also important. The hormones of the adrenal cortex are summarized in Table 10–7.

OVARIES

The **ovaries** are located in the pelvic cavity, one on each side of the uterus. The hormones produced by the ovaries are the steroids estrogen and progesterone. Although their functions will be an integral part of Chapters 20 and 21, we will briefly discuss some of them here.

ESTROGEN

Estrogen is secreted by the follicle cells of the ovary; secretion is stimulated by **FSH** from the anterior pituitary gland. Estrogen promotes the maturation of the ovum in the ovarian follicle and stimulates the growth of blood vessels in the endometrium (lining) of the uterus in preparation for a possible fertilized egg.

The **secondary sex characteristics** in women also develop in response to estrogen. These include growth of the duct system of the mammary glands, growth of the uterus, and the deposition of fat subcutaneously in the hips and thighs. The closure of the epiphyseal discs in long bones is brought about by estrogen, and growth in height stops. Estrogen is also believed to lower blood levels of cholesterol and triglycerides. For women before the age of menopause this is beneficial in that it decreases the risk of atherosclerosis and coronary artery disease.

PROGESTERONE

When a mature ovarian follicle releases an ovum, the follicle becomes the **corpus luteum** and begins to secrete **progesterone** in addition to estrogen. This is stimulated by **LH** from the anterior pituitary gland.

Progesterone promotes the storage of glycogen and the further growth of blood vessels in the endometrium, which thus becomes a potential placenta. The secretory cells of the mammary glands also develop under the influence of progesterone.

Both progesterone and estrogen are secreted by

the placenta during pregnancy; these functions will be covered in Chapter 21.

TESTES

The **testes** are located in the scrotum, a sac of skin between the upper thighs. Two hormones, testosterone and inhibin, are secreted by the testes.

TESTOSTERONE

Testosterone is a steroid hormone secreted by the interstitial cells of the testes; the stimulus for secretion is LH from the anterior pituitary gland.

Testosterone promotes maturation of sperm in the seminiferous tubules of the testes; this process begins at puberty and continues throughout life. At puberty, testosterone stimulates development of the male **secondary sex characteristics.** These include growth of all the reproductive organs, growth of facial and body hair, growth of the larynx and deepening of the voice, and growth (protein synthesis) of the skeletal muscles. Testosterone also brings about closure of the epiphyses of the long bones.

INHIBIN

The hormone **inhibin** is secreted by the sustentacular cells of the testes; the stimulus for secretion is increased testosterone. The function of inhibin is to decrease the secretion of FSH by the anterior pituitary gland. The interaction of inhibin, testosterone, and the anterior pituitary hormones maintains spermatogenesis at a constant rate.

OTHER HORMONES

There are other organs that produce hormones that have only one or a few target organs. For example, the stomach and duodenum produce hormones that regulate aspects of digestion. The thymus gland produces several hormones necessary for the normal functioning of the immune system, and the kidneys produce a hormone that stimulates red blood cell production. All of these will be discussed in later chapters.

PROSTAGLANDINS

Prostaglandins (PG) are made by virtually all cells from the phospholipids of their cell membranes. They differ from other hormones in that they do not circulate in the blood to target organs, but rather exert their effects locally, where they are produced.

There are many types of prostaglandins, designated by the letters A–I, as in PGA, PGB, and so on. The functions of prostaglandins are also many, and we will list only a few of them here. Prostaglandins are known to be involved in inflammation, pain mechanisms, blood clotting, vasoconstriction and vasodilation, contraction of the uterus, reproduction, secretion of digestive glands, and nutrient metabolism. Current research is directed at determining the normal functioning of prostaglandins in the hope that many of them may eventually be used clinically.

One familiar example may illustrate the widespread activity of prostaglandins. For minor pain such as a headache, many people take aspirin. Aspirin inhibits the synthesis of prostaglandins involved in pain mechanisms, and usually relieves the pain. Some people, however, such as those with rheumatoid arthritis, may take large amounts of aspirin to diminish pain and inflammation. These people may bruise easily because blood clotting has been impaired. This too is an effect of aspirin, which blocks the synthesis of prostaglandins necessary for blood clotting.

MECHANISMS OF HORMONE ACTION

Exactly how hormones exert their effects on their target organs involves a number of complex processes, which will be presented simply here.

A hormone must first bond to a **receptor** for it

on or in the target cell. Cells respond to certain hormones and not to others because of the presence of specific receptors, which are proteins. These receptor proteins may be part of the cell membrane or within the cytoplasm or nucleus of the target cells. A hormone will affect only those cells which have its specific receptors. Liver cells, for example, have cell membrane receptors for insulin, glucagon,

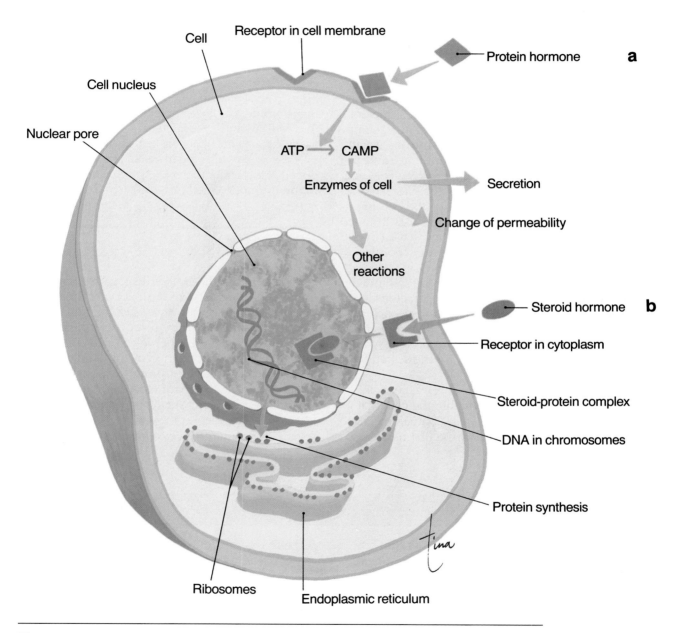

Figure 10–6   Mechanisms of hormone action. **(a)**, Two-messenger model of the action of protein hormones. **(b)**, Action of steroid hormones. See text for description.

growth hormone, and epinephrine; bone cells have receptors for growth hormone, PTH, and calcitonin. Cells of the ovaries and testes do not have receptors for PTH and calcitonin, but do have receptors for FSH and LH, which bone cells and liver cells do not have. Once a hormone has bonded to a receptor on or in its target cell, other reactions will take place.

THE TWO-MESSENGER THEORY— PROTEIN HORMONES

This theory of hormone action involves "messengers" that make something happen, that is, stimulate specific reactions. **Protein hormones** usually bond to receptors of the cell membrane, and the hormone is called the first messenger. The hormone-receptor bonding activates the enzyme adenyl cyclase on the inner surface of the cell membrane. Adenyl cyclase synthesizes a substance called cyclic adenosine monophosphate **(cyclic AMP)** from ATP, and cyclic AMP is the second messenger.

Cyclic AMP activates specific enzymes within the cell, which bring about the cell's characteristic response to the hormone. These responses include a change in the permeability of the cell membrane to a specific substance, an increase in protein synthesis, activation of other enzymes, or the secretion of a cellular product.

In summary, a cell's response to a hormone is determined by the enzymes within the cell, that is, the reactions of which the cell is capable. These reactions are brought about by the first messenger,

the hormone, which stimulates the formation of the second messenger, cyclic AMP. Cyclic AMP then activates the cell's enzymes to elicit a response to the hormone (Fig. 10–6).

ACTION OF STEROID HORMONES

Steroid hormones are soluble in the lipids of the cell membrane, and diffuse easily into a target cell. Once inside the cell, the steroid hormone combines with a protein receptor in the cytoplasm, and this steroid–protein complex enters the nucleus of the cell. Within the nucleus, the steroid–protein complex activates specific genes, which begin the process of **protein synthesis.** The enzymes produced bring about the cell's characteristic response to the hormone (see Fig. 10–6).

SUMMARY

The hormones of endocrine glands are involved in virtually all aspects of normal body functioning. The growth and repair of tissues, the utilization of food to produce energy, responses to stress, the maintenance of the proper levels and pH of body fluids, and the continuance of the human species all depend on hormones. Some of these topics will be discussed in later chapters. As you might expect, you will be reading about the contributions of many of these hormones and reviewing their important roles in the maintenance of homeostasis.

STUDY OUTLINE

Endocrine glands are ductless glands that secrete hormones into the blood. Hormones exert their effects on target organs or tissues.
Chemistry of Hormones
1. Amines—structural variations of the amino acid tyrosine; thyroxine, epinephrine.
2. Proteins—chains of amino acids; peptides are short chains. Insulin, GH, glucagon are proteins; ADH and oxytocin are peptides.

3. Steroids—made from cholesterol; cortisol, aldosterone, estrogen, testosterone.

Regulation of Hormone Secretion
1. Hormones are secreted when there is a need for their effects. Each hormone has a specific stimulus for secretion.
2. The secretion of most hormones is regulated by negative feedback mechanisms: as the hormone

exerts its effects, the stimulus for secretion is reversed, and secretion of the hormone decreases.

Pituitary Gland (Hypophysis)—hangs from hypothalamus by the infundibulum; enclosed by sella turcica of sphenoid bone (see Fig. 10–1)

Posterior Pituitary (Neurohypophysis)—stores hormones produced by the hypothalamus (Figs. 10–2 and 10–3 and Table 10–1).

- ADH—increases water reabsorption by the kidneys. Result: decreases urinary output and increases blood volume. Stimulus: nerve impulses from hypothalamus when body water decreases.
- Oxytocin—stimulates contraction of myometrium of uterus during labor and release of milk from mammary glands. Stimulus: nerve impulses from hypothalamus as cervix is stretched or as infant sucks on nipple.

Anterior Pituitary (Adenohypophysis)—secretions are regulated by releasing hormones from the hypothalamus (Figs. 10–2 and 10–3 and Table 10–2).

- GH—increases amino acid transport into cells and increases protein synthesis; increases rate of mitosis; increases use of fats for energy. Stimulus: GHRH from the hypothalamus.
- TSH—increases secretion of thyroxine and T_3 by the thyroid. Stimulus: TRH from the hypothalamus.
- ACTH—increases secretion of cortisol by the adrenal cortex. Stimulus: CRH from the hypothalamus.
- Prolactin—initiates and maintains milk production by the mammary glands. Stimulus: PRH from the hypothalamus.
- FSH—*In women:* initiates development of ova in ovarian follicles and secretion of estrogen by follicle cells.
  In men: initiates sperm development in the testes. Stimulus: GnRH from the hypothalamus.
- LH—*In women:* stimulates ovulation, transforms mature follicle into corpus luteum and stimulates secretion of progesterone.
  In men: stimulates secretion of testosterone by the testes. Stimulus: GnRH from the hypothalamus.

Thyroid Gland—on front and sides of trachea below the larynx (see Fig. 10–1 and Table 10–3)

- Thyroxine (T_4) and T_3—produced by thyroid follicles. Increase use of all food types for energy and increase protein synthesis. Necessary for normal physical, mental, and sexual development. Stimulus: TSH from the anterior pituitary.
- Calcitonin—produced by parafollicular cells. Decreases reabsorption of calcium from bones and lowers blood calcium level. Stimulus: hypercalcemia.

Parathyroid Glands—four; two on posterior of each lobe of thyroid (see Fig. 10–4 and Table 10–4)

- PTH—increases reabsorption of calcium and phosphate from bones to the blood; increases absorption of calcium and phosphate by the small intestine; increases reabsorption of calcium and excretion of phosphate by the kidneys. Result: raises blood calcium and lowers blood phosphate levels. Stimulus: hypocalcemia.

Pancreas—extends from curve of duodenum to the spleen. Islets of Langerhans consist of alpha cells and beta cells (see Figs. 10–1 and 10–5 and Table 10–5)

- Glucagon—secreted by alpha cells. Stimulates liver to change glycogen to glucose; increases use of fats and amino acids for energy. Result: raises blood glucose level. Stimulus: hypoglycemia
- Insulin—secreted by beta cells. Increases use of glucose by cells to produce energy; stimulates liver and muscles to change glucose to glycogen; increases cellular intake of fatty acids and amino acids to use for synthesis of lipids and proteins. Result: lowers blood glucose level. Stimulus: hyperglycemia.

Adrenal Glands—one on top of each kidney; each has an inner adrenal medulla and an outer adrenal cortex (see Fig. 10–1)

Adrenal Medulla—produces catecholamines in stress situations (Table 10–6).

- Norepinephrine—stimulates vasoconstriction and raises blood pressure.
- Epinephrine—increases heart rate and force, causes vasoconstriction in skin and viscera and vasodilation in skeletal muscles; dilates bronchioles; slows peristalsis; causes liver to change glycogen to glucose; increases use of fats for energy; increases rate of cell respiration. Stimulus: sympathetic impulses from the hypothalamus.

Adrenal Cortex—produces mineralocorticoids, glucocorticoids, and very small amounts of sex hormones (function not known) (Table 10–7).

- Aldosterone—increases reabsorption of sodium and excretion of potassium by the kidneys. Results: hydrogen ions are excreted in exchange for sodium; chloride and bicarbonate ions and water follow sodium back to the blood; maintains normal blood pH, blood volume, and blood pressure. Stimulus: decreased blood sodium or elevated blood potassium; decreased blood volume or blood pressure (activates the renin-angiotensin mechanism of the kidneys).
- Cortisol—increases use of fats and amino acids for energy; decreases use of glucose to conserve glucose for the brain; anti-inflammatory effect: blocks effects of histamine and stabilizes lysosomes to prevent excessive tissue damage. Stimulus: ACTH from hypothalamus during physiological stress.

Ovaries—in pelvic cavity on either side of uterus (see Fig. 10–1)

- Estrogen—produced by follicle cells. Promotes maturation of ovum; stimulates growth of blood vessels in endometrium; stimulates development of secondary sex characteristics: growth of duct system of mammary glands, growth of uterus, fat deposition. Promotes closure of epiphyses of long bones; lowers blood levels of cholesterol and triglycerides. Stimulus: FSH from anterior pituitary.
- Progesterone—produced by the corpus luteum. Promotes storage of glycogen and further growth of blood vessels in the endometrium; promotes growth of secretory cells of mammary glands. Stimulus: LH from anterior pituitary.

Testes—in scrotum between the upper thighs (see Fig. 10–1)

- Testosterone—produced by interstitial cells. Promotes maturation of sperm in testes; stimulates development of secondary sex characteristics: growth of reproductive organs, facial and body hair, larynx, skeletal muscles; promotes closure of epiphyses of long bones. Stimulus: LH from anterior pituitary.
- Inhibin—produced by sustentacular cells. Inhibits secretion of FSH to maintain a constant rate of sperm production. Stimulus: increased testosterone.

Prostaglandins

- Synthesized by cells from the phospholipids of their cell membranes; exert their effects locally. Are involved in inflammation and pain, reproduction, nutrient metabolism, changes in blood vessels, blood clotting.

Mechanisms of Hormone Action (see Fig. 10–6)

- A hormone affects cells that have receptors for it. Receptors are proteins that may be part of the cell membrane, or within the cytoplasm or nucleus of the target cell.
- The Two-Messenger Theory: a protein hormone (1st messenger) bonds to a membrane receptor; stimulates formation of cyclic AMP (2nd messenger), which activates the cell's enzymes to bring about the cell's characteristic response to the hormone.
- Steroid hormones diffuse easily through cell membranes and bond to cytoplasmic receptors. Steroid-protein complex enters the nucleus and activates certain genes which initiate protein synthesis.

REVIEW QUESTIONS

1. Use the following to describe a negative feedback mechanism: TSH, TRF, decreased metabolic rate, thyroxine, and T_3. (pp. 218, 226, 227)

2. Name the two hormones stored in the posterior pituitary gland. Where are these hormones produced? State the functions of each of these hormones. (pp. 220–221)

3. Name the two hormones of the anterior pituitary gland that affect the ovaries or testes, and state their functions. (pp. 224–225)

4. Describe the antagonistic effects of PTH and calcitonin on bones and on blood calcium level. State the other functions of PTH. (p. 227)

5. Describe the antagonistic effects of insulin and glucagon on the liver and on blood glucose level. (pp. 228, 230)

6. Describe how cortisol affects the use of foods for energy. State the anti-inflammatory effects of cortisol. (pp. 232–233)

7. State the effect of aldosterone on the kidneys. Describe the results of this effect on the composition of the blood. (p. 231)

8. When are epinephrine and norepinephrine secreted? Describe the effects of these hormones. (pp. 230–231)

9. Name the hormones necessary for development of egg cells in the ovaries. Name the hormones necessary for development of sperm in the testes. (pp. 224, 233, 234)

10. State what prostaglandins are made from. State three functions of prostaglandins. (p. 234)

11. Name the hormones that promote the growth of the endometrium of the uterus in preparation for a fertilized egg, and state precisely where each hormone is produced. (pp. 233–234)

12. State the functions of thyroxine and T_3. For what aspects of growth are these hormones necessary? (pp. 225–226)

13. Explain the functions of GH as they are related to normal growth. (p. 222)

14. State the direct stimulus for secretion of each of these hormones: (pp. 221, 224, 227, 230, 233)
    a. thyroxine
    b. insulin
    c. cortisol
    d. PTH
    e. aldosterone
    f. calcitonin
    g. GH
    h. glucagon
    i. progesterone
    j. ADH

Chapter 11

Chapter Outline

Student Objectives

- Describe the composition and explain the functions of blood plasma.
- Name the hemopoietic tissues and the kinds of blood cells each produces.
- State the function of red blood cells, including the protein and the mineral involved.
- Name the nutrients necessary for red blood cell production, and state the function of each.
- Explain how hypoxia may change the rate of red blood cell production.
- Describe what happens to red blood cells that have reached the end of their life span; what happens to the hemoglobin?
- Explain the ABO and Rh blood types.
- Name the five kinds of white blood cells and the function of each.
- State what platelets are, and explain how they are involved in hemostasis.
- Describe the three stages of chemical blood clotting.
- Explain how abnormal clotting is prevented in the vascular system.
- State the normal values in a complete blood count (CBC).

Blood

New Terminology

ABO group (A–B–O GROOP)
Albumin (al–**BYOO**–min)
Bilirubin (**BILL**–ee–roo–bin)
Chemical clotting (**KEM**–i–kuhl **KLAH**–ting)
Embolism (**EM**–boh–lizm)
Erythrocytes (e–**RITH**–roh–sites)
Hemoglobin (**HEE**–muh–GLOW–bin)
Hemostasis (HEE–moh–**STAY**–sis)
Heparin (**HEP**–ar–in)
Immunity (im–**YOO**–ni–tee)
Leukocytes (**LOO**–koh–sites)
Macrophage (**MAK**–roh–fahj)
Normoblast (**NOR**–moh–blast)
Reticulocyte (re–**TIK**–yoo–loh–site)
Rh factor (R–H **FAK**–ter)
Stem cells (STEM SELLS)
Thrombocytes (**THROM**–boh–sites)
Thrombus (**THROM**–bus)

Related Clinical Terminology

Anemia (uh–**NEE**–mee–yah)
Differential count (**DIFF**–er–EN–shul KOWNT)
Erythroblastosis fetalis (e–RITH–roh–blass–**TOH**–sis fee–**TAL**–is)
Hematocrit (hee–**MAT**–oh–krit)
Hemophilia (HEE–moh–**FILL**–ee–ah)
Jaundice (**JAWN**–diss)
Leukemia (loo–**KEE**–mee–ah)
Leukocytosis (LOO–koh–sigh–**TOH**–sis)
RhoGam (**ROH**–gam)
Tissue-typing (**TISH**-yoo-**TIGH**–ping)
   Typing and cross-matching (**TIGH**–ping and **KROSS**-match–ing)

Terms that appear in **bold type** in the chapter text are defined in the glossary, which begins on page 549.

One of the simplest and most familiar life-saving medical procedures is a blood transfusion. As you know, however, the blood of one individual is not always compatible with that of another person. The ABO blood types were discovered in the early 1900s by Karl Landsteiner, an Austrian-American. He also contributed to the discovery of the Rh factor in 1940. In the early 1940s, Charles Drew, an African-American, developed techniques for processing and storing blood plasma, which could then be used in transfusions for people with any blood type. When we donate blood today, our blood may be given to a recipient as whole blood, or it may be separated into its component parts, and recipients will then receive only those parts they need, such as red cells, plasma, Factor VIII, or platelets. Each of these parts has a specific function, and all of the functions of blood are essential to our survival.

The general functions of blood are transportation, regulation, and protection. Materials transported by the blood include nutrients, waste products, gases, and hormones. The blood helps regulate fluid–electrolyte balance, acid–base balance, and the body temperature. Protection against pathogens is provided by white blood cells, and the blood clotting mechanism prevents excessive loss of blood after injuries. Each of these functions will be covered in more detail in this chapter.

CHARACTERISTICS OF BLOOD

Blood has distinctive physical characteristics:

Amount—a person has 4 to 6 liters of blood, depending on his or her size. Of the total blood volume in the human body, 38% to 48% is composed of the various blood cells, also called "formed elements." The remaining 52% to 62% of the blood volume is plasma, the liquid portion of blood (Fig. 11–1).

Color—you're probably saying to yourself, "of course, it's red!" Mention is made of this obvious fact, however, since the color does vary. Arterial blood is bright red because it contains high levels of oxygen. Venous blood has given up much of its oxygen in tissues, and has a darker, dull red color. This may be important in the assessment of the source of bleeding. If blood is bright red it is probably from a severed artery, and dark red blood is probably venous blood.

pH—the normal pH range of blood is 7.35 to 7.45, which is slightly alkaline. Venous blood normally has a lower pH than does arterial blood because of the presence of more carbon dioxide.

Viscosity—this means thickness or resistance to flow. Blood is about three to five times thicker than water. Viscosity is increased by the presence of blood cells and the plasma proteins, and this thickness contributes to normal blood pressure.

PLASMA

Plasma is the liquid part of blood and is approximately 91% water. The solvent ability of water enables the plasma to transport many types of substances. Nutrients absorbed in the digestive tract are circulated to all body tissues, and waste products of the tissues circulate through the kidneys and are excreted in urine. Hormones produced by endocrine glands are carried in the plasma to their target organs, and antibodies are also transported in plasma. Most of the carbon dioxide produced by cells is carried in the plasma in the form of bicarbonate ions (HCO_3^-). When the blood reaches the lungs, the CO_2 is reformed, diffuses into the alveoli, and is exhaled.

Also in the plasma are the **plasma proteins.** The clotting factors **prothrombin, fibrinogen,** and others are synthesized by the liver and circulate until activated to form a clot in a ruptured or damaged blood vessel. **Albumin** is the most abundant plasma protein. It too is synthesized by the liver. Albumin contributes to the colloid osmotic pressure of blood, which pulls tissue fluid into capillaries. This is important to maintain normal blood volume and blood pressure. Yet another group of plasma proteins are the **globulins.** Alpha and beta globulins are synthesized by the liver and act as carriers for molecules such as fats. The gamma globulins are antibodies produced by lymphocytes. Antibodies initiate the destruction of pathogens and provide us with immunity.

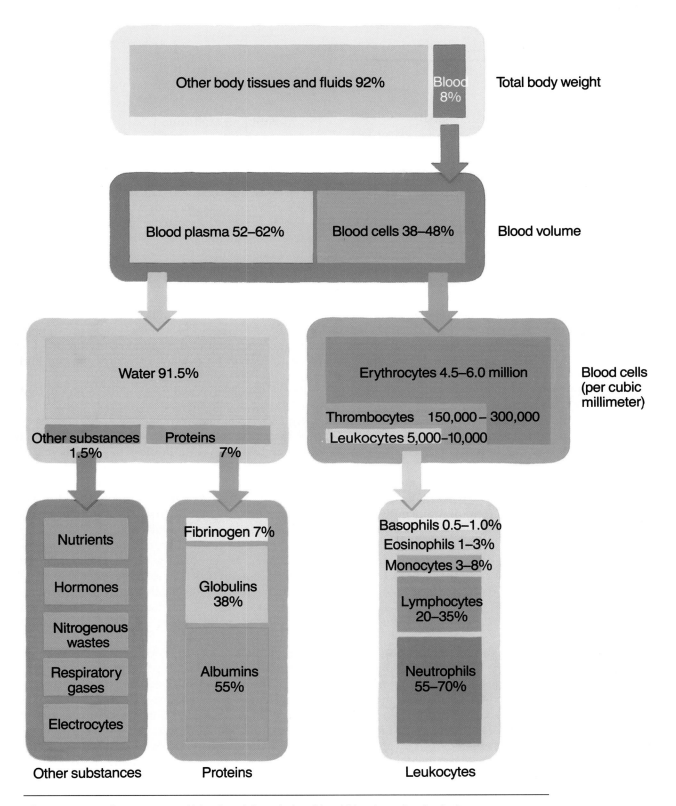

Figure 11–1 Components of blood and the relationship of blood to other body tissues.

BLOOD CELLS

There are three kinds of blood cells: red blood cells, white blood cells, and platelets. Blood cells are produced in **hemopoietic tissues,** of which there are two: **red bone marrow,** found in flat and irregular bones, and **lymphatic tissue,** found in the spleen, lymph nodes, and thymus gland.

RED BLOOD CELLS

Also called **erythrocytes**, red blood cells (RBCs) are biconcave discs, which means their centers are thinner than their edges. You may recall from Chapter 3 that red blood cells are the only human cells without nuclei. Their nuclei disintegrate as the red blood cells mature and are not needed for normal functioning.

A normal RBC count ranges from 4.5 to 6.0 million cells per mm³ of blood (the volume of a cubic millimeter is approximately that of a very small droplet). RBC counts for men are often toward the high end of this range; those for women are often toward the low end. Another way to measure the amount of RBCs is the **hematocrit.** This test involves drawing blood into a thin glass tube called a capillary tube, and centrifuging the tube to force all the cells to one end. The percentages of cells and plasma can then be determined. Since RBCs are by far the most abundant of the blood cells, a normal hematocrit range is just like that of the total blood cells: 38% to 48%. Both RBC count and hematocrit (Hct) are part of a complete blood count (CBC).

Function

Red blood cells contain the protein **hemoglobin** (Hb), which gives them the ability to carry oxygen. In the pulmonary capillaries, RBCs pick up oxygen and oxyhemoglobin is formed. In the systemic capillaries, hemoglobin gives up much of its oxygen and becomes reduced hemoglobin. A determination of hemoglobin level is also part of a CBC; the normal range is 12 to 18 grams per 100 ml of blood. Essential to the formation of hemoglobin is the mineral iron; there are four atoms of iron in each mol-

ecule of hemoglobin. It is the iron that actually bonds to the oxygen and also makes RBCs red.

Production and Maturation

Red blood cells are formed in red bone marrow (RBM) in flat and irregular bones. Within the red bone marrow are precursor cells called **stem cells**, which constantly undergo mitosis to produce all the kinds of blood cells, most of which are RBCs (Fig. 11–2). The rate of production is very rapid (estimated at several million new RBCs per second) and a major regulating factor is oxygen. If the body is in a state of **hypoxia,** or lack of oxygen, the kidneys produce a hormone called **erythropoietin,** which stimulates the red bone marrow to increase the rate of RBC production. This will occur following hemorrhage, or if a person stays for a time at a higher altitude. As a result of the action of erythropoietin, more RBCs will be available to carry oxygen and correct the hypoxic state.

The stem cells that will become RBCs go through a number of developmental stages, only the last two of which we will mention (see Fig. 11–2). The **normoblast** is the last stage with a nucleus, which then disintegrates. The **reticulocyte** has fragments of the endoplasmic reticulum, which are visible when blood smears are stained for microscopic evaluation. These immature cells are usually found in the red bone marrow, although a small number of reticulocytes in the peripheral circulation is considered normal. Large numbers of reticulocytes or normoblasts in the circulating blood mean that the number of mature RBCs is not sufficient to carry the oxygen needed by the body. Such situations include hemorrhage, when mature RBCs have been destroyed, as in Rh disease of the newborn, and malaria.

The maturation of red blood cells requires many nutrients. Protein and iron are necessary for the synthesis of hemoglobin and become part of hemoglobin molecules. The vitamins folic acid and B_{12} are required for DNA synthesis in the stem cells of the red bone marrow. As these cells undergo mitosis they must continually produce new sets of chromosomes. Vitamin B_{12} is also called the **extrinsic factor,** because its source is external, our food. Parietal cells of the stomach lining produce the **in-**

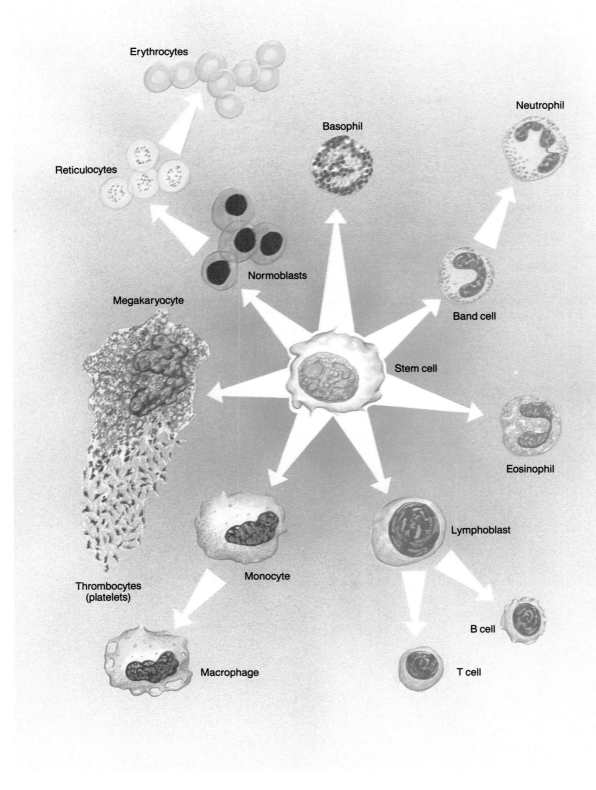

Figure 11–2 Production of blood cells. Stem cells are found in red bone marrow and in lymphatic tissues, and are the precursor cells for all the types of blood cells.

trinsic factor, a chemical that combines with the vitamin B_{12} in food to prevent its digestion and promote its absorption in the small intestine. A deficiency of either vitamin B_{12} or the intrinsic factor results in **pernicious anemia** (see Box 11–1: Anemia).

Life Span

Red blood cells live for approximately 120 days. As they reach this age they become fragile and are removed from circulation by cells of the **tissue macrophage system** (formerly called the reticuloendothelial or RE system). The organs that contain macrophages (literally, "big eaters") are the liver, spleen, and red bone marrow. The old RBCs are phagocytized and digested by macrophages, and the iron they contained is put into the blood to be returned to the red bone marrow to be used for the synthesis of new hemoglobin. If not needed immediately for this purpose, excess iron is stored in the liver. The iron of RBCs is actually recycled over and over again.

Another part of the hemoglobin molecule is the heme portion, which cannot be recycled and is a waste product. The heme is converted to **bilirubin** by macrophages. The liver removes bilirubin from circulation and excretes it into bile; bilirubin is called a bile pigment. Bile is secreted by the liver into the duodenum and passes through the small intestine and colon, so bilirubin is eliminated in feces, and gives feces their characteristic brown color. If bilirubin is not excreted properly, perhaps because of liver disease such as hepatitis, it remains in the blood. This may cause **jaundice,** a condition in which the whites of the eyes appear yellow. This yellow color may also be seen in the skin of light-skinned people (see Box 11–2: Jaundice).

Box 11–1   ANEMIA

Anemia is a deficiency of red blood cells, or insufficient hemoglobin within the red blood cells. There are many different types of anemia.

Iron deficiency anemia is caused by a lack of dietary iron, and there is not enough of this mineral to form sufficient hemoglobin. A person with this type of anemia may have a normal RBC count and a normal hematocrit, but the hemoglobin level will be below normal.

A deficiency of vitamin B_{12}, which is found only in animal foods, leads to **pernicious anemia,** in which the RBCs are large, misshapen, and fragile. Another cause of this form of anemia is lack of the intrinsic factor due to autoimmune destruction of the parietal cells of the stomach lining.

Sickle cell anemia has already been discussed in Chapter 3. It is a genetic disorder of hemoglobin, which causes RBCs to sickle, clog capillaries, and rupture.

Aplastic anemia is suppression of the red bone marrow, with decreased production of RBCs, the granular WBCs, and platelets. This is a very serious disorder which may be caused by exposure to radiation, certain chemicals such as benzene, or some medications. There are several antibiotics that must be used with caution since they may have this potentially fatal side effect.

Hemolytic anemia is any disorder that causes rupture of RBCs before the end of their normal life span. Sickle cell anemia and Rh disease of the newborn are examples. Another example is malaria, in which a protozoan parasite reproduces in RBCs and destroys them. Hemolytic anemias are often characterized by jaundice because of the increased production of bilirubin.

Box 11–2 JAUNDICE

Jaundice is not a disease, but rather a sign caused by excessive accumulation of bilirubin in the blood. Since one of the liver's many functions is the excretion of bilirubin, jaundice may be a sign of liver disease such as hepatitis or cirrhosis. This may be called **hepatic jaundice,** since the problem is with the liver.

Other types of jaundice are prehepatic jaundice and posthepatic jaundice: the name of each tells us where the problem is. Recall that bilirubin is the waste product formed from the heme portion of the hemoglobin of old RBCs. **Prehepatic jaundice** means that the problem is "before" the liver, that is, hemolysis of RBCs is taking place at a more rapid rate. Rapid hemolysis is characteristic of sickle cell anemia, malaria, and Rh disease of the newborn; these are hemolytic anemias. As excessive numbers of RBCs are destroyed, bilirubin is formed at a faster rate than the liver can excrete it. The bilirubin that the liver cannot excrete remains in the blood and causes jaundice. Another name for this type is **hemolytic jaundice.**

Posthepatic jaundice means that the problem is "after" the liver. The liver excretes bilirubin into bile, which is stored in the gall bladder and then moves to the small intestine. If the bile ducts are obstructed, perhaps by gall stones or inflammation of the gall bladder, bile cannot pass to the small intestine and backs up in the liver. Bilirubin may then be reabsorbed back into the blood and cause jaundice. Another name for this type is **obstructive jaundice.**

Blood Types

Our blood types are genetic, that is, we inherit genes from our parents that determine our own types. There are many red blood cell factors or types; we will discuss the two most important ones: the **ABO group** and the **Rh factor.**

The **ABO group** contains four blood types: A, B, AB, and O. The letters A and B represent antigens (protein-oligosaccharides) on the red blood cell membrane. A person with type A blood has the A antigen on the RBCs, and someone with type B blood has the B antigen. Type AB means that both A and B antigens are present, and type O means that neither the A nor the B antigen is present.

In the plasma of each person are natural antibodies for those antigens *not* present on the RBCs. Therefore, a type A person has anti-B antibodies in the plasma; a type B person has anti-A antibodies; a type AB person has neither anti-A nor anti-B antibodies, and a type O person has both anti-A and anti-B antibodies (see Table 11–1 and Fig. 11–3).

These natural antibodies are of great importance for transfusions. If possible, a person should receive blood of his or her own type; only if this type is not available should another type be given. For example, let us say that there is a type A person who needs a transfusion to replace blood lost in hemorrhage. If this person were to receive type B blood, what would happen? The type-A recipient has anti-B antibodies that would bind to the type B antigens of the RBCs of the donated blood. The type-B RBCs would first clump **(agglutination)** then rupture **(hemolysis),** thus defeating the purpose of the

Table 11–1 ABO BLOOD TYPES

Type	Antigens Present on RBCs	Antibodies Present in Plasma
A	A	anti-B
B	B	anti-A
AB	both A and B	neither anti-A nor anti-B
O	neither A nor B	both anti-A and anti-B

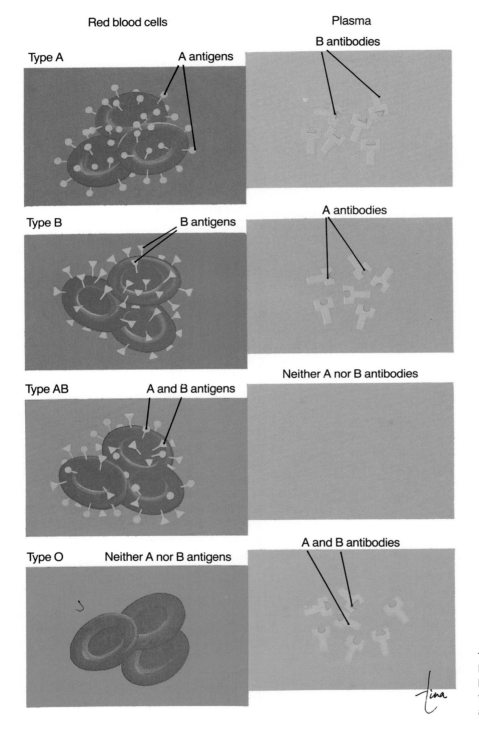

Red blood cells

Plasma

Figure 11–3 (A), The ABO blood types. Schematic representation of antigens on the RBCs and antibodies in the plasma.

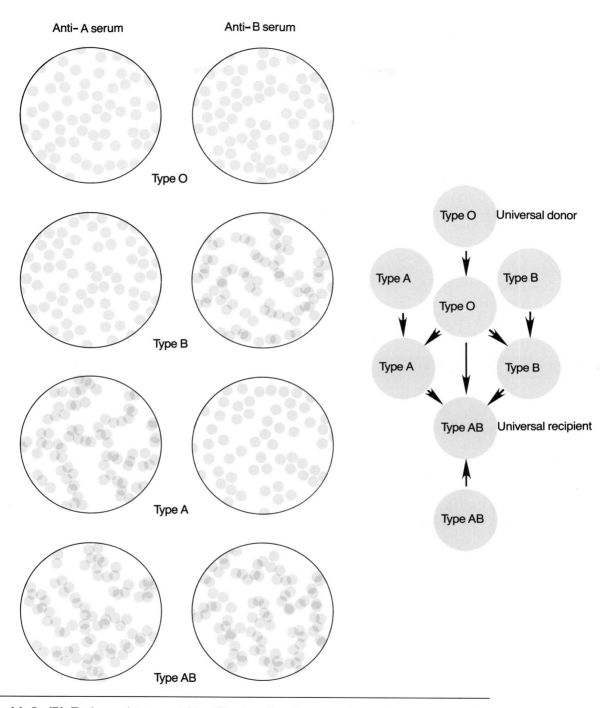

Figure 11–3 (B), Typing and crossmatching. The A or B antiserum causes agglutination of RBCs with the matching antigen. Acceptable transfusions are diagrammed on the right and presuppose compatible Rh factors.

transfusion. An even more serious consequence is that the hemoglobin of the ruptured RBCs, now called free hemoglobin, may clog the capillaries of the kidneys and lead to renal damage or renal failure. You can see why **typing** and **crossmatching** of donor and recipient blood in the hospital laboratory is so important before any transfusion is given (see Fig. 11–3). This procedure helps ensure that donated blood will not bring about a hemolytic transfusion reaction in the recipient.

You may have heard of the concept that type O is the "universal donor." Usually, type O negative blood may be given to people with any other blood type. This is so because type-O RBCs have neither the A nor the B antigens and will not react with whatever antibodies the recipient may have. The term "negative" refers to the Rh factor, which we will now consider.

The **Rh factor** is another antigen (often called D) that may be present on RBCs. People whose RBCs have the Rh antigen are Rh positive; those without the antigen are Rh negative. Rh negative people do not have natural antibodies to the Rh antigen, and for them this antigen is foreign. If an Rh negative person receives Rh positive blood by mistake, antibodies will be formed just as they would be to bacteria or viruses. A first mistaken transfusion often does not cause problems, because antibody production is slow upon the first exposure to Rh positive RBCs. A second transfusion, however, when anti-Rh antibodies are already present, will bring about a transfusion reaction, with hemolysis and possible kidney damage (see also Box 11–3: Rh Disease of the Newborn).

WHITE BLOOD CELLS

White blood cells (WBCs) are also called **leukocytes**. There are five kinds of WBCs; all are larger than RBCs and have nuclei when mature. The nu-

---

Box 11–3   Rh DISEASE OF THE NEWBORN

Rh disease of the newborn may also be called **erythroblastosis fetalis** and is the result of an Rh incompatibility between mother and fetus. During a normal pregnancy, maternal blood and fetal blood do not mix in the placenta. However, during delivery of the placenta (the "afterbirth" that follows the birth of the baby), some fetal blood may enter maternal circulation.

If the woman is Rh negative and her baby is Rh positive, this exposes the woman to Rh positive RBCs. In response, her immune system will now produce anti-Rh antibodies following this first delivery. In a subsequent pregnancy, these maternal antibodies will cross the placenta and enter fetal circulation. If this next fetus is also Rh positive, the maternal antibodies will cause destruction (hemolysis) of the fetal RBCs. In severe cases this may result in the death of the fetus. In less severe cases, the baby will be born anemic and jaundiced from the loss of RBCs. Such an infant may require a gradual exchange transfusion to remove the blood with the maternal antibodies and replace it with Rh negative blood. The baby will continue to produce its own Rh-positive RBCs, which will not be destroyed once the maternal antibodies have been removed.

Much better than treatment, however, is prevention. If an Rh negative woman delivers an Rh positive baby, she should be given **RhoGam** within 72 hours after delivery. RhoGam is an anti-Rh antibody that will destroy any fetal RBCs that have entered the mother's circulation *before* her immune system can respond and produce antibodies. The RhoGam antibodies themselves break down within a few months. The woman's next pregnancy will be like the first, as if she had never been exposed to Rh positive RBCs.

cleus may be in one piece or appear as several lobes. Special staining for microscopic examination gives each kind of WBC a distinctive appearance (see Fig. 11–2).

A normal WBC count (part of a CBC) is 5000 to 10,000 per mm³. Notice that this number is quite small compared to a normal RBC count. Many of our WBCs are not within blood vessels but are carrying out their functions in tissue fluid.

Classification and Sites of Production

The five kinds of white blood cells may be classified in two groups: granular and agranular. The granular leukocytes are produced in the red bone marrow; these are the **neutrophils, eosinophils**, and **basophils**, which have distinctly colored granules when stained. The agranular leukocytes are **lymphocytes** and **monocytes**, which are produced in the lymphatic tissue of the spleen, lymph nodes, and thymus, as well as in the red bone marrow. A **differential WBC count** (part of a CBC) is the percentage of each kind of leukocyte. Normal ranges are listed in Table 11–2, along with other normal values of a CBC.

Functions

White blood cells all contribute to the same general function, which is to protect the body from infectious disease and to provide **immunity** to certain diseases. Each kind of leukocyte has a role in this very important aspect of homeostasis.

Neutrophils and monocytes are capable of the **phagocytosis** of pathogens. Neutrophils are the more abundant phagocytes, but the monocytes are the more efficient phagocytes, since they differentiate into **macrophages,** which also phagocytize dead or damaged tissue at the site of any injury, helping to make tissue repair possible.

Eosinophils are believed to detoxify foreign proteins. This is especially important in allergic reactions and parasitic infections such as trichinosis (a worm parasite). Basophils contain granules of heparin and histamine. **Heparin** is an anticoagulant that helps prevent abnormal clotting within blood vessels. **Histamine,** you may recall, is released as part of the inflammation process, and it makes capillaries more permeable, allowing tissue fluid, proteins, and white blood cells to accumulate in the damaged area.

There are two major kinds of lymphocytes: T cells and B cells. For now we will say that **T cells** (or T lymphocytes) recognize foreign antigens, may directly destroy some foreign antigens, and stop the immune response when the antigen has been destroyed. **B cells** (or B lymphocytes) become plasma cells that produce antibodies to foreign antigens. These T cell and B cell functions will be discussed in the context of the mechanisms of immunity in Chapter 14.

As mentioned earlier, leukocytes function in tissue fluid as well as in the blood. Many WBCs are capable of self-locomotion (ameboid movement) and are able to squeeze between the cells of capillary walls and out into tissue spaces (Fig. 11–4). Macrophages provide a good example of the dual locations of leukocytes. Some macrophages are "fixed," that is, stationary in organs such as the liver, spleen, and red bone marrow (part of the tissue macrophage or RE system) and in the lymph nodes. They phagocytize pathogens that circulate in blood or lymph through these organs (these are the same macrophages that also phagocytize old RBCs). Other "wandering" macrophages move about in tissue fluid, especially in the areolar connective tissue of mucous membranes and below the skin. Pathogens that gain entry into the body through natural

Table 11–2   COMPLETE BLOOD COUNT

Measurement	Normal Range*
Red blood cells	• 4.5–6.0 million/mm³
Hemoglobin	• 12–18 grams/100 ml
Hematocrit	• 38%–48%
Reticulocytes	• 0%–1.5%
White blood cells (total)	• 5000–10,000/mm³
Neutrophils	• 55%–70%
Eosinophils	• 1%–3%
Basophils	• 0.5%–1%
Lymphocytes	• 20%–35%
Monocytes	• 3%–8%
Platelets	• 150,000–300,000/mm³

*The values on hospital lab slips may vary somewhat but will be very similar to the normal ranges given here.

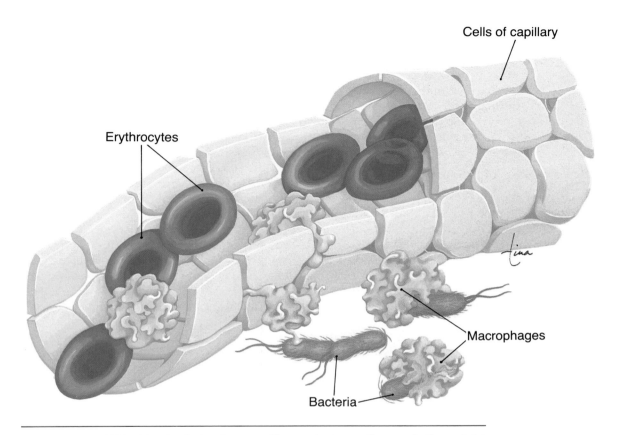

Figure 11–4 White blood cells leaving a capillary to destroy pathogens in tissue fluid.

Box 11–4   LEUKEMIA

Leukemia is the term for malignancy of the blood-forming tissues, the red bone marrow or lymphatic tissue. There are many types of leukemia, which are classified as acute or chronic, by the types of abnormal cells produced, and by either childhood or adult onset.

In general, leukemia is characterized by an overproduction of immature white blood cells. These immature cells cannot perform their normal functions, and the person becomes very susceptible to infection. As a greater proportion of the body's nutrients are used by malignant cells, the production of other blood cells decreases. Severe anemia is a consequence of decreased red blood cell production, and the tendency to hemorrhage is the result of decreased platelets.

Chemotherapy may bring about cure or remission for some forms of leukemia, but other forms remain resistant to treatment and may be fatal within a few months of diagnosis. In such cases, the cause of death is often pneumonia or some other serious infection, because the abnormal white blood cells cannot prevent the growth of pathogens within the body.

openings or through breaks in the skin are usually destroyed by the leukocytes in connective tissue before they can cause serious disease.

A high WBC count, called **leukocytosis**, is often an indication of infection. **Leukopenia** is a low WBC count, which may be present in the early stages of diseases such as tuberculosis. Exposure to radiation or to chemicals such as benzene may de-

---

Box 11–5   WHITE BLOOD CELL TYPES: HLA

Human leukocyte antigens (HLA) are antigens on WBCs that are representative of the antigens present on all the cells of an individual. These are our "self" antigens that identify cells that belong in the body.

Recall that in the ABO blood group of RBCs, there are only two antigens, A and B, and four possible types: A, B, AB, and O. HLA antigens are also given letter names (A, B, C, D), but there are as many as 40 possible antigens in each of these letter categories. Each individual will have two antigens in each category; the HLA types are inherited, just as RBC types are inherited. Members of the same family may have some of the same HLA types, and identical twins have exactly the same HLA type.

The purpose of the HLA types is to provide a "self" comparison for the immune system to use when pathogens enter the body. The T lymphocytes compare the "self" antigens on macrophages to the antigens on bacteria and viruses. Since these antigens do not match ours, they are recognized as foreign; this is the first step in the destruction of a pathogen.

The surgical transplantation of organs has also focused on the HLA. The most serious problem for the recipient of a transplanted heart or kidney is rejection of the organ and its destruction by the immune system. You may be familiar with the term **"tissue-typing."** This involves determining the HLA types of a donated organ to see if one or several will match the HLA types of the potential recipient. If even one HLA type matches, the chance of rejection is significantly lessened. Although all transplant recipients (except corneal) must receive immunosuppressive medications to prevent rejection, such medications make them more susceptible to infection. The closer the HLA match of the donated organ, the lower the dosage of such medications, and the less chance of serious infections. (The chance of finding a perfect HLA match in the general population is estimated at 1 in 20,000.)

There is yet another aspect of the importance of HLA: people with certain HLA types seem to be more likely to develop certain non-infectious diseases. For example, insulin-dependent diabetes mellitus is often found in people with HLA DR3 or DR4, and a form of arthritis of the spine called ankylosing spondylitis is often found in those with HLA B27. These are *not* genes for these diseases, but may be predisposing factors. What may happen is this: a virus enters the body and stimulates the immune system to produce antibodies. The virus is destroyed, but one of the person's own antigens is so similar to the viral antigen that the immune system continues its activity and begins to destroy this similar part of the body. Another possibility is that a virus damages a self-antigen to the extent that it is now so different that it will be perceived as foreign. These are two theories of how autoimmune diseases are triggered, which is the focus of much research in the field of immunology.

stroy WBCs and lower the total count. Such a person is then very susceptible to infection. **Leukemia,** or malignancy of leukocyte-forming tissues, is discussed in Box 11–4: Leukemia.

The white blood cell types (analagous to RBC types such as the ABO group) are called **human leukocyte antigens (HLAs)** and are discussed in Box 11–5: White Blood Cell Types: HLA.

PLATELETS

The more formal name for platelets is **thrombocytes,** which are not whole cells but rather fragments or pieces of cells. A normal platelet count (part of a CBC) is 150,000 to 300,000/mm³ (the high end of the range may be extended to 500,000). **Thrombocytopenia** is the term for a low platelet count.

Site of Production

Some of the stem cells in the red bone marrow differentiate into large cells called **megakaryocytes** (see Fig. 11–2), which break up into small pieces that enter circulation. These circulating pieces are platelets, which may survive for 5 to 9 days, if not utilized before that.

Function

Platelets are necessary for **hemostasis,** which means prevention of blood loss. There are three mechanisms, and platelets are involved in each. Two of these mechanisms are shown in Fig. 11–5.

1. **Vascular spasm**—when a large vessel such as an artery or vein is severed, the smooth muscle in its wall contracts in response to the damage (called the myogenic response). Platelets in the area of the rupture release serotonin, which also brings about vasoconstriction. The diameter of the vessel is thereby made smaller, and the smaller opening may then be blocked by a blood clot. If the vessel did not constrict first, the clot that forms would quickly be washed out by the force of the blood pressure.

2. **Platelet plugs**—when capillaries rupture, the

damage is too slight to initiate the formation of a blood clot. The rough surface, however, causes platelets to become sticky and stick to the edges of the break and to each other. The platelets form a mechanical barrier or wall to close off the break in the capillary. Capillary ruptures are quite frequent, and platelet plugs, although small, are all that is needed to seal them.

Would platelet plugs be effective for breaks in larger vessels? No, they are too small, and would be washed away as fast as they form. Would vascular spasm be effective for capillaries? Again, the answer is no, because capillaries have no smooth muscle and cannot constrict at all.

3. **Chemical clotting**—The stimulus for clotting is a rough surface within a vessel, or a break in the vessel, which also creates a rough surface. The more damage there is, the faster clotting begins, usually within 15 to 120 seconds.

The clotting mechanism is a series of reactions involving chemicals that normally circulate in the blood and others that are released when a vessel is damaged.

The chemicals involved in clotting include platelet factors, chemicals released by damaged tissues, calcium ions, and the plasma proteins prothrombin, fibrinogen, Factor 8, and others synthesized by the liver. Vitamin K is necessary for the liver to synthesize prothrombin and several other clotting factors (Factors 7, 9, and 10). Most of our vitamin K is produced by the bacteria that live in the colon; the vitamin is absorbed as the colon absorbs water.

Chemical clotting is usually described in three stages, which are shown in Table 11–3. As you follow the pathway, notice that the product of stage 1, prothrombin activator, brings about the stage 2 reaction. The product of stage 2, thrombin, brings about the stage 3 reaction (see Box 11–6: Hemophilia).

The clot itself is made of **fibrin,** the product of stage 3. Fibrin is a thread-like protein. Many strands of fibrin form a mesh that traps RBCs and creates a wall across the break in the vessel.

Once the clot has formed and bleeding has stopped, **clot retraction** and **fibrinolysis** occur. Clot retraction requires platelets, ATP, and Factor

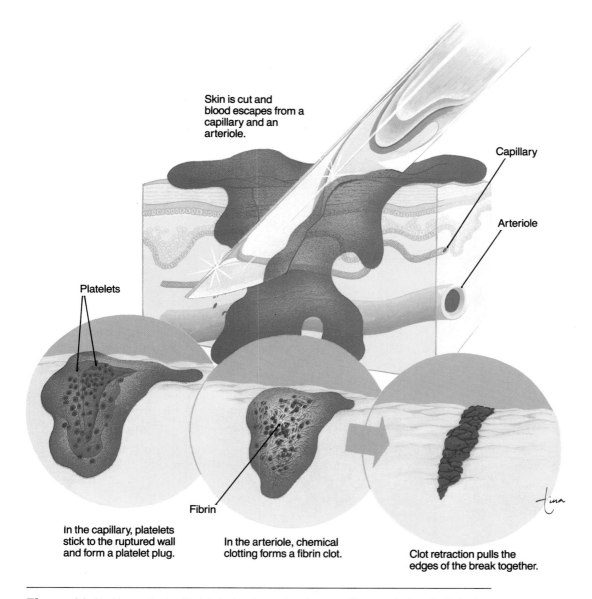

Skin is cut and
blood escapes from a
capillary and an
arteriole.

Capillary

Arteriole

Platelets

Fibrin

In the capillary, platelets
stick to the ruptured wall
and form a platelet plug.

In the arteriole, chemical
clotting forms a fibrin clot.

Clot retraction pulls the
edges of the break together.

Figure 11–5 Hemostasis. Platelet plug formation in a capillary and chemical clotting
and clot retraction in an arteriole.

13 and involves folding of the fibrin threads to pull
the edges of the rupture in the vessel wall closer
together. This will make the area to be repaired
smaller. As repair begins, the clot is dissolved, a
process called fibrinolysis.

Prevention of Abnormal Clotting

Clotting should take place to stop bleeding, but
too much clotting would obstruct vessels and interfere with normal circulation of blood. Clots do not
usually form in intact vessels because the simple

Table 11–3 CHEMICAL CLOTTING

Clotting Stage	Factors Needed	Reaction
Stage 1	• Platelet factors • Chemicals from damaged tissue (tissue thromboplastin) • Factors 5,7,8,9,10,11,12 • Calcium ions	Platelet factors + tissue thromboplastin + other clotting factors + calcium ions form prothrombin activator
Stage 2	• Prothrombin activator from stage 1 • Prothrombin • Calcium ions	Prothrombin activator converts prothrombin to thrombin
Stage 3	• Thrombin from stage 2 • Fibrinogen • Calcium ions • Factor 13 (fibrin stabilizing factor)	Thrombin converts fibrinogen to fibrin

squamous epithelial lining is very smooth and repels the platelets and clotting factors. If the lining becomes roughened, as happens with the lipid deposits of atherosclerosis, a clot will form.

Heparin, produced by basophils, is a natural anticoagulant that inhibits the clotting process. The liver produces a globulin called **antithrombin,** which combines with and inactivates excess thrombin. This usually limits the fibrin formed to what is

Box 11–6 HEMOPHILIA

There are three forms of **hemophilia;** all are genetic and are characterized by the inability of the blood to clot properly. Hemophilia A is the most common form and involves a deficiency of clotting Factor 8. The gene for hemophilia A is located on the X chromosome, so this is a **sex-linked trait,** with the same pattern of inheritance as red-green colorblindness and Duchenne's muscular dystrophy.

Without Factor 8, the first stage of chemical clotting cannot be completed, and prothrombin activator is not formed. Without treatment, a hemophiliac experiences prolonged bleeding after even minor injuries and extensive internal bleeding, especially in joints subjected to the stresses of weight bearing. In recent years, treatment (but not cure) has become possible with Factor 8 obtained from blood donors. The Factor 8 is extracted from the plasma of donated blood and administered in concentrated form to hemophiliacs, enabling them to live normal lives.

In what is perhaps the most tragic irony of medical progress, many hemophiliacs were inadvertently infected with HIV, the virus that causes AIDS. Prior to 1985, there was no test to detect HIV in donated blood, and the virus was passed to hemophiliacs in the very blood product that was meant to control their disease and prolong their lives. Today, all donated blood and blood products are tested for HIV, and the risk of AIDS transmission to hemophiliacs is now very small.

Box 11–7   DISSOLVING CLOTS

Abnormal clots may cause serious problems in coronary arteries, pulmonary arteries, cerebral vessels, and even veins in the legs. However, if clots can be dissolved before they cause death of tissue, normal circulation and tissue functioning may be restored.

One of the first substances used to dissolve clots in coronary arteries was **streptokinase,** which is actually a bacterial toxin produced by some members of the genus *Streptococcus.* Streptokinase does indeed dissolve clots, but its use creates the possibility of clot destruction throughout the body, with hemorrhage a potential consequence.

Recently, a natural enzyme has been produced by genetic engineering; this is tissue plasminogen activator, or t–PA. When a blood clot forms, a plasma protein called plasminogen is incorporated into the clot. Plasminogen is converted to plasmin by a substance present in the blood vessel lining and lysosomes, and by thrombin. Plasmin then begins to dissolve the fibrin of which the clot is made. The synthetic t–PA functions in a similar way: it converts the plasminogen in an abnormal clot to plasmin, which dissolves the clot and permits blood to flow through the vessel. In a case of coronary thrombosis, if t–PA can be administered within a few hours, the clot may be dissolved and permanent heart damage prevented.

needed to create a useful clot but not an obstructive one.

Thrombosis refers to clotting in an intact vessel; the clot itself is called a **thrombus.** Coronary thrombosis, for example, is abnormal clotting in a coronary artery, which will decrease the blood (oxygen) supply to part of the heart muscle. An **embolism** is a clot or other tissue transported from elsewhere that lodges in and obstructs a vessel (see Box 11–7: Dissolving Clots).

SUMMARY

All the functions of blood described in this chapter contribute to the homeostasis of the body as a whole. However, these functions could not be carried out if the blood did not circulate properly. The circulation of blood throughout the blood vessels is dependent upon the proper functioning of the heart, the pump of the circulatory system.

STUDY OUTLINE

The general functions of blood are transportation, regulation, and protection.

Characteristics of Blood
1. Amount—4 to 6 liters; 38% to 48% is cells; 52% to 62% is plasma (Fig. 11–1).
2. Color—arterial blood has a high oxygen content and is bright red; venous blood has less oxygen and is dark red.
3. pH—7.35 to 7.45; venous blood has more CO_2 and a lower pH than arterial blood.
4. Viscosity—thickness or resistance to flow; due to the presence of cells and plasma proteins; contributes to normal blood pressure.

Plasma—the liquid portion of blood
1. 91% water.
2. Plasma transports nutrients, wastes, hormones, antibodies, CO_2 as HCO_3^-.

3. Plasma proteins: clotting factors are synthesized by the liver; albumin is synthesized by the liver and provides colloid osmotic pressure which pulls tissue fluid into capillaries to maintain normal blood volume and blood pressure; alpha and beta globulins are synthesized by the liver and are carriers for fats and other substances in the blood; gamma globulins are antibodies produced by lymphocytes.

Blood Cells

1. Formed elements are RBCs, WBCs, and platelets (Fig. 11–2).
2. The hemopoietic tissues are red bone marrow (RBM) and lymphatic tissue of the spleen, lymph nodes, and thymus.

Red Blood Cells—erythrocytes (see Table 11–2 for normal values)

1. Biconcave discs; no nuclei when mature.
2. RBCs carry O_2 bonded to the iron in hemoglobin.
3. RBCs are formed in the RBM from stem cells (precursor cells).
4. Hypoxia stimulates the kidneys to produce the hormone erythropoietin, which increases the rate of RBC production in the RBM.
5. Immature RBCs: normoblasts (have nuclei) and reticulocytes (large numbers in peripheral circulation indicate a need for more RBCs to carry oxygen).
6. Vitamin B_{12} is the extrinsic factor, needed for DNA synthesis (mitosis) in stem cells in the RBM. Intrinsic factor is produced by the parietal cells of the stomach lining; it combines with B_{12} to prevent its digestion and promote its absorption.
7. RBCs live for 120 days and are then phagocytized by macrophages in the liver, spleen, and RBM. The iron is returned to the RBM or stored in the liver. The heme of the hemoglobin is converted to bilirubin, which the liver excretes into bile to be eliminated in feces. Jaundice is the accumulation of bilirubin in the blood, perhaps due to liver disease.
8. ABO blood types are hereditary. The type indicates the antigen(s) on the RBCs (see Table 11–1 and Fig. 11–3); antibodies in plasma are for those antigens not present on the RBCs and are important for transfusions.
9. The Rh type is also hereditary. Rh positive means that the D antigen is present on the RBCs; Rh negative means that the D antigen is not present on the RBCs. Rh negative people do not have natural antibodies but will produce them if given Rh positive blood.

White Blood Cells—leukocytes (see Table 11–2 for normal values)

1. Larger than RBCs; have nuclei when mature (Fig. 11–2).
2. Granular WBCs are the neutrophils, eosinophils, and basophils and are produced in the RBM.
3. Agranular WBCs are the lymphocytes and monocytes and are produced in lymphatic tissue, as well as in RBM.
4. Neutrophils and monocytes phagocytize pathogens; monocytes become macrophages which also phagocytize dead tissue.
5. Eosinophils detoxify foreign proteins during allergic reactions and parasitic infections.
6. Basophils contain the anticoagulant heparin and histamine, which contributes to inflammation.
7. Lymphocytes: T cells and B cells. T cells recognize foreign antigens and stop the immune response once the antigen has been destroyed. B cells become plasma cells which produce antibodies to foreign antigens.
8. WBCs carry out their functions in tissue fluid as well as in the blood.

Platelets—thrombocytes (see Table 11–2 for normal values)

1. Platelets are formed in the RBM and are fragments of megakaryocytes.
2. Platelets are involved in all mechanisms of hemostasis (prevention of blood loss) (Fig. 11–5).
3. Vascular spasm—large vessels constrict when damaged, the myogenic response. Platelets release serotonin, which also causes vasoconstriction. The break in the vessel is made smaller and may be closed with a blood clot.
4. Platelet plugs—rupture of a capillary creates a rough surface to which platelets stick and form a barrier over the break.

5. Chemical clotting involves platelet factors, chemicals from damaged tissue, prothrombin, fibrinogen and other clotting factors synthesized by the liver, and calcium ions. See Table 11–3 for the three stages of chemical clotting. The clot is formed of fibrin threads that form a mesh over the break in the vessel.

6. Clot retraction is the folding of the fibrin threads to pull the cut edges of the vessel closer together.

Fibrinolysis is the dissolving of the clot once it has served its purpose.

7. Abnormal clotting (thrombosis) is prevented by the very smooth simple squamous epithelium (endothelium) that lines blood vessels; heparin, which inhibits the clotting process; and antithrombin (synthesized by the liver) which inactivates excess thrombin.

REVIEW QUESTIONS

1. Name four different kinds of substances transported in blood plasma. (p. 242)

2. Name the precursor cell of all blood cells. Name the two types of hemopoietic tissue, their locations, and the types of blood cells produced by each. (pp. 244, 251, 254)

3. State the normal values (CBC) for RBCs, WBCs, platelets, hemoglobin, hematocrit. (p. 251)

4. State the function of RBCs; include the protein and mineral needed. (p. 244)

5. Explain why iron, protein, vitamin B_{12}, and the intrinsic factor are needed for RBC production. (pp. 244, 246)

6. Explain how bilirubin is formed and excreted. (p. 246)

7. Explain what will happen if a person with type O positive blood receives a transfusion of type A negative blood. (pp. 247, 249)

8. Name the WBC with each of the following functions: (p. 251)
   a. become macrophages and phagocytize dead tissue
   b. produce antibodies
   c. detoxify foreign proteins
   d. phagocytize pathogens
   e. contain the anticoagulant heparin
   f. recognize antigens as foreign
   g. secrete histamine during inflammation

9. Explain how and why platelet plugs form in ruptured capillaries. (p. 254)

10. Explain how vascular spasm prevents excessive blood loss when a large vessel is severed. (p. 254)

11. With respect to chemical blood clotting: (pp. 254, 256)
    a. name the mineral necessary
    b. name the organ that produces many of the clotting factors
    c. name the vitamin necessary for prothrombin synthesis
    d. state what the clot itself is made of

12. Explain what is meant by clot retraction and fibrinolysis. (p. 254, 255)

13. State two ways abnormal clotting is prevented in the vascular system. (pp. 255–256)

14. Explain what is meant by blood viscosity, the factors that contribute, and why viscosity is important. (p. 242)

15. State the normal pH range of blood. What gas has an effect on blood pH? (p. 242)

16. Define anemia, leukocytosis, thrombocytopenia. (pp. 246, 253–254)

Chapter 12
The Heart

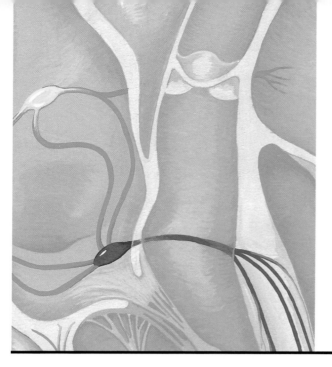

Chapter 12

Chapter Outline

Student Objectives

- Describe the location of the heart and the pericardial membranes.
- Name the chambers of the heart and the vessels that enter or leave each.
- Name the valves of the heart, and explain their functions.
- Describe coronary circulation, and explain its purpose.
- Describe the cardiac cycle.
- Explain how heart sounds are created.
- Name the parts of the cardiac conduction pathway, and explain why it is the SA node that initiates each beat.
- Explain stroke volume, cardiac output, and Starling's Law of the Heart.
- Explain how the nervous system regulates heart rate and force of contraction.

The Heart

New Terminology

Aorta (ay–**OR**–tah)
Atrium (**AY**–tree–um)
Cardiac cycle (**KAR**–dee–yak **SIGH**–kuhl)
Cardiac output (**KAR**–dee–yak **OUT**–put)
Coronary arteries (**KOR**–uh–na–ree **AR**–tuh–rees)
Diastole (dye–**AS**–tuh–lee)
Endocardium (EN–doh–**KAR**–dee–um)
Epicardium (EP–ee–**KAR**–dee–um)
Mediastinum (ME–dee–ah–**STYE**–num)
Mitral valve (**MYE**–truhl VALV)
Myocardium (MY–oh–**KAR**–dee–um)
Sinoatrial (SA) node (**SIGH**–noh–AY–tree–al
  NOHD)
Stroke volume (STROHK **VAHL**–yoom)
Systole (**SIS**–tuh–lee)
Tricuspid valve (try–**KUSS**–pid VALV)
Venous return (**VEE**–nus ree–**TURN**)
Ventricle (**VEN**–tri–kuhl)

Related Clinical Terminology

Arrhythmias (uh–**RITH**–me–yahs)
Ectopic focus (ek–**TOP**–ik **FOH**–kus)
Electrocardiogram (ECG) (ee–LEK–troh–**KAR**–
  dee–oh–GRAM)
Fibrillation (fi–bri–**LAY**–shun)
Heart murmur (HART **MUR**–mur)
Ischemic (iss–**KEY**–mik)
Myocardial infarction (MY–oh–**KAR**–dee–yuhl in–
  FARK–shun)
Pulse (**PULS**)
Stenosis (ste–**NO**–sis)

Terms that appear in **bold type** in the chapter text are defined in the glossary, which begins on page 549.

In the embryo, the heart begins to beat at 4 weeks of age, even before its nerve supply has been established. If a person lives to be 80 years old, his or her heart would continue to beat an average of 100,000 times a day, every day for each of those 80 years. Imagine trying to squeeze a tennis ball 70 times a minute. After a few minutes, your arm muscles would begin to tire. Then imagine increasing your squeezing rate to 120 times a minute. Most of us could not keep that up very long, but that is what the heart does during exercise. A healthy heart can increase its rate and force of contraction to meet the body's need for more oxygen, then return to its resting rate and keep on beating as if nothing very extraordinary had happened. In fact, it isn't extraordinary at all; this is the job the heart is meant to do.

The primary function of the heart is to pump blood through the arteries, capillaries, and veins. As you learned in the last chapter, blood transports oxygen and nutrients and has other important functions, as well. The heart is the pump that keeps blood circulating properly.

LOCATION AND PERICARDIAL MEMBRANES

The heart is located in the thoracic cavity between the lungs. This area is called the **mediastinum.** The cone-shaped heart has its tip (apex) just above the diaphragm to the left of the midline. This is why we may think of the heart as being on the left side, since the strongest beat can be heard or felt here.

The heart is enclosed in the **pericardial membranes,** of which there are three (Fig. 12–1). The outermost is the **fibrous pericardium,** a loose-fitting sac of fibrous connective tissue that extends inferiorly over the diaphragm and superiorly over the bases of the large vessels that enter and leave the heart. Lining the fibrous pericardium is the **parietal pericardium,** a serous membrane. On the surface of the heart muscle is the **visceral pericardium,** also called the **epicardium,** another serous membrane. Between the parietal and visceral pericardial membranes is **serous fluid,** which prevents friction as the heart beats.

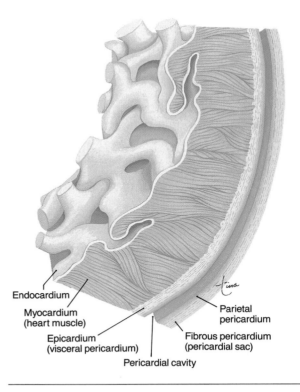

Endocardium
Myocardium (heart muscle)
Epicardium (visceral pericardium)
Pericardial cavity
Parietal pericardium
Fibrous pericardium (pericardial sac)

Figure 12–1 Layers of the wall of the heart and the pericardial membranes. The endocardium is the lining of the chambers of the heart. The fibrous pericardium is the outermost layer.

CHAMBERS—VESSELS AND VALVES

The walls of the four chambers of the heart are made of cardiac muscle called the **myocardium.** The chambers are lined with **endocardium,** simple squamous epithelium that also covers the valves of the heart and continues into the vessels as their lining. The important physical characteristic of the endocardium is not its thinness, but rather its smoothness. This very smooth tissue prevents abnormal blood clotting, since clotting would be initiated by contact of blood with a rough surface.

The upper chambers of the heart are the right and left **atria** (singular: **atrium**), which have relatively thin walls and are separated by a common wall of myocardium called the **interatrial septum.** The lower chambers are the right and left **ventricles,** which have thicker walls and are separated by the **interventricular septum** (Fig. 12–2). As you will see, the atria receive blood, either from the body or the lungs, and the ventricles pump blood, either to the lungs or the body.

RIGHT ATRIUM

Two large veins return blood from the body to the right atrium (see Fig. 12–2). The **superior vena cava** carries blood from the upper body, and the **inferior vena cava** carries blood from the lower body. From the right atrium, blood will flow through the right **atrioventricular (AV) valve,** or **tricuspid valve,** into the right ventricle.

The tricuspid valve is made of three flaps (or cusps) of endocardium reinforced with connective tissue. The general purpose of all valves in the circulatory system is to prevent backflow of blood. The specific purpose of the tricuspid valve is to prevent backflow of blood from the right ventricle to the right atrium when the right ventricle contracts. As the ventricle contracts, blood is forced behind the three valve flaps, forcing them upward and together to close the valve.

LEFT ATRIUM

The left atrium receives blood from the lungs, by way of four **pulmonary veins.** This blood will then flow into the left ventricle through the left atrioventricular (AV) valve, also called the **mitral valve** or **bicuspid** (two flaps) valve. The mitral valve prevents backflow of blood from the left ventricle to the left atrium when the left ventricle contracts.

A recently discovered function of the atria is the production of a hormone involved in blood pressure maintenance. When the walls of the atria are stretched, as by increased blood volume or blood pressure, the cells produce **Atrial Natriuretic Hormone** (ANH). ANH decreases the reabsorption of sodium ions by the kidneys, so that more sodium ions are excreted in urine, which in turn increases the elimination of water. The loss of water lowers blood volume and blood pressure. You may have

noticed that ANH is an antagonist to the hormone aldosterone, which raises blood pressure.

RIGHT VENTRICLE

When the right ventricle contracts, the tricuspid valve closes, and the blood is pumped to the lungs through the pulmonary artery (or trunk). At the junction of this large artery and the right ventricle is the **pulmonary semilunar valve.** Its three flaps are forced open when the right ventricle contracts and pumps blood into the pulmonary artery. When the right ventricle relaxes, blood tends to come back, but this fills the valve flaps and closes the pulmonary semilunar valve to prevent backflow of blood into the right ventricle.

Projecting into the lower part of the right ventricle are columns of myocardium called **papillary muscles** (see Fig. 12–2). Strands of fibrous connective tissue, the **chordae tendineae,** extend from the papillary muscles to the flaps of the tricuspid valve. When the right ventricle contracts, the papillary muscles also contract and pull on the chordae tendineae to prevent inversion of the tricuspid valve. If you have ever had your umbrella blown inside out by a strong wind, you can see what would happen if the flaps of the tricuspid valve were not anchored by the chordae tendineae and papillary muscles.

LEFT VENTRICLE

The walls of the left ventricle are thicker than those of the right ventricle, which enables the left ventricle to contract more forcefully. The left ventricle pumps blood to the body through the **aorta,** the largest artery of the body. At the junction of the aorta and the left ventricle is the **aortic semilunar valve** (see Fig. 12–2). This valve is opened by the force of contraction of the left ventricle, which also closes the mitral valve. The aortic semilunar valve closes when the left ventricle relaxes, to prevent backflow of blood from the aorta to the left ventricle. When the mitral (left AV) valve closes, it prevents backflow of blood to the left atrium; the flaps of the mitral valve are also anchored by chordae tendineae and papillary muscles. All valves are shown in Fig. 12–3.

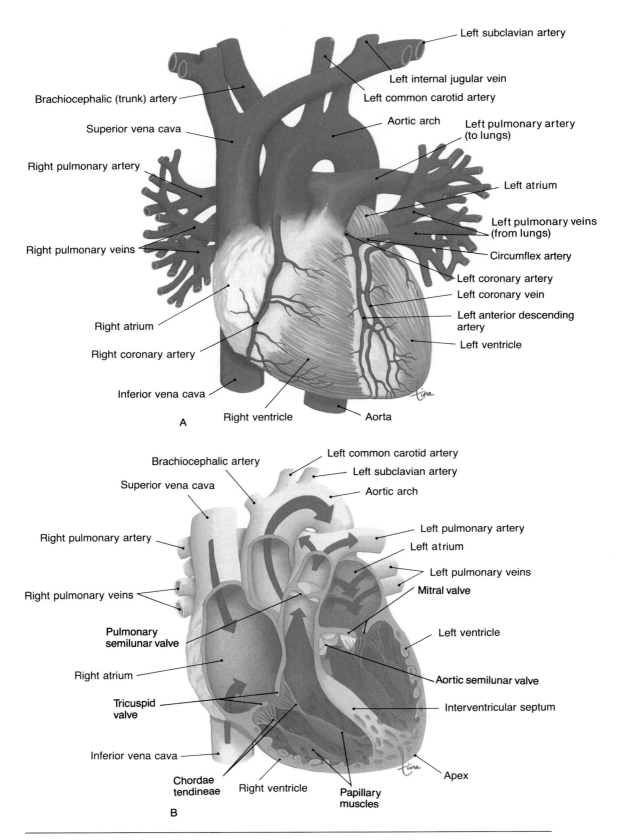

Figure 12–2 **(A)**, Anterior view of the heart and major blood vessels. **(B)**, Frontal section of the heart in anterior view, showing internal structures.

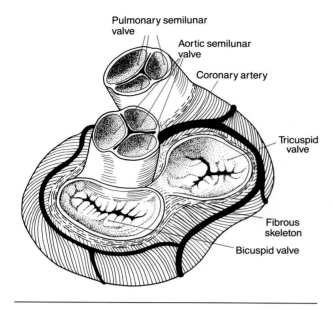

Pulmonary semilunar valve

Aortic semilunar valve

Coronary artery

Tricuspid valve

Fibrous skeleton

Bicuspid valve

Figure 12–3 Heart valves in superior view. The atria have been removed. The fibrous skeleton of the heart is fibrous connective tissue that anchors the valve flaps and prevents enlargement of the valve openings.

As you can see from this description of the chambers and their vessels, the heart is really a double, or two-sided, pump. The right side of the heart receives deoxygenated blood from the body and pumps it to the lungs to pick up oxygen and release carbon dioxide. The left side of the heart receives oxygenated blood from the lungs and pumps it to the body. Both pumps work simultaneously, that is, both atria contract together, followed by the contraction of both ventricles.

CORONARY VESSELS

The right and left **coronary arteries** are the first branches of the ascending aorta, just beyond the aortic semilunar valve (Fig. 12–4). The two arteries branch into smaller arteries and arterioles, then to capillaries. The coronary capillaries merge to form coronary veins, which empty blood into a large coronary sinus that returns blood to the right atrium.

The purpose of the coronary vessels is to supply blood to the myocardium itself, because oxygen is essential for normal myocardial contraction. If a coronary artery becomes obstructed, by a blood clot for example, part of the myocardium becomes **ischemic,** that is, deprived of its blood supply. Prolonged ischemia will create an **infarct,** an area of necrotic (dead) tissue. This is a myocardial infarction, commonly called a heart attack (see also Box 12–1: Risk Factors for Heart Disease).

CARDIAC CYCLE AND HEART SOUNDS

The **cardiac cycle** is the sequence of events in one heartbeat. In its simplest form, the cardiac cycle is the simultaneous contraction of the two atria, followed a fraction of a second later by the simultaneous contraction of the two ventricles. **Systole** is another term for contraction. The term for relaxation is **diastole.** You are probably familiar with these terms as they apply to blood pressure readings. If we apply them to the cardiac cycle, we can say that atrial systole is followed by ventricular systole. There is, however, a significant difference between the movement of blood from the atria to the ventricles and the movement of blood from the ventricles to the arteries. Refer to Fig. 12–5 as you read the following events of the cardiac cycle.

Blood is constantly flowing from the veins into both atria. As more blood accumulates, its pressure forces open the right and left AV valves. Two thirds of the atrial blood flows passively into the ventricles; the atria then contract to pump the remaining blood into the ventricles.

Following their contraction, the atria relax and the ventricles begin to contract. Ventricular contraction forces blood against the flaps of the right and left AV valves and closes them; the force of blood also opens the aortic and pulmonary semilunar valves. As the ventricles continue to contract, they pump blood into the arteries. Notice that blood that enters the arteries must all be pumped. The ventricles then relax, and at the same time blood continues to flow into the atria, and the cycle will begin again.

The important distinction here is that most blood

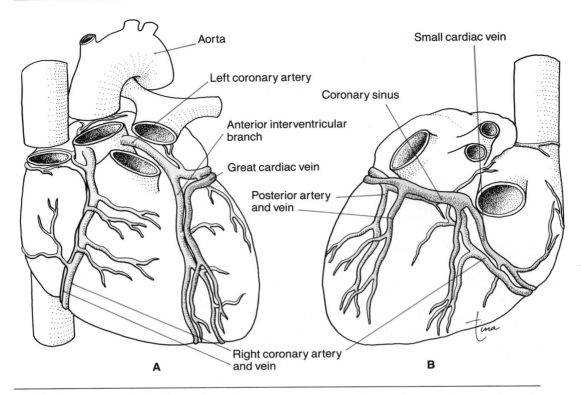

Figure 12–4 **(A)**, Coronary vessels in anterior view. The pulmonary artery has been cut to show the left coronary artery emerging from the ascending aorta. **(B)**, Coronary vessels in posterior view. The coronary sinus empties blood into the right atrium.

flows passively from atria to ventricles, but *all* blood to the arteries is actively pumped by the ventricles. For this reason, the proper functioning of the ventricles is much more crucial to survival than is atrial functioning.

You may be asking: all this in one heartbeat? The answer is yes. The cardiac cycle is this precise sequence of events that keeps blood moving from the veins, through the heart, and into the arteries.

The cardiac cycle also creates the **heart sounds:** each heartbeat produces two sounds, often called lub-dup, that can be heard with a stethoscope. The first sound, the loudest and longest, is caused by ventricular systole closing the AV valves. The second sound is caused by the closure of the aortic and pulmonary semilunar valves. If any of the valves do not close properly, an extra sound called a **heart murmur** may be heard (see Box 12–2: Heart Murmur).

CARDIAC CONDUCTION PATHWAY

The cardiac cycle is a sequence of mechanical events that is regulated by the electrical activity of the myocardium. Cardiac muscle cells have the ability to contract spontaneously, that is, nerve impulses are not required to cause contraction. The heart generates its own beat, and the electrical impulses follow a very specific route throughout the myocardium. You may find it helpful to refer to Fig. 12–6 as you read the following.

The natural pacemaker of the heart is the **sinoatrial (SA) node,** a specialized group of cardiac muscle cells located in the wall of the right atrium. The SA node is considered specialized because it has the most rapid rate of contraction, that is, it de-

Box 12–1   RISK FACTORS FOR HEART DISEASE

Coronary artery disease results in decreased blood flow to the myocardium. If blood flow is diminished but not completely obstructed, the person may experience angina, which is chest pain caused by lack of oxygen to part of the heart muscle. If blood flow is completely blocked, however, the result is a myocardial infarction (necrosis of cardiac muscle).

The most common cause of coronary artery disease is **atherosclerosis**. Plaques of cholesterol form in the walls of a coronary artery; this narrows the lumen (cavity) and creates a rough surface where a clot (thrombus) may form. Predisposing factors for atherosclerosis include cigarette smoking, diabetes mellitus, and high blood pressure. Any one of these may cause damage to the lining of coronary arteries, which is the first step in the abnormal deposition of cholesterol. A diet high in cholesterol and saturated fats and high blood levels of these lipids will increase the rate of cholesterol deposition.

Another predisposing factor is a family history of coronary artery disease. There is no "gene for heart attacks," but we do have genes for the enzymes involved in cholesterol metabolism. Many of these are liver enzymes that regulate the transport of cholesterol in the blood in the form of lipoproteins and regulate the liver's excretion of excess cholesterol in bile. Some people, therefore, have a greater tendency than others to have higher blood levels of cholesterol and certain lipoproteins. In women before menopause, estrogen is believed to exert a protective effect by lowering blood lipid levels. This is why heart attacks in the 30- to 50-year-old age range are far more frequent in men than in women.

What can an individual do to minimize the risk of coronary artery disease? There is no way to change a hereditary predisposition, but other sensible steps can be taken. First, don't smoke. Second, maintain a diet low in cholesterol and saturated fats. Third, maintain normal body weight through proper diet and exercise. Fourth, have regular blood pressure checks and, if hypertension develops, follow the recommendations of a physician to maintain normal blood pressure.

When coronary artery disease may be life-threatening, coronary artery **by-pass surgery** may be performed. In this procedure, a synthetic vessel or a vein (such as the saphenous vein of the leg) is grafted around the obstructed coronary vessel to restore blood flow to the myocardium. This is not a cure, for atherosclerosis may occur in a grafted vein or at other sites in the coronary arteries. The person who has had such surgery should follow the guidelines described above.

polarizes more rapidly than any other part of the myocardium. As you may recall, depolarization is the rapid entry of Na$^+$ ions and the reversal of charges on either side of the cell membrane. The cells of the SA node are more permeable to Na$^+$ ions than are other cardiac muscle cells. Therefore, they depolarize more rapidly, then contract and initiate each heartbeat.

From the SA node, impulses for contraction travel to the **atrioventricular (AV) node,** located in the lower interatrial septum. The transmission of impulses from the SA node to the AV node and to the rest of the atrial myocardium brings about atrial systole.

Within the upper interventricular septum is the **Bundle of His** (AV bundle), which receives im-

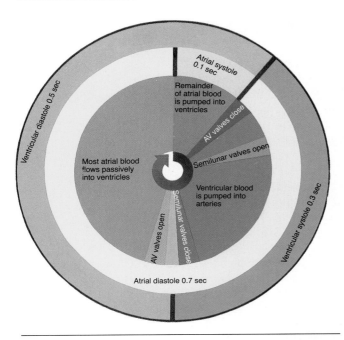

Atrial systole 0.1 sec

Remainder of atrial blood is pumped into ventricles

Ventricular diastole 0.5 sec

Most atrial blood flows passively into ventricles

AV valves close

Semilunar valves open

Ventricular blood is pumped into arteries

Semilunar valves close

AV valves open

Ventricular systole 0.3 sec

Atrial diastole 0.7 sec

Figure 12–5 The cardiac cycle depicted in one heart-beat (pulse: 75). The outer circle represents the ventricles, the middle circle the atria, and the inner circle the movement of blood and its effect on the heart valves. See text for description.

pulses from the AV node and transmits them to the right and left **bundle branches.** From the bundle branches, impulses travel along **Purkinje fibers** to the rest of the ventricular myocardium and bring about ventricular systole. The electrical activity of the atria and ventricles is depicted by an electro-cardiogram (ECG); this is discussed in Box 12–3: Electrocardiogram.

If the SA node does not function properly, the AV node will initiate the heartbeat, but at a slower rate (50 to 60 beats per minute). The Bundle of His is also capable of generating the beat of the ventricles, but at a much slower rate (15 to 40 beats per minute). This may occur in certain kinds of heart disease in which transmission of impulses from the atria to the ventricles is blocked.

Arrhythmias are irregular heartbeats; their effects range from harmless to life-threatening. Nearly everyone experiences heart **palpitations** (becoming aware of an irregular beat) from time to time. These are usually not serious and may be the result of too much caffeine, nicotine, or alcohol. Much more serious is ventricular **fibrillation,** a very rapid and uncoordinated ventricular beat that is totally ineffective for pumping blood (see Box 12–4: Arrhythmias).

Box 12–2   HEART MURMUR

A heart murmur is an abnormal or extra heart sound caused by a malfunctioning heart valve. The function of heart valves is to prevent backflow of blood, and when a valve does not close properly, blood will regurgitate (go backward), creating turbulence that may be heard with a stethoscope.

Rheumatic heart disease is a now uncommon complication of a streptococcal infection. In rheumatic fever, the heart valves are damaged by an abnormal response by the immune system. Erosion of the valves makes them "leaky" and inefficient, and a murmur of backflowing blood will be heard. Mitral valve regurgitation, for example, will be heard as a systolic murmur, because this valve is meant to close and prevent backflow during ventricular systole.

Some valve defects involve a narrowing **(stenosis)** and are congenital, that is, the child is born with an abnormally narrow valve. In aortic stenosis, for example, blood cannot easily pass from the left ventricle to the aorta. The ventricle must then work harder to pump blood through the narrow valve to the arteries, and the turbulence created is also heard as a systolic murmur.

Children sometimes have heart murmurs that are called "functional" because no structural cause can be found. These murmurs usually disappear with no adverse effects on the child.

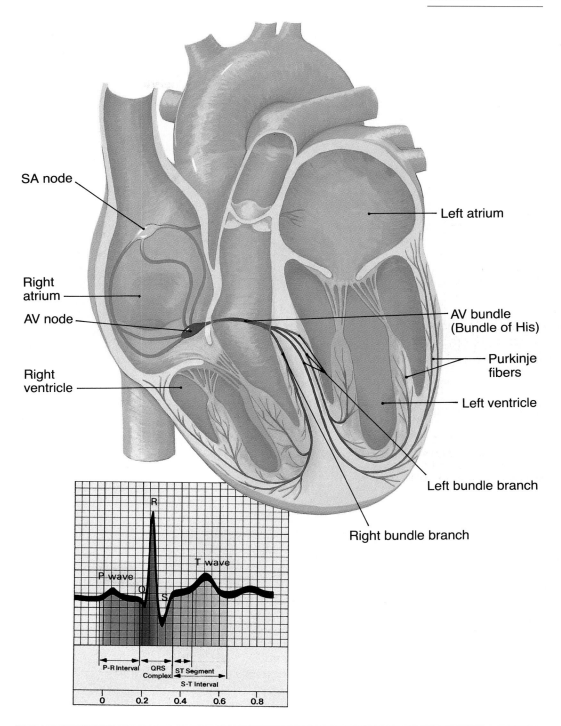

Figure 12–6 Conduction pathway of the heart. Anterior view of the interior of the heart. The electrocardiogram tracing is of one normal heartbeat. See text and Box 12–3 for description.

Box 12–3 ELECTROCARDIOGRAM

A heartbeat is a series of electrical events, and the electrical charges generated by the myocardium can be recorded by placing electrodes on the body surface. Such a recording is called an **electrocardiogram (ECG)** (see Fig. 12–6).

A typical ECG consists of three distinguishable waves or deflections: the P wave, the QRS complex, and the T wave. Each represents a specific electrical event; all are shown at right in a normal ECG tracing.

The P wave represents depolarization of the atria, that is, the transmission of electrical impulses from the SA node throughout the atrial myocardium.

The QRS complex represents depolarization of the ventricles as the electrical impulses spread throughout the ventricular myocardium. The T wave represents repolarization of the ventricles (atrial repolarization does not appear as a separate wave because it is masked by the QRS complex).

Detailed interpretation of abnormal ECGs is beyond the scope of this book, but in general, the length of each wave and the time intervals between waves are noted. An ECG may be helpful in the diagnosis of coronary atherosclerosis which deprives the myocardium of oxygen, or of rheumatic fever or other valve disorders that result in enlargement of a chamber of the heart and prolong a specific wave of an ECG. For example, the enlargement of the left ventricle that is often a consequence of hypertension may be indicated by an abnormal QRS complex.

Box 12–4 ARRHYTHMIAS

Arrhythmias are irregular heartbeats caused by damage to part of the conduction pathway, or by an **ectopic focus,** which is a beat generated in part of the myocardium other than the SA node.

Flutter is a very rapid but fairly regular heartbeat. In atrial flutter, the atria may contract up to 300 times per minute. Since atrial pumping is not crucial, however, blood flow to the ventricles may be maintained for a time, and flutter may not be immediately life-threatening. Ventricular flutter is usually only a brief transition between ventricular tachycardia and fibrillation.

Fibrillation is very rapid and uncoordinated contractions. Ventricular fibrillation is a medical emergency that must be rapidly corrected to prevent death. Normal contraction of the ventricles is necessary to pump blood into the arteries, but fibrillating ventricles are not pumping, and cardiac output decreases sharply.

Ventricular fibrillation may follow a non-fatal heart attack (myocardial infarction). Damaged cardiac muscle cells may not be able to maintain a normal state of polarization, and they depolarize spontaneously and rapidly. From this ectopic focus, impulses spread to other parts of the ventricular myocardium in a rapid and haphazard pattern, and the ventricles quiver rather than contract as a unit.

It is often possible to correct ventricular fibrillation with the use of an electrical defibrillator. This instrument delivers an electric shock to the heart, which causes the entire myocardium to depolarize and contract, then relax. If the first part of the heart to recover is the SA node (which usually has the most rapid rate of contraction), a normal heartbeat may be restored.

HEART RATE

A healthy adult has a resting heart rate **(pulse)** of 60 to 80 beats per minute, which is the rate of depolarization of the SA node. A child's normal heart rate may be as high as 100 beats per minute, that of an infant as high as 120, and that of a near-term fetus as high as 140 beats per minute. These higher rates are not related to age, but rather to size: the smaller the individual, the higher the metabolic rate and the faster the heart rate. Parallels may be found among animals of different sizes; the heart rate of a mouse is about 200 beats per minute and that of an elephant about 30 beats per minute.

Let us return to the adult heart rate and consider the person who is in excellent physical condition. As you may know, well-conditioned athletes have low resting pulse rates. Those of basketball players are often around 50 beats per minute, and the pulse of a marathon runner often ranges from 35 to 40 beats per minute. To understand why this is so, remember that the heart is a muscle. When our skeletal muscles are exercised, they become stronger and more efficient. The same is true for the heart; consistent exercise makes it a more efficient pump, as you will see in the next discussion.

CARDIAC OUTPUT

Cardiac output is the amount of blood pumped by a ventricle in 1 minute. A certain level of cardiac output is needed at all times to transport oxygen to tissues and to remove waste products. During exercise, cardiac output must increase to meet the body's need for more oxygen. We will return to exercise after first considering resting cardiac output.

In order to calculate cardiac output, we must know the pulse rate and how much blood is pumped per beat. **Stroke volume** is the term for the amount of blood pumped by a ventricle per beat; an average resting stroke volume is 60 to 80 ml per beat. A simple formula then enables us to determine cardiac output:

Cardiac output = stroke volume × pulse (heart rate)

Let us put into this formula an average resting stroke volume, 70 ml, and an average resting pulse, 70 beats per minute (bpm):

Cardiac output = 70 ml × 70 bpm
Cardiac output = 4900 ml per minute
(approximately 5 liters)

Naturally, cardiac ouput varies with the size of the person, but the average resting cardiac output is 5 to 6 liters per minute.

If we now reconsider the athlete, you will be able to see precisely why the athlete has a low resting pulse. In our formula, we will use an average resting cardiac output (5 liters) and an athlete's pulse rate (50):

Cardiac output = stroke volume × pulse
5000 ml = stroke volume × 50 bpm
$\frac{5000}{50}$ = stroke volume
100 ml = stroke volume

Notice that the athlete's resting stroke volume is significantly higher than the average. The athlete's more efficient heart pumps more blood with each beat and so can maintain a normal resting cardiac output with fewer beats.

Now let us see how the heart responds to exercise. Heart rate (pulse) increases during exercise, and so does stroke volume. The increase in stroke volume is the result of **Starling's Law of the Heart,** which states that the more the cardiac muscle fibers are stretched, the more forcefully they contract. During exercise, more blood returns to the heart; this is called **venous return.** Increased venous return stretches the myocardium of the ventricles, which contract more forcefully and pump more blood, thereby increasing stroke volume. Therefore, during exercise, our formula might be the following:

Cardiac output = stroke volume × pulse
Cardiac output = 100 ml × 100 bpm
Cardiac output = 10,000 ml (10 liters)

This exercise cardiac output is twice the resting cardiac output we first calculated, which should not be considered unusual. The cardiac output of a healthy young person may increase up to four times the resting level during strenuous exercise. The marathon runner's cardiac output may increase six times or more compared to the resting level; this is the result of the marathoner's extremely efficient heart.

REGULATION OF HEART RATE

Although the heart generates and maintains its own beat, the rate of contraction can be changed to adapt to different situations. The nervous system can and does bring about necessary changes in heart rate as well as in force of contraction.

The **medulla** of the brain contains the two cardiac centers, the **accelerator center** and the **inhibitory center.** These centers send impulses to the heart along autonomic nerves. Recall from Chapter 8 that the autonomic nervous system has two divisions: sympathetic and parasympathetic. Sympathetic impulses from the accelerator center along sympathetic nerves increase heart rate and force of contraction. Parasympathetic impulses from the inhibitory center along the vagus nerves decrease the heart rate.

Our next question might be: what information is received by the medulla to initiate changes? Since the heart pumps blood, it is essential to maintain normal blood pressure. Blood contains oxygen, which all tissues must receive continuously. Therefore, changes in blood pressure and oxygen level of the blood are stimuli for changes in heart rate.

You may also recall from Chapter 9 that pressoreceptors and chemoreceptors are located in the carotid arteries and aortic arch. **Pressoreceptors** in the carotid sinuses and aortic sinus detect changes in blood pressure. **Chemoreceptors** in the carotid bodies and aortic body detect changes in the oxygen content of the blood. The sensory nerves for the carotid receptors are the glossopharyngeal (9th cranial) nerves; the sensory nerves for the aortic arch receptors are the vagus (10th cranial) nerves. If we now put all these facts together in a specific example, you will see that the regulation of heart rate is a reflex. Fig. 12–7 depicts all the structures mentioned above.

A person who stands up suddenly from a lying position may feel light-headed or dizzy for a few moments, because blood pressure to the brain has decreased abruptly. The drop in blood pressure is detected by pressoreceptors in the carotid sinuses—notice that they are "on the way" to the brain, a very strategic location. Impulses generated by the pressoreceptors travel along the glossopharyngeal nerves to the medulla and stimulate the accelerator center. The accelerator center generates impulses that are carried by sympathetic nerves to the SA node, AV node, and the ventricular myocardium. As heart rate and force increase, blood pressure to the brain is raised to normal, and the sensation of light-headedness passes. When blood pressure to the brain is restored to normal, the heart receives more parasympathetic impulses from the inhibitory center along the vagus nerves to the SA node and AV node. These parasympathetic impulses slow the heart rate to a normal resting pace.

The heart will also be the effector in a reflex stimulated by a decrease in the oxygen content of the blood. The aortic receptors are strategically located so as to detect such an important change as soon as blood leaves the heart. The reflex arc in this situation would be: (1) aortic chemoreceptors, (2) vagus nerves (sensory), (3) accelerator center in the medulla, (4) sympathetic nerves, and (5) the heart, which will increase its rate and force of contraction to circulate more oxygen to correct the hypoxia.

SUMMARY

As you can see, the nervous system regulates the functioning of the heart based on what the heart is supposed to do. The pumping of the heart maintains normal blood pressure and proper oxygenation of tissues, and the nervous system ensures that the heart will be able to meet these demands in different situations.

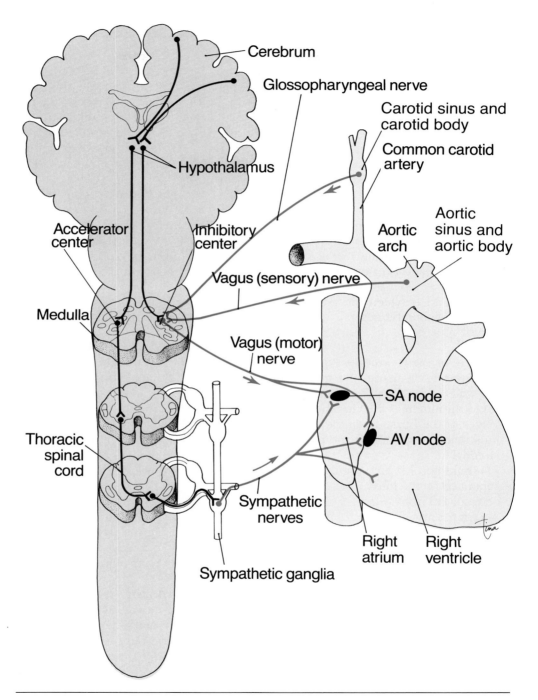

Figure 12-7 Nervous regulation of the heart. The brain and spinal cord are shown on the left. The heart and major blood vessels are shown on the right.

STUDY OUTLINE

The heart pumps blood, which creates blood pressure, and circulates oxygen, nutrients, and other substances. The heart is located in the mediastinum, the area between the lungs in the thoracic cavity.

Pericardial Membranes—three layers that enclose the heart (see Fig. 12–1)
1. The outer, fibrous pericardium, made of fibrous connective tissue, is a loose-fitting sac that surrounds the heart and extends over the diaphragm and the bases of the great vessels.
2. The parietal pericardium is a serous membrane that lines the fibrous pericardium.
3. The visceral pericardium, or epicardium, is a serous membrane on the surface of the myocardium.
4. Serous fluid between the parietal and visceral pericardial membranes prevents friction as the heart beats.

Chambers of the Heart (see Fig. 12–2)
1. Cardiac muscle tissue, the myocardium, forms the walls of the four chambers of the heart.
2. Endocardium lines the chambers and covers the valves of the heart; is simple squamous epithelium that is very smooth and prevents abnormal clotting.
3. The right and left atria are the upper chambers, separated by the interatrial septum. The atria receive blood from veins.
4. The right and left ventricles are the lower chambers, separated by the interventricular septum. The ventricles pump blood into arteries.

Right Atrium
1. Receives blood from the upper body by way of the superior vena cava and receives blood from the lower body by way of the inferior vena cava.
2. The tricuspid (right AV) valve prevents backflow of blood from the right ventricle to the right atrium when the right ventricle contracts.

Left Atrium
1. Receives blood from the lungs by way of four pulmonary veins.

2. The mitral (left AV or bicuspid) valve prevents backflow of blood from the left ventricle to the left atrium when the left ventricle contracts.
   • The walls of the atria produce Atrial Natriuretic Hormone when stretched by increased blood volume or BP. ANH increases the loss of Na^+ ions and water in urine, which decreases blood volume and BP to normal.

Right Ventricle—has relatively thin walls
1. Pumps blood to the lungs through the pulmonary artery.
2. The pulmonary semilunar valve prevents backflow of blood from the pulmonary artery to the right ventricle when the right ventricle relaxes.
3. Papillary muscles and chordae tendineae prevent inversion of the right AV valve when the right ventricle contracts.

Left Ventricle—has thicker walls than does the right ventricle
1. Pumps blood to the body through the aorta.
2. The aortic semilunar valve prevents backflow of blood from the aorta to the left ventricle when the left ventricle relaxes.
3. Papillary muscles and chordae tendineae prevent inversion of the left AV valve when the left ventricle contracts.
   • The heart is a double pump: the right heart receives deoxygenated blood from the body and pumps it to the lungs; the left heart receives oxygenated blood from the lungs and pumps it to the body. Both sides of the heart work simultaneously.

Coronary Vessels (see Fig. 12–4)
1. Pathway: ascending aorta to right and left coronary arteries, to smaller arteries, to capillaries, to coronary veins, to the coronary sinus, to the right atrium.
2. Coronary circulation supplies oxygenated blood to the myocardium.
3. Obstruction of a coronary artery causes a myocardial infarction: death of an area of myocardium due to lack of oxygen.

Cardiac Cycle—the sequence of events in one heartbeat (see Fig. 12–5)

1. The atria continually receive blood from the veins; as pressure within the atria increases, the AV valves are opened.
2. Two thirds of the atrial blood flows passively into the ventricles; atrial contraction pumps the remaining blood into the ventricles; the atria then relax.
3. The ventricles contract, which closes the AV valves and opens the aortic and pulmonary semilunar valves.
4. Ventricular contraction pumps all blood into the arteries. The ventricles then relax. Meanwhile, blood is filling the atria, and the cycle begins again.
5. Systole means contraction; diastole means relaxation. In the cardiac cycle, atrial systole is followed by ventricular systole. When the ventricles are in systole, the atria are in diastole.
6. The mechanical events of the cardiac cycle keep blood moving from the veins through the heart and into the arteries.

Heart Sounds—two sounds per heartbeat: lub-dup

1. The first sound is created by closure of the AV valves during ventricular systole.
2. The second sound is created by closure of the aortic and pulmonary semilunar valves.
3. Improper closing of a valve results in a heart murmur.

Cardiac Conduction Pathway—the pathway of impulses during the cardiac cycle (see Fig. 12–6)

1. The SA node in the wall of the right atrium initiates each heartbeat; the cells of the SA node are more permeable to Na^+ ions and depolarize more rapidly than any other part of the myocardium.
2. The AV node is in the lower interatrial septum. Depolarization of the SA node spreads to the AV node and to the atrial myocardium and brings about atrial systole.
3. The Bundle of His is in the upper interventricular septum; the first part of the ventricles to depolarize.

4. The right and left bundle branches in the interventricular septum transmit impulses to the Purkinje fibers in the ventricular myocardium, which complete ventricular systole.
5. An electrocardiogram (ECG) depicts the electrical activity of the heart (see Fig. 12–6).
6. If part of the conduction pathway does not function properly, the next part will initiate contraction, but at a slower rate.
7. Arrhythmias are irregular heartbeats; their effects range from harmless to life-threatening.

Heart Rate

1. Healthy adult: 60 to 80 beats per minute (heart rate equals pulse); children and infants have faster pulses because of their smaller size and higher metabolic rate.
2. A person in excellent physical condition has a slow resting pulse because the heart is a more efficient pump and pumps more blood per beat.

Cardiac Output

1. Cardiac output is the amount of blood pumped by a ventricle in 1 minute.
2. Stroke volume is the amount of blood pumped by a ventricle in one beat; average is 60 to 80 ml.
3. Cardiac output equals stroke volume × pulse; average resting cardiac output is 5 to 6 liters.
4. Starling's Law of the Heart—the more that cardiac muscle fibers are stretched, the more forcefully they contract.
5. During exercise, stroke volume increases as venous return increases and stretches the myocardium of the ventricles (Starling's Law).
6. During exercise, the increase in stroke volume and the increase in pulse result in an increase in cardiac output: two to four times the resting level.

Regulation of Heart Rate (see Fig. 12–7)

1. The heart generates its own beat, but the nervous system brings about changes to adapt to different situations.
2. The medulla contains the cardiac centers: the accelerator center and the inhibitory center.
3. Sympathetic impulses to the heart increase rate and force of contraction; parasympathetic im-

pulses (vagus nerves) to the heart decrease heart rate.

4. Pressoreceptors in the carotid and aortic sinuses detect changes in blood pressure.

5. Chemoreceptors in the carotid and aortic bodies detect changes in the oxygen content of the blood.

6. The glossopharyngeal nerves are sensory for the carotid receptors. The vagus nerves are sensory for the aortic receptors.

7. If blood pressure to the brain decreases, pressoreceptors in the carotid sinuses detect this decrease and send sensory impulses along the glossopharyngeal nerves to the medulla. The accelerator center dominates and sends motor impulses along sympathetic nerves to increase heart rate and force to restore blood pressure to normal.

8. A similar reflex is activated by hypoxia.

REVIEW QUESTIONS

1. Describe the location of the heart with respect to the lungs and to the diaphragm. (p. 264)

2. Name the three pericardial membranes. Where is serous fluid found and what is its function? (p. 264)

3. Describe the location and explain the function of endocardium. (p. 264)

4. Name the veins that enter the right atrium; name those that enter the left atrium. For each, where does the blood come from? (p. 265)

5. Name the artery that leaves the right ventricle; name the artery that leaves the left ventricle. For each, where is the blood going? (p. 265)

6. Explain the purpose of the right and left AV valves and the purpose of the aortic and pulmonary semilunar valves. (p. 265)

7. Describe the coronary system of vessels and explain the purpose of coronary circulation. (p. 267)

8. Define: systole, diastole, cardiac cycle. (p. 267)

9. Explain how movement of blood from atria to ventricles differs from movement of blood from ventricles to arteries. (pp. 267–268)

10. Explain why the heart is considered a double pump. Trace the path of blood from the right atrium back to the right atrium, naming the chambers of the heart and their vessels that the blood passes through. (pp. 265, 267)

11. Name the parts, in order, of the cardiac conduction pathway. Explain why it is the SA node that generates each heartbeat. State a normal range of heart rate for a healthy adult. (pp. 268–270)

12. Calculate cardiac output if stroke volume is 75 ml and pulse is 75 bpm. Using the cardiac output you just calculated as a resting normal, what is the stroke volume of a marathoner whose resting pulse is 40 bpm? (p. 273)

13. Name the two cardiac centers and state their location. Sympathetic impulses to the

heart have what effect? Parasympathetic impulses to the heart have what effect? Name the parasympathetic nerves to the heart. (p. 274)

14. State the locations of arterial pressoreceptors and chemoreceptors; what they detect, and their sensory nerves. (p. 274)

15. Describe the reflex arc to increase heart rate and force when blood pressure to the brain decreases. (p. 274)

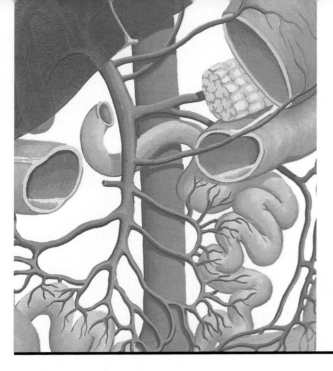

Chapter 13

Student Objectives

- Describe the structure of arteries and veins, and relate their structure to function.
- Explain the purpose of arterial and venous anastomoses.
- Describe the structure of capillaries, and explain the exchange processes that take place in capillaries.
- Describe the pathway and purpose of pulmonary circulation.
- Name the branches of the aorta and their distributions.
- Name the major systemic veins, and the parts of the body they drain of blood.
- Describe the pathway and purpose of hepatic portal circulation.
- Describe the modifications of fetal circulation, and explain the purpose of each.
- Define blood pressure, and state the normal ranges for systemic and pulmonary blood pressure.
- Explain the factors that maintain systemic blood pressure.
- Explain how the heart and kidneys are involved in the regulation of blood pressure.
- Explain how the medulla and the autonomic nervous system regulate the diameter of blood vessels.

The Vascular System

New Terminology

Anastomosis (a–NAS–ti–**MOH**–sis)

Arteriole (ar–**TIR**–ee–ohl)

Circle of Willis (**SIR**–kuhl of **WILL**–iss)

Ductus arteriosus (**DUK**–tus ar–TIR–ee–**OH**–sis)

Endothelium (EN–doh–**THEEL**–ee–um)

Foramen ovale (for–**RAY**–men oh–**VAHL**–ee)

Hepatic portal (hep–**PAT**–ik **POOR**–tuhl)

Peripheral resistance (puh–**RIFF**–uh–ruhl ree–**ZIS**–tense)

Placenta (pluh–**SEN**–tah)

Precapillary sphincter (pre–**KAP**–i–lar–ee **SFINK**–ter)

Sinusoid (**SIGH**–nuh–soyd)

Umbilical arteries (uhm–**BILL**–i–kull **AR**–tuh–rees)

Umbilical vein (uhm–**BILL**–i–kull VAIN)

Venule (**VEN**–yool)

Terms that appear in **bold type** in the chapter text are defined in the glossary, which begins on page 549.

Related Clinical Terminology

Anaphylactic (AN–uh–fi–**LAK**–tik)

Aneurysm (**AN**–yur–izm)

Arteriosclerosis (ar–TIR–ee–oh–skle–**ROH**–sis)

Hypertension (HIGH–per–**TEN**–shun)

Hypovolemic (HIGH–poh–voh–**LEEM**–ik)

Phlebitis (fle–**BY**–tis)

Pulse deficit (PULS **DEF**–i–sit)

Septic shock (**SEP**–tik SHAHK)

Varicose veins (**VAR**–i–kohs VAINS).

The role of blood vessels in the circulation of blood has been known since 1628, when William Harvey, an English anatomist, demonstrated that blood in veins always flowed toward the heart. Before that time, it was believed that blood was static or stationary, some of it within the vessels but the rest sort of in puddles throughout the body. Harvey showed that blood indeed does move, and only in the blood vessels. In the centuries that followed, the active (rather than merely passive) roles of the vascular system were discovered, and all contribute to homeostasis.

The vascular system consists of the arteries, capillaries, and veins through which the heart pumps blood throughout the body. As you will see, the major "business" of the vascular system, which is the exchange of materials between the blood and tissues, takes place in the capillaries. The arteries and veins, however, are just as important, transporting blood between the capillaries and the heart.

Another important topic of this chapter will be blood pressure (BP), which is the force the blood exerts against the walls of the vessels. Normal blood pressure is essential for circulation and for some of the material exchanges that take place in capillaries.

ARTERIES

Arteries carry blood from the heart to capillaries; smaller arteries are called **arterioles.** If we look at an artery in cross-section, we find three layers (or tunics) of tissues, each with different functions (Fig. 13–1).

The innermost layer, the **tunica intima,** is simple squamous epithelium called **endothelium.** This is the same type of tissue that forms the endocardium, the lining of the chambers of the heart. As you might guess, its function is also the same: its extreme smoothness prevents abnormal blood clotting. The **tunica media,** or middle layer, is made of smooth muscle and elastic connective tissue. Both these tissues are involved in the maintenance of normal blood pressure, especially diastolic blood pressure when the heart is relaxed. Fibrous connective tissue forms the outer layer, the **tunica externa.** This tissue is very strong, which is important

to prevent the rupture or bursting of the larger arteries that carry blood under high pressure (see Box 13–1: Disorders of Arteries).

The outer and middle layers of large arteries are quite thick. In the arterioles, only individual smooth muscle cells encircle the tunica intima. The smooth muscle layer enables arteries to constrict or dilate. This is regulated by the medulla and autonomic nervous system and will be discussed in a later section on blood pressure.

VEINS

Veins carry blood from capillaries back to the heart; the smaller veins are called **venules.** The same three tissue layers are present in veins as in the walls of arteries, but there are some differences when compared to the arterial layers. The inner layer of veins is smooth endothelium, but at intervals this lining is folded to form **valves** (see Fig. 13–1). Valves prevent backflow of blood and are most numerous in veins of the legs, where blood must often return to the heart against the force of gravity.

The middle layer of veins is a thin layer of smooth muscle. It is thin because veins do not regulate blood pressure and blood flow into capillaries as arteries do. Veins can constrict extensively, however, and this function becomes very important in certain situations such as severe hemorrhage. The outer layer of veins is also thin; not as much fibrous connective tissue is necessary because blood pressure in veins is very low.

ANASTOMOSES

An **anastomosis** is a connection, or joining, of vessels, that is, artery to artery or vein to vein. The general purpose of these connections is to provide alternate pathways for the flow of blood if one vessel becomes obstructed.

An arterial anastomosis helps ensure that blood will get to the capillaries of an organ to deliver oxygen and nutrients, and to remove waste products. There are arterial anastomoses, for example, between some of the coronary arteries that supply blood to the myocardium.

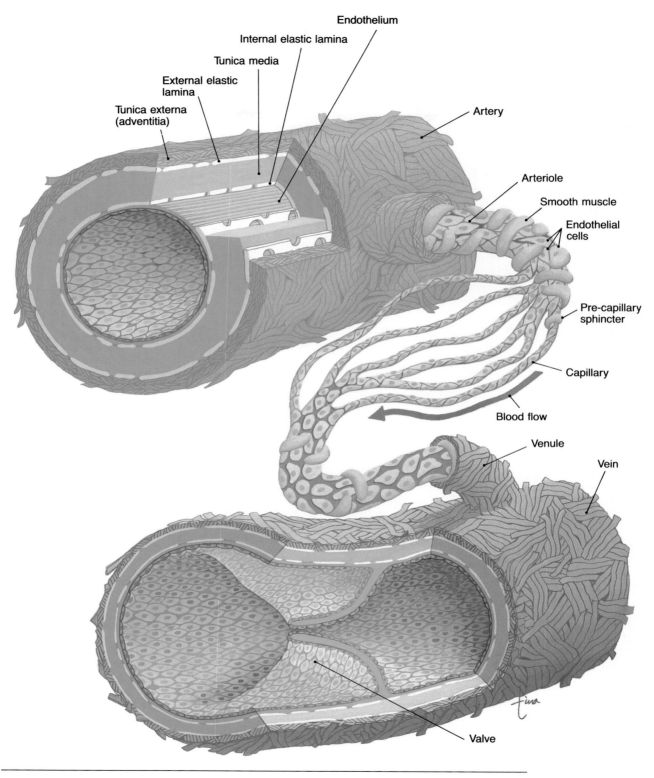

Figure 13–1 Structure of an artery, arteriole, capillary network, venule, and vein. See text for description.

Box 13–1 DISORDERS OF ARTERIES

Arteriosclerosis—although commonly called "hardening of the arteries," arteriosclerosis really means that the arteries lose their elasticity, and their walls become weakened. Arteries carry blood under high pressure, so deterioration of their walls is part of the aging process.

Aneurysm—a weak portion of an arterial wall may bulge out, forming a sac or bubble called an aneurysm. Arteriosclerosis is a possible cause, but some aneurysms are congenital. An aneurysm may be present for many years without any symptoms and may only be discovered during diagnostic procedures for some other purpose.

The most common sites for aneurysm formation are the cerebral arteries and the aorta. Rupture of a cerebral aneurysm is a possible cause of a cerebrovascular accident (CVA). Rupture of an aortic aneurysm is life-threatening and requires immediate corrective surgery. The damaged portion of the artery is removed and replaced with a graft. Such surgery may also be performed when an aneurysm is found before it ruptures.

Atherosclerosis—this condition has been mentioned previously; see Chapter 12.

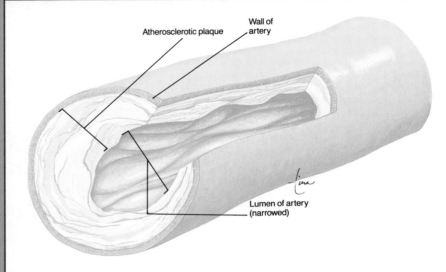

Box Figure 13–A Atherosclerosis obstructing blood flow in an artery.

A venous anastomosis helps ensure that blood will be able to return to the heart in order to be pumped again. Venous anastomoses are most numerous among the veins of the legs, where the possibility of obstruction increases as a person gets older (see Box 13–2: Disorders of Veins).

CAPILLARIES

Capillaries carry blood from arterioles to venules. Their walls are only one cell in thickness; capillaries are actually the extension of just the lining of arteries and veins (see Fig. 13–1). Some tissues do not have capillaries; these are the epidermis, cartilage, and the lens and cornea of the eye.

Most tissues, however, have extensive capillary networks. Blood flow into these networks is regulated by smooth muscle cells called **precapillary sphincters,** found at the beginning of each network (see Fig. 13–1). Precapillary sphincters are not regulated by the nervous system but rather constrict or dilate depending on the needs of the tissues. Since there is not enough blood in the body to fill all the capillaries at once, precapillary sphincters are usually slightly constricted. In an active tissue that requires more oxygen, such as exercising muscle,

Box 13–2 DISORDERS OF VEINS

Phlebitis—inflammation of a vein. This condition is most common in the veins of the legs because they are subjected to great pressure as the blood is returned to the heart against the force of gravity. Often no specific cause can be determined, but advancing age, obesity, and blood disorders may be predisposing factors.

If a superficial vein is affected, the area may be tender or painful, but blood flow is usually maintained because there are so many anastomoses among these veins. Deep vein phlebitis is potentially more serious, with the possibility of clot formation (thrombophlebitis) and subsequent dislodging of the clot to form an embolism.

Varicose Veins—swollen and distended veins that occur most often in the superficial veins of the legs. This condition may develop in people who must sit or stand in one place for long periods of time. Without contraction of the leg muscles, blood tends to pool in the leg veins, stretching their walls. If the veins become overly stretched, the valves within them no longer close properly. These incompetent valves no longer prevent backflow of blood, leading to further pooling and even further stretching of the walls of the veins. Varicose veins may cause discomfort and cramping in the legs, or become even more painful. Severe varicosities may be removed surgically.

This condition may also develop during pregnancy, when the enlarged uterus presses against the iliac veins and slows blood flow into the inferior vena cava. Varicose veins of the anal canal are called **hemorrhoids,** which may also be a result of pregnancy or of chronic constipation and straining to defecate. Hemorrhoids that cause discomfort or pain may also be removed surgically. The recent developments in laser surgery have made this a simpler procedure than it was in the past.

the precapillary sphincters dilate to increase blood flow. These automatic responses ensure that blood, the volume of which is constant, will circulate where it is needed most.

Some organs have another type of capillary called **sinusoids,** which are larger and more permeable than are other capillaries. The permeability of sinusoids permits large substances such as proteins and blood cells to enter or leave the blood. Sinusoids are found in hemopoietic tissues such as the red bone marrow and spleen and in organs such as the liver and pituitary gland, which produce and secrete proteins into the blood.

EXCHANGES IN CAPILLARIES

Capillaries are the sites of exchanges of materials between the blood and the tissue fluid surrounding cells. Some of these substances move from the blood to tissue fluid, and others move from tissue

fluid to the blood. The processes by which these substances are exchanged are illustrated in Fig. 13–2.

Gases move by **diffusion,** that is, from their area of greater concentration to their area of lesser concentration. Oxygen, therefore, diffuses from the blood in systemic capillaries to the tissue fluid, and carbon dioxide diffuses from tissue fluid to the blood to be brought to the lungs and exhaled.

Let us now look at the blood pressure as blood enters capillaries from the arterioles. Blood pressure here is about 30 to 35 mmHg, and the pressure of the surrounding tissue fluid is much lower, about 2 mmHg. Since the capillary blood pressure is higher, the process of **filtration** occurs, which forces plasma and dissolved nutrients out of the capillaries and into tissue fluid. This is how nutrients such as glucose, amino acids, and vitamins are brought to cells.

Blood pressure decreases as blood reaches the

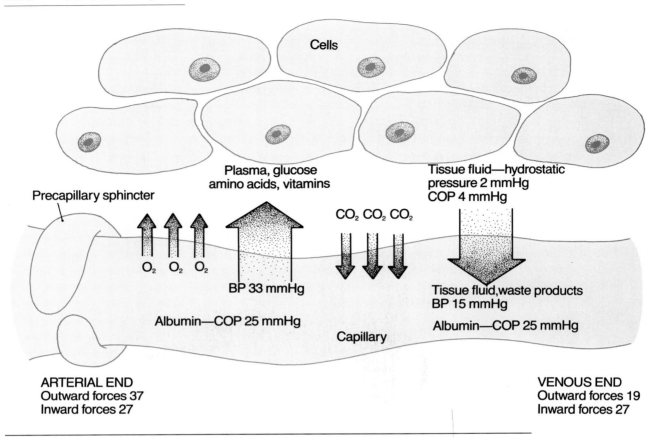

Figure 13–2 Exchanges between blood in a systemic capillary and the surrounding tissue fluid. Arrows depict the direction of movement. Filtration takes place at the arterial end of the capillary. Osmosis takes place at the venous end. Gases are exchanged by diffusion.

venous end of capillaries, but notice that proteins such as albumin have remained in the blood. Albumin contributes to the **colloid osmotic pressure** of blood; this is a "pulling" rather than a "pushing" pressure. At the venous end of capillaries, the presence of albumin in the blood pulls tissue fluid into the capillaries, which also brings into the blood the waste products produced by cells. The tissue fluid that returns to the blood also helps maintain normal blood volume and blood pressure.

The amount of tissue fluid formed is slightly greater than the amount returned to the capillaries. If this were to continue, blood volume would be gradually depleted. The excess tissue fluid, how-

ever, enters lymph capillaries. Now called lymph, it will be returned to the blood to be recycled again as plasma, thus maintaining blood volume. This is discussed further in Chapter 14.

PATHWAYS OF CIRCULATION

The two major pathways of circulation are pulmonary and systemic. Pulmonary circulation begins at the right ventricle, and systemic circulation begins at the left ventricle. Hepatic portal circulation is a special segment of systemic circulation that will be covered separately. Fetal circulation involves pathways that are present only before birth and will also be discussed separately.

PULMONARY CIRCULATION

The right ventricle pumps blood into the pulmonary artery (or trunk), which divides into the right and left pulmonary arteries, one to each lung. Within the lungs each artery branches extensively into smaller arteries and arterioles, then to capillaries. The pulmonary capillaries surround the alveoli of the lungs; it is here that exchanges of oxygen and carbon dioxide take place. The capillaries unite to form venules, which merge into veins, and finally into the two pulmonary veins from each lung that return blood to the left atrium. This oxygenated blood will then travel through the systemic circulation. (Notice that the pulmonary veins contain oxygenated blood; these are the only veins that carry blood with a high oxygen content. The blood in systemic veins has a low oxygen content; it is systemic arteries that carry oxygenated blood.)

SYSTEMIC CIRCULATION

The left ventricle pumps blood into the aorta, the largest artery of the body. We will return to the aorta and its branches in a moment, but first we will summarize the rest of systemic circulation. The branches of the aorta take blood into arterioles and capillary networks throughout the body. Capillaries merge to form venules and veins. The veins from the lower body take blood to the inferior vena cava; veins from the upper body take blood to the superior vena cava. These two caval veins return blood to the right atrium. The major arteries and veins are shown in Figs. 13–3 to 13–5, and their functions are listed in Tables 13–1 and 13–2.

The aorta is a continuous vessel, but for the sake of precise description is divided into sections that are named anatomically: ascending aorta, aortic arch, thoracic aorta, and abdominal aorta. The ascending aorta is the first inch that emerges from the top of the left ventricle. The arch of the aorta curves posteriorly over the heart and turns downward. The thoracic aorta continues down through the chest cavity and through the diaphragm. Below the level of the diaphragm, the abdominal aorta continues to the level of the 4th lumbar vertebra, where it divides into the two common iliac arteries. Along its course, the aorta has many branches through which blood travels to specific organs and parts of the body.

The ascending aorta has only two branches: the right and left coronary arteries, which supply blood to the myocardium. This pathway of circulation was described previously in Chapter 12.

The aortic arch has three branches that supply blood to the head and arms: the brachiocephalic artery, left common carotid artery, and left subclavian artery. The brachiocephalic (literally: "arm-head") artery is very short and divides into the right common carotid artery and right subclavian artery. The right and left common carotid arteries extend into the neck, where each divides into an internal carotid artery and external carotid artery, which supply the head. The right and left subclavian arteries are in the shoulders behind the clavicles and continue into the arms. The branches of the carotid and subclavian arteries are diagrammed in Figs. 13–3 and 13–5. As you look at these diagrams, keep in mind that the name of the vessel often tells us where it is. The ulnar artery, for example, is found in the forearm along the ulna.

Some of these vessels contribute to an important arterial anastomosis, the **Circle of Willis** (or cerebral arterial circle), which is a "circle" of arteries around the pituitary gland (Fig. 13–6). The Circle of Willis is formed by the right and left internal carotid arteries and the basilar artery, which is the union of the right and left vertebral arteries (branches of the subclavians). The brain must have a constant flow of blood to supply oxygen and remove waste products, and there are four vessels that bring blood to the Circle of Willis. From this anastomosis, several paired arteries extend into the brain itself.

The thoracic aorta and its branches supply the chest wall and the organs within the thoracic cavity. These vessels are listed in Table 13–1.

The abdominal aorta gives rise to arteries that supply the abdominal wall and organs and to the common iliac arteries which continue into the legs. These vessels are also listed in Table 13–1 (see Box 13–3: Pulse Sites).

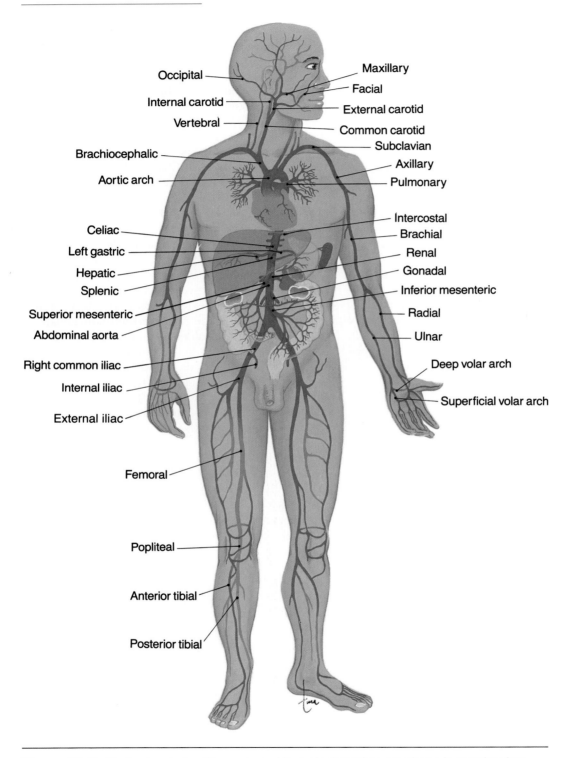

Occipital

Maxillary

Facial

Internal carotid

External carotid

Vertebral

Common carotid

Subclavian

Brachiocephalic

Axillary

Aortic arch

Pulmonary

Intercostal

Celiac

Brachial

Left gastric

Renal

Hepatic

Gonadal

Splenic

Inferior mesenteric

Superior mesenteric

Radial

Abdominal aorta

Ulnar

Right common iliac

Internal iliac

Deep volar arch

External iliac

Superficial volar arch

Femoral

Popliteal

Anterior tibial

Posterior tibial

Figure 13–3 Systemic arteries. The aorta and its major branches are shown in anterior view.

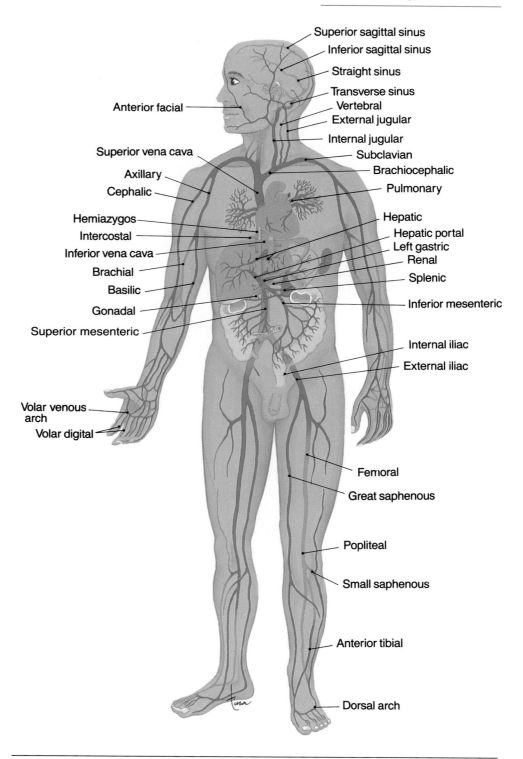

Superior sagittal sinus
Inferior sagittal sinus
Straight sinus
Transverse sinus
Vertebral
External jugular
Internal jugular
Subclavian
Brachiocephalic
Pulmonary
Hepatic
Hepatic portal
Left gastric
Renal
Splenic
Inferior mesenteric
Internal iliac
External iliac
Femoral
Great saphenous
Popliteal
Small saphenous
Anterior tibial
Dorsal arch

Anterior facial
Superior vena cava
Axillary
Cephalic
Hemiazygos
Intercostal
Inferior vena cava
Brachial
Basilic
Gonadal
Superior mesenteric
Volar venous arch
Volar digital

Figure 13–4 Systemic veins shown in anterior view.

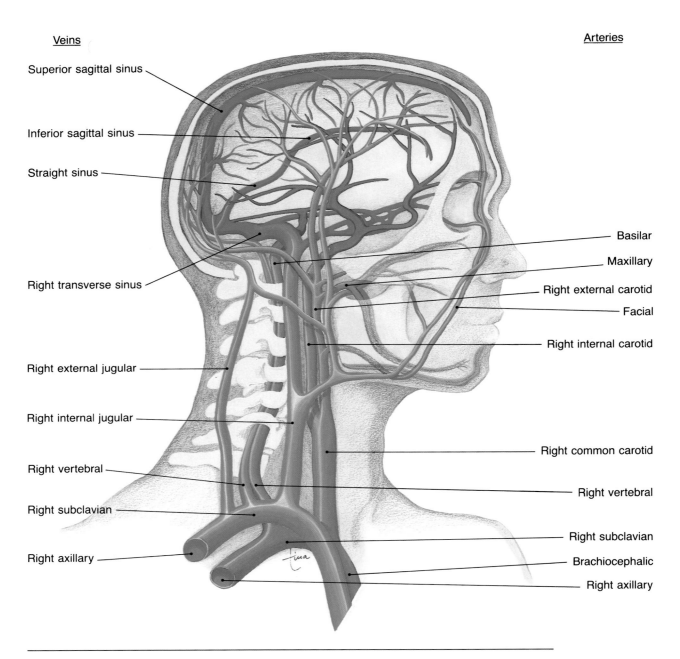

<u>Veins</u>

Superior sagittal sinus

Inferior sagittal sinus

Straight sinus

Right transverse sinus

Right external jugular

Right internal jugular

Right vertebral

Right subclavian

Right axillary

<u>Arteries</u>

Basilar

Maxillary

Right external carotid

Facial

Right internal carotid

Right common carotid

Right vertebral

Right subclavian

Brachiocephalic

Right axillary

Figure 13–5 Arteries and veins of the head and neck shown in right lateral view. Veins are labeled on the left. Arteries are labeled on the right.

Table 13–1 MAJOR SYSTEMIC ARTERIES

A. Branches of the Ascending Aorta and Aortic Arch

Artery	Branch of	Region Supplied
Coronary a.	Ascending aorta	• Myocardium
Brachiocephalic a.	Aortic arch	• Right arm and head
Right common carotid a.	Brachiocephalic a.	• Right side of head
Right subclavian a.	Brachiocephalic a.	• Right shoulder and arm
Left common carotid a.	Aortic arch	• Left side of head
Left subclavian a.	Aortic arch	• Left shoulder and arm
External carotid a.	Common carotid a.	• Superficial head
Superficial temporal a.	External carotid a.	• Scalp
Internal carotid a.	Common carotid a.	• Brain (Circle of Willis)
Ophthalmic a.	Internal carotid a.	• Eye
Vertebral a.	Subclavian a.	• Cervical vertebrae and Circle of Willis
Axillary a.	Subclavian a.	• Armpit
Brachial a.	Axillary a.	• Upper arm
Radial a.	Brachial a.	• Forearm
Ulnar a.	Brachial a.	• Forearm
Volar arch	Radial and Ulnar a.	• Hand

B. Branches of the Thoracic Aorta

Artery	Region Supplied
Intercostal a. (9 pairs)	• Skin, muscles, bones of trunk
Superior phrenic a.	• Diaphragm
Pericardial a.	• Pericardium
Esophageal a.	• Esophagus
Bronchial a.	• Bronchioles and connective tissue of the lungs

C. Branches of the Abdominal Aorta

Artery	Region Supplied
Inferior phrenic a.	• Diaphragm
Lumbar a.	• Lumbar area of back
Middle sacral a.	• Sacrum, coccyx, buttocks
Celiac a.	• (see branches)
Hepatic a.	• Liver
Left gastric a.	• Stomach
Splenic a.	• Spleen, pancreas
Superior mesenteric a.	• Small intestine, part of colon
Suprarenal a.	• Adrenal glands
Renal a.	• Kidneys
Inferior mesenteric a.	• Most of colon and rectum
Testicular or ovarian a.	• Testes or ovaries

Table 13–1   MAJOR SYSTEMIC ARTERIES (*Continued*)

C. Branches of the Abdominal Aorta (*Continued*)

Artery	Region Supplied
Common iliac a.	• The two large vessels that receive blood from the Abdominal Aorta; each branches as follows below:
Internal iliac a.	• Bladder, rectum, reproductive organs
External iliac a.	• Lower pelvis to leg
Femoral a.	• Thigh
Popliteal a.	• Back of knee
Anterior tibial a.	• Front of lower leg
Dorsalis pedis	• Top of ankle and foot
Plantar arches	• Foot
Posterior tibial a.	• Back of lower leg
Peroneal a.	• Medial lower leg
Plantar arches	• Foot

The systemic veins drain blood from organs or parts of the body and often parallel their corresponding arteries. The most important veins are diagrammed in Fig. 13–4 and listed in Table 13–2.

HEPATIC PORTAL CIRCULATION

Hepatic portal circulation is a subdivision of systemic circulation in which blood from the abdominal digestive organs and spleen circulates through the liver before returning to the heart.

Blood from the capillaries of the stomach, small intestine, colon, pancreas, and spleen flows into two large veins, the superior mesenteric vein and the splenic vein, which unite to form the portal vein (Fig. 13–7). The portal vein takes blood into the liver, where it branches extensively and empties blood into the sinusoids, the capillaries of the liver (see also Fig. 16–6). From the sinusoids, blood flows into hepatic veins, to the inferior vena cava and back to the right atrium. Notice that in this pathway there are two sets of capillaries, and keep in mind that it is in capillaries that exchanges take place. Let us use some specific examples to show the purpose and importance of portal circulation.

Glucose from carbohydrate digestion is absorbed into the capillaries of the small intestine; after a big meal this may greatly increase the blood glucose level. If this blood were to go directly back to the heart and then circulate through the kidneys, some of the glucose might be lost in urine. However, blood from the small intestine passes first through the liver sinusoids, and the liver cells remove the excess glucose and store it as glycogen. The blood that returns to the heart will then have a blood glucose level in the normal range.

Another example: alcohol is absorbed into the capillaries of the stomach. If it were to circulate directly throughout the body, the alcohol would rapidly impair the functioning of the brain. Portal circulation, however, takes blood from the stomach to the liver, the organ that can detoxify the alcohol and prevent its detrimental effects on the brain. Of course, if alcohol consumption continues, the blood alcohol level rises faster than the liver's capacity to detoxify, and the well-known signs of alcohol intoxication appear.

As you can see, this portal circulation pathway enables the liver to modify the blood from the digestive organs and spleen. Some nutrients may be stored or changed, bilirubin from the spleen is excreted into bile, and potential poisons are detoxified before the blood returns to the heart and the rest of the body.

Table 13–2   MAJOR SYSTEMIC VEINS

Vein	Vein Joined	Region Drained
Head and Neck		
Cranial venous sinuses	Internal jugular v.	• Brain, including reabsorbed CSF
Internal jugular v.	Brachiocephalic v.	• Face and neck
External jugular v.	Subclavian v.	• Superficial face and neck
Subclavian v.	Brachiocephalic v.	• Shoulder
Brachiocephalic v.	Superior vena cava	• Upper body
Superior vena cava	Right atrium	• Upper body
Arm and Shoulder		
Radial v.	Brachial v.	• Forearm and hand
Ulnar v.	Brachial v.	• Forearm and hand
Cephalic v.	Axillary v.	• Superficial arm and forearm
Basilic v.	Axillary v.	• Superficial upper arm
Brachial v.	Axillary v.	• Upper arm
Axillary v.	Subclavian v.	• Armpit
Subclavian v.	Brachiocephalic v.	• Shoulder
Trunk		
Brachiocephalic v.	Superior vena cava	• Upper body
Azygos v.	Superior vena cava	• Deep structures of chest and abdomen; links inferior vena cava to superior vena cava
Hepatic v.	Inferior vena cava	• Liver
Renal v.	Inferior vena cava	• Kidney
Testicular or Ovarian v.	Inferior vena cava and left renal v.	• Testes or ovaries
Internal iliac v.	Common iliac v.	• Rectum, bladder, reproductive organs
External iliac v.	Common iliac v.	• Leg and abdominal wall
Common iliac v.	Inferior vena cava	• Leg and lower abdomen
Leg		
Anterior and posterior tibial v.	Popliteal v.	• Lower leg and foot
Popliteal v.	Femoral v.	• Knee
Small saphenous v.	Popliteal v.	• Superficial leg and foot
Great saphenous v.	Femoral v.	• Superficial foot, leg, and thigh
Femoral v.	External iliac v.	• Thigh
External iliac v.	Common iliac v.	• Leg and abdominal wall
Common iliac v.	Inferior vena cava	• Leg and lower abdomen
Inferior vena cava	Right atrium	• Lower body

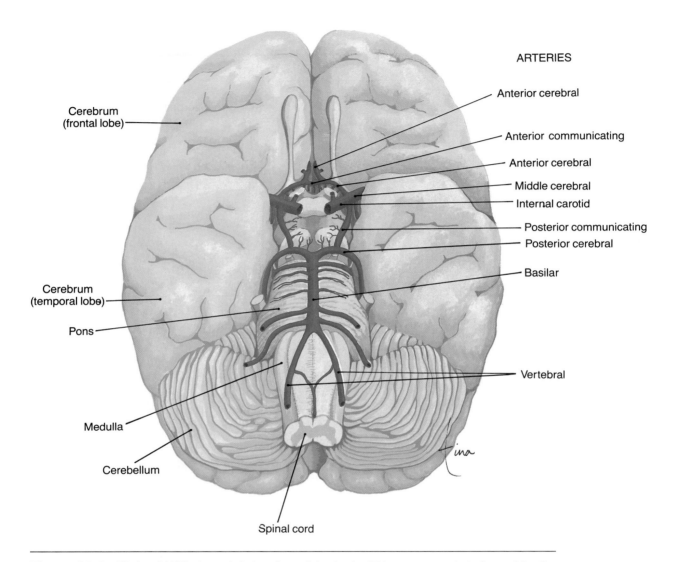

ARTERIES

Anterior cerebral

Anterior communicating

Anterior cerebral

Middle cerebral

Internal carotid

Posterior communicating

Posterior cerebral

Basilar

Vertebral

Cerebrum
(frontal lobe)

Cerebrum
(temporal lobe)

Pons

Medulla

Cerebellum

Spinal cord

Figure 13–6   Circle of Willis in an inferior view of the brain. This anastomosis is formed by the following arteries: internal carotid, anterior communicating, posterior communicating, and basilar. The cerebral arteries extend from the Circle of Willis into the brain.

Box 13–3 PULSE SITES

A pulse is the heartbeat that is felt at an arterial site. What is felt is not actually the force exerted by the blood, but the force of ventricular contraction transmitted through the walls of the arteries. This is why pulses are not felt in veins; they are too far from the heart for the force to be detectable.

The most commonly used pulse sites are:

Radial—the radial artery on the thumb side of the wrist.

Carotid—the carotid artery lateral to the larynx in the neck.

Temporal—the temporal artery just in front of the ear.

Femoral—the femoral artery at the top of the thigh.

Popliteal—the popliteal artery at the back of the knee.

Dorsalis pedis—the dorsalis pedis artery on the top of the foot (commonly called the pedal pulse).

Pulse rate is, of course, the heart rate. However, if the heart is beating weakly, a radial pulse may be lower than an **apical pulse** (listening to the heart itself with a stethoscope). This is called a **pulse deficit** and indicates heart disease of some kind.

When taking a pulse, the careful observer also notes the rhythm and force of the pulse. Abnormal rhythms may reflect cardiac arrhythmias, and the force of the pulse (strong or weak) is helpful in assessing the general condition of the heart and arteries.

FETAL CIRCULATION

The fetus depends upon the mother for oxygen and nutrients and for the removal of carbon dioxide and other waste products. The site of exchange between fetus and mother is the **placenta,** which contains fetal and maternal blood vessels that are very close to one another (see Fig. 13–8 and 21–5). The blood of the fetus does not mix with the blood of the mother; substances are exchanged by diffusion and active transport mechanisms.

The fetus is connected to the placenta by the umbilical cord, which contains two umbilical arteries and one umbilical vein (see Fig. 13–8). The **umbilical arteries** are branches of the fetal internal iliac arteries; they carry blood from the fetus to the placenta. In the placenta, carbon dioxide and waste products in the fetal blood enter maternal circulation, and oxygen and nutrients from the mother's blood enter fetal circulation.

The **umbilical vein** carries this oxygenated blood from the placenta to the fetus. Within the body of the fetus, the umbilical vein branches: one branch takes some blood to the fetal liver, but most of the blood passes through the **ductus venosus** to the inferior vena cava, to the right atrium. After birth, when the umbilical cord is cut, the remnants of these fetal vessels constrict and become nonfunctional.

The other modifications of fetal circulation concern the fetal heart and large arteries (also shown in Fig. 13–8). Since the fetal lungs are deflated and do not provide for gas exchange, blood is shunted away from the lungs and to the body. The **foramen ovale** is an opening in the interatrial septum that permits some blood to flow from the right atrium to the left atrium, not, as usual, to the right ventricle. The blood that does enter the right ventricle is pumped into the pulmonary artery. The **ductus arteriosus** is a short vessel that diverts most of the blood in the pulmonary artery to the aorta, to the body. Both the foramen ovale and the ductus

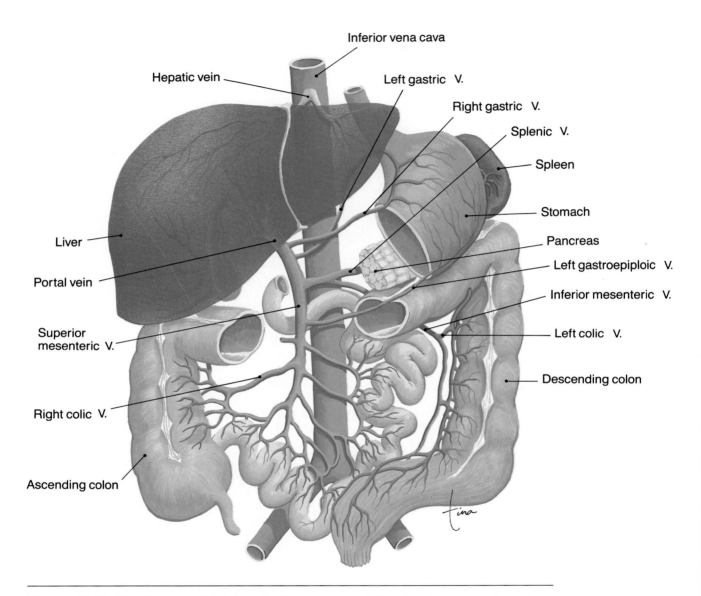

Figure 13–7 Hepatic portal circulation. Portions of some of the digestive organs have been removed to show the veins that unite to form the portal vein.

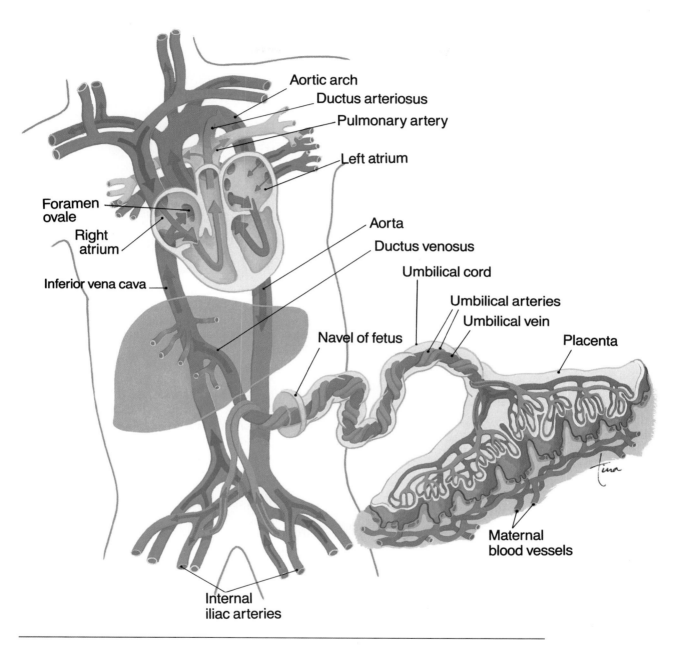

Aortic arch
Ductus arteriosus
Pulmonary artery
Left atrium
Foramen ovale
Right atrium
Inferior vena cava
Aorta
Ductus venosus
Umbilical cord
Umbilical arteries
Umbilical vein
Placenta
Navel of fetus
Maternal blood vessels
Internal iliac arteries

Figure 13–8   Fetal circulation. Fetal heart and blood vessels are shown on the left. Arrows depict the direction of blood flow. The placenta and umbilical blood vessels are shown on the right.

arteriosus permit blood to bypass the fetal lungs.

Just after birth, the baby breathes and expands its lungs, which pulls more blood into the pulmonary circulation. More blood then returns to the left atrium, and a flap on the left side of the foramen ovale is closed. The ductus arteriosus constricts, probably in response to the higher oxygen content of the blood, and pulmonary circulation becomes fully functional within a few days.

BLOOD PRESSURE

Blood pressure is the force the blood exerts against the walls of the blood vessels. Filtration in capillaries depends upon blood pressure; filtration brings nutrients to tissues, and as you will see in Chapter 18, is the first step in the formation of urine. Blood pressure is one of the "vital signs" often measured, and indeed a normal blood pressure is essential to life.

The pumping of the ventricles creates blood pressure, which is measured in mmHg (millimeters of mercury). When a systemic blood pressure reading is taken, two numbers are obtained: systolic and diastolic, as in 110/70 mmHg. **Systolic** pressure is always the higher of the two and represents the blood pressure when the left ventricle is contracting. The lower number is the **diastolic** pressure, when the left ventricle is relaxed and does not exert force. Diastolic pressure is maintained by the arteries and arterioles and is discussed in a later section.

Systemic blood pressure is highest in the aorta, which receives all the blood pumped by the left ventricle. As blood travels further away from the heart, blood pressure decreases (Fig. 13–9). The brachial artery is most often used to take a blood pressure reading; here a normal systolic range is 90 to 135 mmHg, and a normal diastolic range is 60 to 85 mmHg. In the arterioles, blood pressure decreases further, and systolic and diastolic pressures merge into one pressure. At the arterial end of capillary networks, blood pressure is about 30 to 35 mmHg, decreasing to 12 to 15 mmHg at the venous end of capillaries. This is high enough to permit filtration but low enough to prevent rupture of the

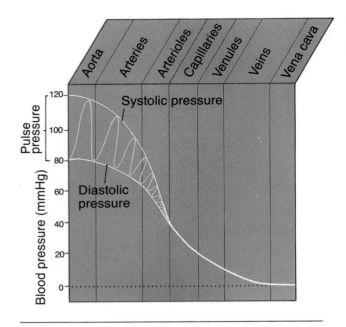

Figure 13–9 Systemic blood pressure changes throughout the vascular system. Notice that systolic and diastolic pressures become one pressure as blood enters the capillaries.

capillaries. As blood flows through veins, the pressure decreases further, and in the caval veins, blood pressure approaches zero as blood enters the right atrium (see also Box 13–4: Hypertension).

Pulmonary blood pressure is created by the right ventricle, which has relatively thin walls and thus exerts about one sixth the force of the left ventricle. The result is that pulmonary arterial pressure is always low: 20 to 25/8 to 10 mmHg, and in pulmonary capillaries is lower still. This is important to *prevent* filtration in pulmonary capillaries, to prevent tissue fluid from accumulating in the alveoli of the lungs.

MAINTENANCE OF SYSTEMIC BLOOD PRESSURE

Since blood pressure is so important, there are many factors and physiological processes that interact to keep blood pressure within normal limits.

1. **Venous return**—the amount of blood that returns to the heart by way of the veins. Venous

Box 13–4   HYPERTENSION

Hypertension is high blood pressure, that is, a resting systemic pressure consistently above 140/90. The term "essential hypertension" means that no specific cause can be determined; most cases are in this category. For some people, however, an overproduction of renin by the kidneys is the cause of their hypertension. Excess renin increases the production of angiotensin II, which raises blood pressure. Although hypertension often produces no symptoms, the long-term consequences may be very serious. Chronic hypertension has its greatest effects on the arteries and on the heart.

Although the walls of arteries are strong, hypertension weakens them and contributes to arteriosclerosis. Such weakened arteries may rupture or develop aneurysms, which may in turn lead to a CVA or kidney damage.

Hypertension affects the heart because the left ventricle must now pump blood against the higher arterial pressure. The left ventricle works harder and, like any other muscle, enlarges as more work is demanded; this is called **left ventricular hypertrophy.** This abnormal growth of the myocardium, however, is not accompanied by a corresponding growth in coronary capillaries, and the blood supply of the left ventricle may not be adequate for all situations. Exercise, for example, puts further demands on the heart, and the person may experience angina due to a lack of oxygen or a myocardial infarction if there is a severe oxygen deficiency.

Although there are several different kinds of medications used to treat hypertension, people with moderate hypertension may limit their dependence on medications by following certain guidelines.

1. *Don't smoke,* because nicotine stimulates vasoconstriction, which raises BP.
2. *Lose weight* if overweight. A weight loss of as little as 10 pounds can lower BP.
3. *Cut salt intake* in half. Although salt consumption may not be the *cause* of hypertension, reducing salt intake may help lower blood pressure by decreasing blood volume.
4. *Exercise* on a regular basis. A moderate amount of aerobic exercise (such as a half hour walk every day) is beneficial for the entire cardiovascular system and may also contribute to weight loss.

return is important because the heart can pump only the blood it receives. If venous return decreases, the cardiac muscle fibers will not be stretched, the force of ventricular systole will decrease (Starling's Law), and blood pressure will decrease. This is what might happen following a severe hemorrhage.

When the body is horizontal, venous return can be maintained fairly easily, but when the body is vertical, gravity must be overcome to return blood from the lower body to the heart. There are three mechanisms that help promote venous return: constriction of veins, the skeletal muscle pump, and the respiratory pump.

Veins contain smooth muscle, which enables them to constrict and force blood toward the heart; the valves prevent backflow of blood. The second mechanism is the **skeletal muscle pump,** which is especially effective for the deep veins of the legs. These veins are surrounded by skeletal muscles that contract and relax during normal activities such as walking. Contractions of the leg muscles squeeze the veins to force blood toward the heart. The third mechanism is the **respiratory pump,** which affects veins that pass through the chest cavity. The pressure changes of inhalation and exhalation alternately expand and compress the veins, and blood is returned to the heart.

2. **Heart rate and force**—in general, if heart rate and force increase, blood pressure increases; this is what happens during exercise. However, if the heart is beating extremely rapidly, the ventricles may not fill completely between beats, and cardiac output and blood pressure will decrease.

3. **Peripheral resistance**—this term refers to the resistance the vessels offer to the flow of blood. The arteries and veins are usually slightly constricted, which maintains normal diastolic blood pressure. It may be helpful to think of the vessels as the "container" for the blood. If a person's body has 5 liters of blood, the "container" must be smaller in order for the blood to exert a pressure against its walls. This is what normal vasoconstriction does: it makes the container (the vessels) smaller than the volume of blood so that the blood will exert pressure even when the left ventricle is relaxed.

   If more vasoconstriction occurs, blood pressure will increase (the container has become even smaller). This is what happens in a stress situation, when greater vasoconstriction is brought about by sympathetic impulses. If vasodilation occurs, blood pressure will decrease (the container is larger). After eating a large meal, for example, there is extensive vasodilation in the digestive tract to supply more oxygenated blood for all the digestive activities. To keep blood pressure within the normal range, vasoconstriction must, and does, occur elsewhere in the body. This is why strenuous exercise should be avoided right after eating; there is not enough blood to completely supply oxygen to exercising muscles and an active digestive tract at the same time.

4. **Elasticity of the large arteries**—when the left ventricle contracts, the blood that enters the large arteries stretches their walls. The arterial walls are elastic and absorb some of the force. When the left ventricle relaxes, the arterial walls recoil or snap back, which helps keep diastolic pressure within the normal range. Normal elasticity, therefore, lowers systolic pressure, raises diastolic pressure, and maintains a normal pulse pressure. (Pulse pressure is the difference between systolic and diastolic pressure. The usual ratio of systolic to diastolic to pulse pressure is approximately 3:2:1. For example, with a blood pressure of 120/80, the pulse pressure is 40, and the ratio is 120:80:40, or 3:2:1.)

5. **Viscosity of the blood**—normal blood viscosity depends upon the presence of red blood cells and plasma proteins, especially albumin. Having too many red blood cells is rare but does occur in the disorder called polycythemia vera and in people who are heavy smokers. This will increase blood viscosity and blood pressure.

   Decreased red blood cells, as in severe anemia, or decreased albumin, as may occur in liver disease or kidney disease, will decrease blood viscosity and blood pressure. In these situations, other mechanisms such as vasoconstriction will maintain blood pressure as close to normal as is possible.

6. **Loss of blood**—a small loss of blood, as when donating a pint of blood, will cause a temporary drop in blood pressure followed by rapid compensation in the form of more rapid heart rate and greater vasoconstriction. After a severe hemorrhage, however, these compensating mechanisms may not be sufficient to maintain normal blood pressure and blood flow to the brain. Although a person may survive blood losses of 50% of the total blood, the possibility of brain damage increases as more blood is lost and not rapidly replaced.

7. **Hormones**—there are several hormones that have effects on blood pressure. You may recall them from Chapters 10 and 12, but let us summarize them here. The adrenal medulla secretes norepinephrine and epinephrine in stress situations. Norepinephrine stimulates vasoconstriction, which raises blood pressure. Epinephrine also causes vasoconstriction, and

increases heart rate and force of contraction, which increase blood pressure.

Antidiuretic hormone (ADH) is secreted by the posterior pituitary gland when the water content of the body decreases. ADH increases the reabsorption of water by the kidneys to prevent further loss of water in urine and a further decrease in blood pressure.

Aldosterone, a hormone from the adrenal cortex, has a similar effect on blood volume. When blood pressure decreases, secretion of aldosterone stimulates the reabsorption of Na$^+$ ions by the kidneys. Water follows sodium back to the blood, which maintains blood volume to prevent a further drop in blood pressure.

Atrial Natriuretic hormone (ANH), secreted by the atria of the heart, functions in opposition to aldosterone. ANH increases the excretion of Na$^+$ ions and water by the kidneys, which decreases blood volume and lowers blood pressure.

REGULATION OF BLOOD PRESSURE

The mechanisms that regulate blood pressure may be divided into two types: intrinsic mechanisms and nervous mechanisms. The nervous mechanisms involve the nervous system, and the intrinsic mechanisms do not require nerve impulses.

INTRINSIC MECHANISMS

The term "intrinsic" means "within." Intrinsic mechanisms work because of the internal characteristics of certain organs. The first such organ is the heart. When venous return increases, cardiac muscle fibers are stretched, and the ventricles pump more forcefully (Starling's Law). Thus, cardiac ouput and blood pressure increase. This is what happens during exercise, when a higher blood pressure is needed. When exercise ends and venous return decreases, the heart pumps less force-

Table 13–3   THE RENIN–ANGIOTENSIN MECHANISM

1. Decreased blood pressure stimulates the kidneys to secrete renin.
2. Renin splits the plasma protein angiotensinogen (synthesized by the liver) to angiotensin I.
3. Angiotensin I is converted to angiotensin II by an enzyme (called converting enzyme) found primarily in lung tissue.
4. Angiotensin II:
   • causes vasoconstriction
   • stimulates the adrenal cortex to secrete aldosterone

fully, which helps return blood pressure to a normal resting level.

The second intrinsic mechanism involves the kidneys. When blood flow through the kidneys decreases, the process of filtration decreases and less urine is formed. This decrease in urinary output preserves blood volume so that it does not decrease further. Following severe hemorrhage or any other type of dehydration, this is very important to maintain blood pressure.

The kidneys are also involved in the **renin-angiotensin mechanism.** When blood pressure decreases, the kidneys secrete the enzyme **renin,** which initiates a series of reactions that result in the formation of **angiotensin II.** These reactions are shown in Table 13–3. Angiotensin II causes vasoconstriction and stimulates secretion of aldosterone by the adrenal cortex, both of which will increase blood pressure.

NERVOUS MECHANISMS

The medulla and the autonomic nervous system are directly involved in the regulation of blood pressure. The first of these nervous mechanisms concerns the heart; this was described previously, so we will not review it here but refer you to Chapter 12.

The second nervous mechanism involves peripheral resistance, that is, the degree of constriction of the arteries and arterioles, and to a lesser extent, the veins. The medulla contains the **vasomotor center,** which consists of a vasoconstrictor area and a

Box 13–5   CIRCULATORY SHOCK

Circulatory shock is any condition in which cardiac output decreases to the extent that tissues are deprived of oxygen and waste products accumulate.

Causes of Shock

Cardiogenic shock occurs most often after a severe myocardial infarction but may also be the result of ventricular fibrillation. In either case, the heart is no longer an efficient pump, and cardiac output decreases.

Hypovolemic shock is the result of decreased blood volume, often due to severe hemorrhage. Other possible causes are extreme sweating (heat stroke) or extreme loss of water through the kidneys (diuresis) or intestines (diarrhea). In these situations, the heart simply does not have enough blood to pump, and cardiac output decreases. Anaphylactic shock, also in this category, is a massive allergic reaction in which great amounts of histamine increase capillary permeability and vasodilation throughout the body. Much plasma is then lost to tissue spaces, which decreases blood volume, blood pressure, and cardiac output.

Septic shock is the result of septicemia, the presence of bacteria in the blood. The bacteria and damaged tissues release inflammatory chemicals that cause vasodilation and extensive loss of plasma into tissue spaces.

Stages of Shock

Compensated shock—the responses by the body maintain cardiac output. Following a small hemorrhage, for example, the heart rate increases, the blood vessels constrict, and the kidneys decrease urinary output to conserve water. These responses help preserve blood volume and maintain blood pressure, cardiac output, and blood flow to tissues.

Progressive shock—the state of shock leads to more shock. Following a severe hemorrhage, cardiac output decreases, and the myocardium itself is deprived of blood. The heart weakens, which further decreases cardiac output. Arteries that are deprived of their blood supply cannot remain constricted. As the arteries dilate, venous return decreases, which in turn decreases cardiac output. Progressive shock is a series of such vicious cycles, and medical intervention is required to restore cardiac output to normal.

Irreversible shock—no amount of medical assistance can restore cardiac output to normal. The usual cause of death is that the heart has been damaged too much to recover. A severe myocardial infarction, massive hemorrhage, or septicemia may all be fatal despite medical treatment.

vasodilator area. The vasodilator area may depress the vasoconstrictor area to bring about vasodilation, which will decrease blood pressure. The vasoconstrictor area may bring about more vasoconstriction by way of the sympathetic division of the autonomic nervous system.

Sympathetic vasoconstrictor fibers innervate the smooth muscle of all arteries and veins, and several impulses per second along these fibers maintain normal vasoconstriction. More impulses per second bring about greater vasoconstriction, and fewer impulses per second cause vasodilation. The medulla receives the information to make such changes from the pressoreceptors in the carotid sinuses and the aortic sinus. The inability to maintain normal blood pressure is one aspect of circulatory shock (see Box 13–5: Circulatory Shock).

SUMMARY

Although the vascular system does form passageways for the blood, you can readily see that the blood vessels are not simply pipes through which the blood flows. The vessels are not passive tubes, but rather active contributors to homeostasis. The arteries and veins help maintain blood pressure, and the capillaries provide sites for the exchanges of materials between the blood and the tissues. Some very important sites of exchange are discussed in the following chapters: the lungs, the digestive tract, and the kidneys.

STUDY OUTLINE

The vascular system consists of the arteries, capillaries, and veins through which blood travels.

Arteries (and arterioles)
1. Carry blood from the heart to capillaries; three layers in their walls.
2. Inner layer (tunica intima): simple squamous epithelial tissue (endothelium), very smooth to prevent abnormal blood clotting.
3. Middle layer (tunica media): smooth muscle and elastic connective tissue; contributes to maintenance of diastolic blood pressure (BP).
4. Outer layer (tunica externa): fibrous connective tissue to prevent rupture.
5. Constriction or dilation is regulated by the autonomic nervous system.

Veins (and venules)
1. Carry blood from capillaries to the heart; three layers in walls.
2. Inner layer: endothelium folded into valves to prevent the backflow of blood.
3. Middle layer: thin smooth muscle, since veins are not as important in the maintenance of BP.
4. Outer layer: thin fibrous connective tissue since veins do not carry blood under high pressure.

Anastomoses—connections between vessels of the same type
1. Provide alternate pathways for blood flow if one vessel is blocked.
2. Arterial anastomoses provide for blood flow to the capillaries of an organ (e.g., Circle of Willis to the brain).
3. Venous anastomoses provide for return of blood to the heart and are most numerous in veins of the legs.

Capillaries
1. Carry blood from arterioles to venules.
2. Walls are one cell thick (simple squamous epithelial tissue) to permit exchanges between blood and tissue fluid.
3. Oxygen and carbon dioxide are exchanged by diffusion.
4. BP in capillaries brings nutrients to tissues and forms tissue fluid in the process of filtration.
5. Albumin in the blood provides colloid osmotic pressure, which pulls waste products and tissue fluid into capillaries. The return of tissue fluid maintains blood volume and BP.
6. Precapillary sphincters regulate blood flow into capillary networks based on tissue needs; in ac-

tive tissues they dilate; in less active tissue they constrict.

7. Sinusoids are very permeable capillaries found in the liver, spleen, pituitary gland, and RBM to permit proteins and blood cells to enter or leave the blood.

Pathways of Circulation

1. Pulmonary: Right ventricle → pulmonary artery → pulmonary capillaries (exchange of gases) → pulmonary veins → left atrium.
2. Systemic: left ventricle → aorta → capillaries in body tissues → superior and inferior caval veins → right atrium (see Table 13–1 for systemic arteries and Table 13–2 for systemic veins).
3. Hepatic Portal Circulation: blood from the digestive organs and spleen flows through the portal vein to the liver before returning to the heart. Purpose: the liver stores some nutrients or regulates their blood levels and detoxifies potential poisons before blood enters the rest of peripheral circulation.

Fetal Circulation—the fetus depends on the mother for oxygen and nutrients and for the removal of waste products

1. The placenta is the site of exchange between fetal blood and maternal blood.
2. Umbilical arteries (two) carry blood from the fetus to the placenta, where CO_2 and waste products enter maternal circulation.
3. The umbilical vein carries blood with O_2 and nutrients from the placenta to the fetus.
4. The umbilical vein branches: some blood flows through the fetal liver; most blood flows through the ductus venosus to the fetal inferior vena cava.
5. The foramen ovale permits blood to flow from the right atrium to the left atrium to bypass the fetal lungs.
6. The ductus arteriosus permits blood to flow from the pulmonary artery to the aorta to bypass the fetal lungs.
7. These fetal structures become nonfunctional after birth, when the umbilical cord is cut and breathing takes place.

Blood Pressure (BP)—the force exerted by the blood against the walls of the blood vessels

1. BP is measured in mmHg: systolic/diastolic. Systolic pressure is during ventricular contraction; diastolic pressure is during ventricular relaxation.
2. Normal range of systemic arterial BP: 90 to 135/60 to 85 mmHg.
3. BP in capillaries is 30 to 35 mmHg at the arterial end and 12 to 15 mmHg at the venous end; high enough to permit filtration but low enough to prevent rupture of the capillaries.
4. BP decreases in the veins and approaches zero in the caval veins.
5. Pulmonary BP is always low (the right ventricle pumps with less force): 20 to 25/8 to 10 mmHg. This low BP prevents filtration and accumulation of tissue fluid in the alveoli.

Maintenance of Systemic BP

1. Venous return—the amount of blood that returns to the heart. If venous return decreases, the heart contracts less forcefully (Starling's Law) and BP decreases. The mechanisms that maintain venous return when the body is vertical are:
   • constriction of veins with the valves preventing backflow of blood
   • skeletal muscle pump—contraction of skeletal muscles, especially in the legs, squeezes the deep veins
   • respiratory pump—the pressure changes of inhalation and exhalation expand and compress the veins in the chest cavity
2. Heart rate and force—if heart rate and force increase, BP increases.
3. Peripheral resistance—the resistance of the arteries and arterioles to the flow of blood. These vessels are usually slightly constricted to maintain normal diastolic BP. Greater vasoconstriction will increase BP; vasodilation will decrease BP. In the body, vasodilation in one area requires vasoconstriction in another area to maintain normal BP.
4. Elasticity of the large arteries—ventricular systole stretches the walls of large arteries, which recoil during ventricular diastole. Normal elastic-

ity lowers systolic BP, raises diastolic BP, and maintains normal pulse pressure.

5. Viscosity of blood—depends on RBCs and plasma proteins, especially albumin. Severe anemia tends to decrease BP. Deficiency of albumin as in liver or kidney disease tends to decrease BP. In these cases, compensation such as greater vasoconstriction will keep BP close to normal.

6. Loss of blood—a small loss will be rapidly compensated for by faster heart rate and greater vasoconstriction. After severe hemorrhage, these mechanisms may not be sufficient to maintain normal BP.

7. Hormones—(a) Norepinephrine stimulates vasoconstriction which raises BP; (b) Epinephrine increases cardiac output and raises BP; (c) ADH increases water reabsorption by the kidneys, which increases blood volume and BP; (d) Aldosterone increases reabsorption of Na^+ ions by the kidneys; water follows Na^+ and increases blood volume and BP; (e) ANH increases excretion of Na^+ ions and water by the kidneys, which decreases blood volume and BP.

Regulation of Blood Pressure—intrinsic mechanisms and nervous mechanisms

Intrinsic Mechanisms

1. The heart—responds to increased venous return by pumping more forcefully (Starling's Law), which increases cardiac output and BP.

2. The kidneys—decreased blood flow decreases filtration, which decreases urinary output to preserve blood volume. Decreased BP stimulates the kidneys to secrete renin, which initiates the renin-angiotensin mechanism (Table 13–3) that results in the formation of angiotensin II, which causes vasoconstriction and stimulates secretion of aldosterone.

Nervous Mechanisms

1. Heart Rate and Force—see Chapter 12.
2. Peripheral Resistance—the medulla contains the vasomotor center, which consists of a vasoconstrictor area and a vasodilator area. The vasodilator area brings about vasodilation by suppressing the vasoconstrictor area. The vasoconstrictor area maintains normal vasoconstriction by generating several impulses per second along sympathetic vasoconstrictor fibers to all arteries and veins. More impulses per second increase vasoconstriction and raise BP; fewer impulses per second bring about vasodilation and a drop in BP.

REVIEW QUESTIONS

1. Describe the structure of the three layers in the walls of arteries, and state the function of each layer. Describe the structural differences in these layers in veins, and explain the reason for each difference. (p. 282)

2. Describe the structure and purpose of anastomoses, and give a specific example. (pp. 282, 284)

3. Describe the structure of capillaries. State the process by which each of the following is exchanged between capillaries and tissue fluid: nutrients, oxygen, waste products, CO_2. (pp. 284–286)

4. State the part of the body supplied by each of the following arteries: (pp. 291–292)
   a. bronchial
   b. femoral
   c. hepatic
   d. brachial
   e. inferior mesenteric
   f. internal carotid
   g. subclavian
   h. intercostal

5. Describe the pathway of blood flow in hepatic portal circulation. Use a specific example to explain the purpose of portal circulation. (p. 292)

6. Begin at the right ventricle and describe the pathway of pulmonary circulation. Explain the purpose of this pathway. (p. 287)

7. Name the fetal structure with each of the following functions: (pp. 295, 298)
   a. permits blood to flow from the right atrium to the left atrium
   b. carries blood from the placenta to the fetus
   c. permits blood to flow from the pulmonary artery to the aorta
   d. carry blood from the fetus to the placenta
   e. carries blood from the umbilical vein to the inferior vena cava

8. Describe the three mechanisms that promote venous return when the body is vertical. (pp. 298–300)

9. Explain how the normal elasticity of the large arteries affects both systolic and diastolic blood pressure. (p. 300)

10. Explain how Starling's Law of the Heart is involved in the maintenance of blood pressure. (p. 301)

11. Name two hormones involved in the maintenance of blood pressure, and state the function of each. (pp. 300–301)

12. Describe two different ways the kidneys respond to decreased blood flow and blood pressure. (p. 301)

13. State two compensations that will maintain blood pressure after a small loss of blood. (p. 300)

14. State the location of the vasomotor center and name its two parts. Name the division of the autonomic nervous system that carries impulses to blood vessels. Which blood vessels? Which tissue in these vessels? Explain why normal vasoconstriction is important. Explain how greater vasoconstriction is brought about. Explain how vasodilation is brought about. How will each of these changes affect blood pressure? (pp. 301, 303)

Chapter 14
The Lymphatic System and Immunity

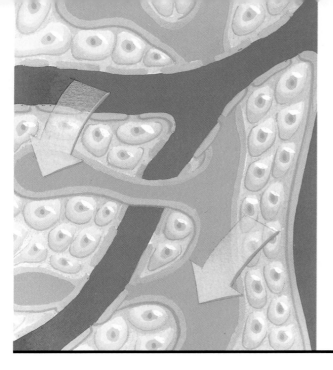

Chapter 14

Student Objectives

- Describe the functions of the lymphatic system.
- Describe how lymph is formed.
- Describe the system of lymph vessels, and explain how lymph is returned to the blood.
- State the locations and functions of the lymph nodes and nodules.
- State the location and functions of the spleen.
- Explain the role of the thymus in immunity.
- Explain what is meant by immunity.
- Describe humoral immunity and cell-mediated immunity.
- Describe the responses to a first and second exposure to a pathogen.
- Explain the difference between genetic immunity and acquired immunity.
- Explain the difference between passive acquired immunity and active acquired immunity.
- Explain how vaccines work.

The Lymphatic System and Immunity

New Terminology

Acquired immunity (uh–**KWHY**–erd)
Active immunity (**AK**–tiv)
Antibody (**AN**–ti–BAH-dee)
Antigen (**AN**–ti–jen)
B cells (B SELLS)
Cell–mediated immunity (SELL **ME**–dee–ay–ted)
Genetic immunity (je–**NET**–ik)
Humoral immunity (**HYOO**–mohr–uhl)
Lymph (LIMF)
Lymphokine (**LIMF**–oh–kine)
Lymph nodes (LIMF NOHDS)
Lymph nodules (LIMF **NAHD**–yools)
Opsonization (OP–sah–ni–**ZAY**–shun)
Passive immunity (**PASS**–iv)
Plasma cell (**PLAZ**–mah SELL)
Spleen (**SPLEEN**)
T cells (T SELLS)
Thymus (**THIGH**–mus)
Tonsils (**TAHN**–sills)

Related Clinical Terminology

AIDS (AYDS)
Allergy (**AL**–er–jee)
Antibody titer (**AN**–ti–BAH–dee **TIGH**–ter)
Attenuated (uh–**TEN**–yoo–AY–ted)
Complement fixation test (**KOM**–ple–ment fik–
    SAY–shun)
Fluorescent antibody test (floor–**ESS**–ent)
Hodgkin's disease (**HODJ**–kinz)
Interferon (in–ter–**FEER**–on)
Tonsillectomy (TAHN–si–**LEK**–toh–mee)
Toxoid (**TOCK**–soid)
Vaccine (vak–**SEEN**)

Terms that appear in **bold type** in the chapter text are defined in the glossary, which begins on page 549.

A child falls and scrapes her knee. Is this likely to be a life-threatening injury? Probably not, even though the breaks in the skin have permitted the entry of thousands or even millions of bacteria. Those bacteria, however, will be quickly destroyed by the cells and organs of the lymphatic system.

Although the lymphatic system may be considered part of the circulatory system, we will consider it separately because its functions are so different from those of the heart and blood vessels. Keep in mind, however, that all of these functions are interdependent. The lymphatic system is responsible for returning tissue fluid to the blood and for protecting the body against foreign material. The parts of the lymphatic system are the lymph, the system of lymph vessels, lymph nodes and nodules, the spleen, and the thymus gland.

LYMPH

Lymph is the name for tissue fluid that enters lymph capillaries. As you may recall from Chapter 13, filtration in capillaries creates tissue fluid, most of which returns almost immediately to the blood in the capillaries by osmosis. Some tissue fluid, however, remains in interstitial spaces and must be returned to the blood by way of the lymphatic vessels. Without this return, blood volume and blood pressure would very soon decrease. The relationship of the lymphatic vessels to the cardiovascular system is depicted in Fig. 14–1.

LYMPH VESSELS

The system of lymph vessels begins as dead-end **lymph capillaries** found in most tissue spaces (Fig. 14–2). Lymph capillaries are very permeable and collect tissue fluid and proteins. **Lacteals** are specialized lymph capillaries in the villi of the small intestine; they absorb the fat-soluble end products of digestion, such as fatty acids and vitamin A.

Lymph capillaries unite to form larger lymph vessels, whose structure is very much like that of veins. There is no pump for lymph (as the heart is the pump for blood), but the lymph is kept moving within lymph vessels by the same mechanisms that promote venous return. The smooth muscle layer of the larger lymph vessels constricts, and the one-way valves (just like those of veins) prevent backflow of lymph. Lymph vessels in the extremities are compressed by the skeletal muscles that surround them; this is the **skeletal muscle pump.** The **respiratory pump** alternately expands and compresses the lymph vessels in the chest cavity and keeps the lymph moving.

Where is the lymph going? Back to the blood to become plasma again. Refer to Fig. 14–3 as you read the following. The lymph vessels from the lower body unite in front of the lumbar vertebrae to form a vessel called the **cisterna chyli,** which continues upward in front of the backbone as the **thoracic duct.** Lymph vessels from the upper left quadrant of the body join the thoracic duct, which empties lymph into the left subclavian vein. Lymph vessels from the upper right quadrant of the body unite to form the right lymphatic duct, which empties lymph into the right subclavian vein. Flaps in both subclavian veins permit the entry of lymph but prevent blood from flowing into the lymph vessels.

LYMPH NODES AND NODULES

Lymph nodes and **nodules** are masses of lymphatic tissue. Recall that lymphatic tissue is one of the hemopoietic tissues and that one of its functions is to produce lymphocytes and monocytes. Nodes and nodules differ with respect to size and location. Nodes are usually larger, 10 to 20 mm in length; nodules range from a fraction of a millimeter to several millimeters in length.

1. **Lymph Nodes**—Lymph nodes are found in groups along the pathways of lymph vessels, and lymph flows through these nodes on its way to the subclavian veins. Lymph enters a node through several afferent lymph vessels

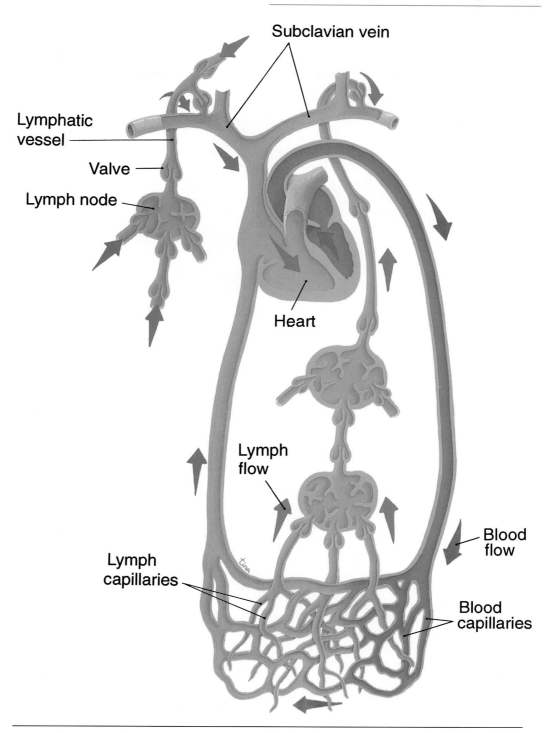

Figure 14–1 Relationship of lymphatic vessels to the cardiovascular system. Lymph capillaries collect tissue fluid, which is returned to the blood. The arrows indicate the direction of flow of the blood and lymph.

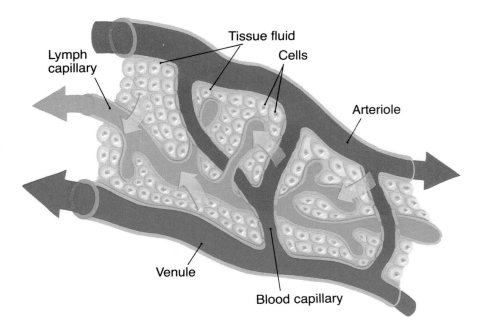

Lymph capillary

Tissue fluid

Cells

Arteriole

Venule

Blood capillary

Figure 14–2 Dead-end lymph capillaries found in tissue spaces. Arrows indicate the movement of plasma, lymph, and tissue fluid.

and leaves through one or two efferent vessels (Fig. 14–4). As lymph passes through a lymph node, bacteria and other foreign materials are phagocytized by fixed (stationary) **macrophages.** Fixed **plasma cells** (from lymphocytes) produce antibodies to any pathogens in the lymph; these antibodies, as well as lymphocytes and monocytes, will eventually reach the blood.

There are many groups of lymph nodes along all the lymph vessels throughout the body, but three paired groups deserve mention because of their strategic locations. These are the **cervical, axillary,** and **inguinal** lymph nodes (see Fig. 14–3). Notice that these are at the junctions of the head and extremities with the trunk of the body. Breaks in the skin, with entry of pathogens, are much more likely to occur in the arms or legs or head rather than in the trunk. If these pathogens get to the lymph, they will be destroyed by the lymph nodes before they get to the trunk, before the lymph is returned to the blood in the subclavian veins.

You may be familiar with the expression "swollen glands," as when a child has a strep throat (an inflammation of the pharynx caused by *Streptococcus* bacteria). These "glands" are the cervical lymph nodes that have enlarged as their macrophages attempt to destroy the bacteria in the lymph from the pharynx (see Box 14–1: Hodgkin's Disease).

2. **Lymph Nodules**—Lymph nodules are small masses of lymphatic tissue found just beneath the epithelium of all **mucous membranes.** The body systems lined with mucous membranes are those that have openings to the environment: the respiratory, digestive, urinary, and reproductive tracts. You can probably see that these are also strategic locations for lymph nodules, since any natural body opening is a possible portal of entry for pathogens. For example, if bacteria in inhaled air get through the epithelium of the trachea, lymph nodules with their macrophages are in position to destroy these bacteria before they get to the blood.

Some of the lymph nodules have specific names. Those of the small intestine are called

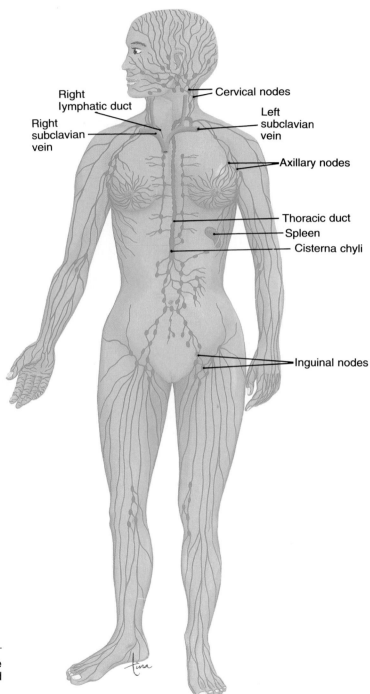

Right lymphatic duct

Right subclavian vein

Cervical nodes

Left subclavian vein

Axillary nodes

Thoracic duct

Spleen

Cisterna chyli

Inguinal nodes

Figure 14–3 System of lymph vessels and the major groups of lymph nodes. Lymph is returned to the blood in the right and left subclavian veins.

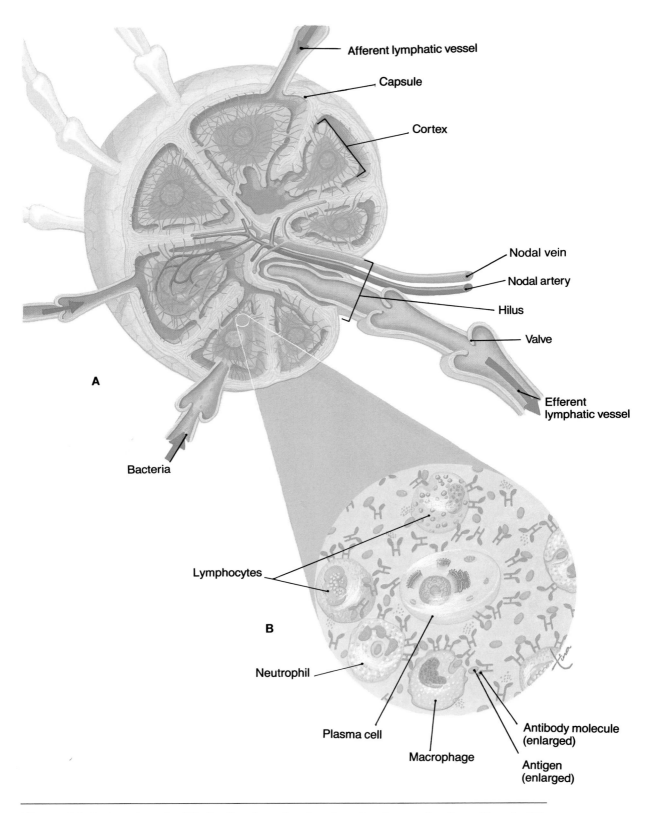

Figure 14–4 Lymph node. **(A)**, Section through a lymph node, showing the flow of lymph. **(B)**, Microscopic detail of bacteria being destroyed within the lymph node.

Box 14–1   HODGKIN'S DISEASE

Hodgkin's disease is a malignant disorder of the lymph nodes; the cause is not known. The first symptom is usually a swollen but painless lymph node, often in the cervical region. The individual is prompted to seek medical attention by other symptoms: chronic fever, fatigue, and weight loss. The diagnosis involves biopsy of the lymph node and the finding of characteristic cells.

Treatment of Hodgkin's disease requires chemotherapy, radiation, or both. With early diagnosis and proper treatment, this malignancy is very often curable.

Peyer's patches, and those of the pharynx are called **tonsils.** The palatine tonsils are on the lateral walls of the pharynx, the adenoid (pharyngeal tonsil) is on the posterior wall, and the lingual tonsils are on the base of the tongue. The tonsils, therefore, form a ring of lymphatic tissue around the pharynx, which is a common pathway for food and air and for the pathogens they contain. A **tonsillectomy** is the surgical removal of the palatine tonsils and the adenoid and may be performed if the tonsils are chronically inflamed and swollen, as may happen in children. As mentioned earlier, the body has redundant structures to help ensure survival if one structure is lost or seriously impaired. Thus, there are many other lymph nodules in the pharynx to take over the function of the surgically removed tonsils.

SPLEEN

The **spleen** is located in the upper left quadrant of the abdominal cavity, just below the diaphragm, behind the stomach. The lower rib cage protects the spleen from physical trauma (see Fig. 16–1).

In the fetus, the spleen produces red blood cells, a function assumed by the red bone marrow after birth.

The functions of the spleen after birth are:

1. Produces lymphocytes and monocytes, which enter the blood.
2. Contains fixed plasma cells that produce antibodies to foreign antigens.
3. Contains fixed macrophages (RE cells) that

phagocytize pathogens or other foreign material in the blood. The macrophages of the spleen also phagocytize old red blood cells and form bilirubin. By way of portal circulation, the bilirubin is sent to the liver for excretion in bile.

The spleen is not considered a vital organ, because other organs compensate for its functions if the spleen must be removed. The liver and red bone marrow will remove old red blood cells from circulation, and the many lymph nodes and nodules will produce lymphocytes and monocytes and phagocytize pathogens (as will the liver). Despite this redundancy, a person without a spleen is somewhat more susceptible to certain bacterial infections such as pneumonia and meningitis.

THYMUS

The **thymus** is located inferior to the thyroid gland. In the fetus and infant, the thymus is large and extends under the sternum (Fig. 14–5). With increasing age, the thymus shrinks, and very little thymus tissue is found in adults.

The lymphocytes produced by the thymus are called T lymphocytes or **T cells;** their functions will be discussed in the next section. Thymosin and other thymic hormones are necessary for what may be called "immunological competence." To be competent means to be able to do something well. The thymic hormones enable the T cells to participate in the recognition of foreign antigens and to provide immunity. This capability of T cells is established early in life and then is perpetuated by the

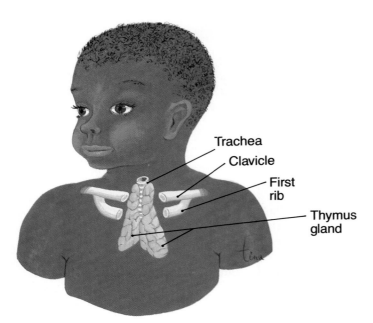

Trachea

Clavicle

First rib

Thymus gland

Figure 14–5 Location of the thymus in a young child.

lymphocytes themselves. The newborn's immune system is not yet fully mature, and infants are more susceptible to certain infections than are older children and adults. Usually by the age of 2 years, the immune system matures and becomes fully functional. This is why some vaccines, such as the measles vaccine, are not recommended for infants younger than 15 to 18 months of age. Their immune systems are not mature enough to respond strongly to the vaccine, and the protection provided by the vaccine may be incomplete.

IMMUNITY

Immunity may be defined as the body's ability to destroy pathogens or other foreign material and to prevent further cases of certain infectious diseases. This ability is of vital importance because the body is exposed to pathogens from the moment of birth.

Malignant cells, which may be formed within the body as a result of mutations of normal cells, are also recognized as foreign and are usually de-stroyed before they can establish themselves and cause cancer. Unfortunately, organ transplants are also foreign tissue, and the immune system may reject (destroy) a transplanted kidney or heart. Sometimes the immune system mistakenly reacts to part of the body itself and causes an autoimmune disease; several of these were mentioned in previous chapters. Most often, however, the immune mechanisms function to protect the body from the microorganisms around us and within us.

LYMPHOCYTES

There are two major types of lymphocytes: T lymphocytes and B lymphocytes, or, more simply, **T cells** and **B cells.** In the embryo, T cells are produced in the bone marrow and thymus. They must pass through the thymus, where the thymic hormones bring about their maturation. The T cells then migrate to the spleen, lymph nodes, and lymph nodules, where they are found after birth.

Produced in the embryo bone marrow, B cells then migrate directly to the spleen and lymph nodes and nodules. When activated during an immune response, some B cells will become plasma cells that produce antibodies to a specific foreign antigen.

ANTIGENS AND ANTIBODIES

Antigens are chemical markers that identify cells. Human cells have their own antigens that identify all the cells in an individual as "self" (recall the HLA types mentioned in Chapter 11). When antigens are foreign, or "non-self," they may be recognized as such and destroyed. Bacteria, viruses, fungi, protozoa, malignant cells, and organ transplants are all foreign antigens that activate immune responses.

Antibodies, also called **immune globulins** or **gamma globulins,** are proteins produced by plasma cells in response to foreign antigens. Antibodies do not themselves destroy foreign antigens, but rather become attached to such antigens to "label" them for destruction. Each antibody produced is specific for only one antigen. Since there are so many different pathogens, you might think that the immune system would have to be capable of producing many different antibodies, and in fact this is so. It is estimated that as many as 1 million different antigen-specific antibodies can be produced, should there be a need for them.

The structure of antibodies is shown in Fig. 14–6, and the five classes of antibodies are described in Table 14–1.

MECHANISMS OF IMMUNITY

The first step in the destruction of a pathogen or foreign cell is the recognition of its antigens as foreign. This is accomplished by macrophages and a specialized group of T lymphocytes called **helper T cells.** The foreign antigen is first phagocytized by a macrophage, and parts of it are "presented" on the macrophage's cell membrane. Also on the macrophage membrane are "self" antigens that are representative of the antigens found on all of the cells of the individual. Therefore, the helper T cell that encounters this macrophage is presented not only with the foreign antigen but also with "self" antigens for comparison. The helper T cell now becomes sensitized to and specific for the foreign antigen, the one that does not belong in the body (see Box 14–2: AIDS).

Once an antigen has been recognized as foreign, the helper T cells initiate one or both of the mechanisms of immunity. These are **cell-mediated immunity,** in which T cells and macrophages participate, and **humoral immunity,** which involves T cells, B cells, and macrophages.

Cell-Mediated Immunity

This mechanism of immunity does not result in the production of antibodies, but it is effective against intracellular pathogens (such as viruses), fungi, malignant cells, and grafts of foreign tissue. As mentioned above, the first step is the recognition of the foreign antigen by macrophages and helper T cells, which now become activated and specific (you may find it helpful to refer to Fig. 14–7 as you read the following).

These activated T cells, which are now antigen specific, divide many times, forming **memory T cells** and **killer (cytotoxic) T cells.** The memory T cells will remember the specific foreign antigen and become active if it enters the body again. Killer T cells are able to chemically destroy foreign antigens by disrupting cell membranes. This is how killer T cells destroy cells infected with viruses, and prevent the viruses from reproducing. These T cells also produce **lymphokines,** which are chemicals that attract macrophages to the area and activate them to phagocytize the foreign antigen.

Other activated T cells become **suppressor T cells,** which will stop the immune response once the foreign antigen has been destroyed. The memory T cells, however, will quickly initiate the cell-mediated immune response should there be a future exposure to the antigen.

Humoral Immunity

This mechanism of immunity does involve the production of antibodies and is diagrammed in Fig. 14–8. Again, the first step is the recognition of the foreign antigen by macrophages and helper T cells. The sensitized helper T cell then presents the foreign antigen to B cells, which in turn become activated and specific. The activated B cells begin to divide many times, and two types of cells are formed. Some of the new B cells produced are

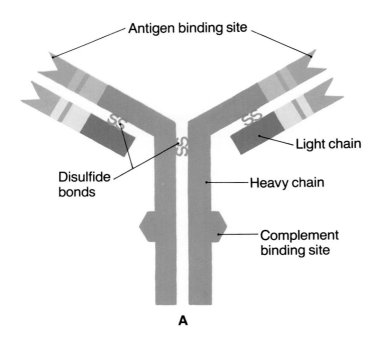

A

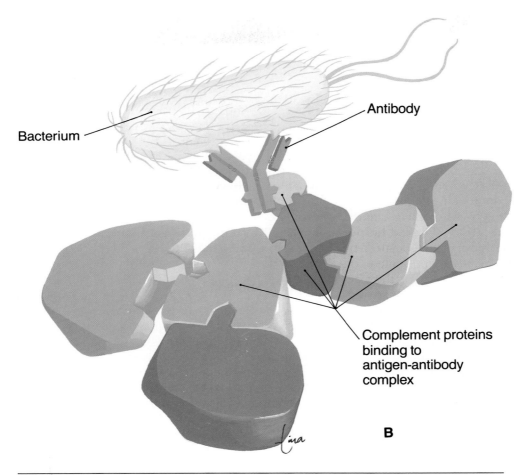

B

Figure 14–6 See legend on facing page.

Table 14–1 CLASSES OF ANTIBODIES

Name	Location	Functions
IgG	Blood Extracellular fluid	• Crosses the placenta to provide passive immunity for newborns • Provides long-term immunity following recovery or a vaccine
IgA	External secretions (tears, saliva, etc.)	• Present in breast milk to provide passive immunity for breast-fed infants • Found in secretions of all mucous membranes
IgM	Blood	• Produced first by the maturing immune system of infants • Produced first during an infection (IgG production follows)
IgD	B lymphocytes	• Receptors on B lymphocytes
IgE	Mast cells or basophils	• Important in allergic reactions (mast cells release histamine)

memory B cells, which will remember the specific antigen. Other B cells become **plasma cells** that produce antibodies specific for this one foreign antigen.

The antibodies then bond to the antigen, forming an antigen-antibody complex. This complex results in **opsonization,** which means that the antigen is now "labeled" for phagocytosis by macrophages or neutrophils. The antigen-antibody complex also stimulates the process of complement fixation (see Box 14–3: Diagnostic Tests).

Complement is a group of about 20 plasma proteins that circulate in the blood until activated, or fixed, by an antigen-antibody complex. Complement fixation may be complete or partial. If the foreign antigen is cellular, the complement proteins bond to the antigen-antibody complex, then to one another, forming an enzymatic ring that punches a hole in the cell to bring about death of the cell. This is complete (or entire) complement fixation and is what happens to bacterial cells (it is also the cause of hemolysis in a transfusion reaction).

If the foreign antigen is not a cell, a virus for example, partial complement fixation takes place, in which some of the complement proteins bond to the antigen-antibody complex. This is a chemotaxic factor. Chemotaxis means "chemical movement" and is actually another label that attracts macrophages to engulf and destroy the foreign antigen.

When the foreign antigen has been destroyed, suppressor T cells that have been sensitized to it stop the immune response. This is important to limit antibody production to just what is necessary to eliminate the pathogen without triggering an autoimmune response.

ANTIBODY RESPONSES

The first exposure to a foreign antigen does stimulate antibody production, but antibodies are produced slowly and in small amounts. Let us take as a specific example the measles virus. On a person's first exposure to this virus, antibody production is usually too slow to prevent the disease itself, and

Figure 14–6 Antibody structure and activity. **(A)**, Structure of an antibody molecule. An antibody is a protein made of four polypeptide chains linked by disulfide bonds. The antigen-binding site is specific for one foreign antigen. **(B)**, Complement fixation.

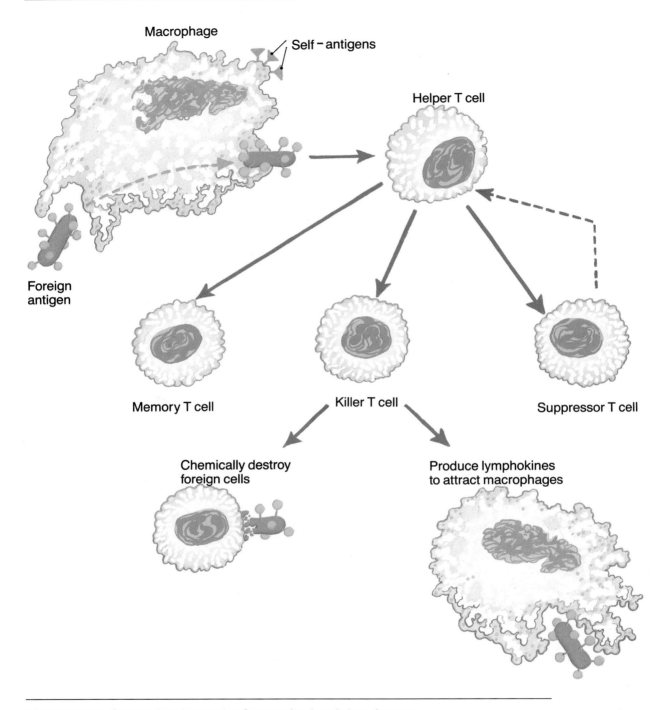

Macrophage

Self – antigens

Helper T cell

Foreign antigen

Memory T cell

Killer T cell

Suppressor T cell

Chemically destroy foreign cells

Produce lymphokines to attract macrophages

Figure 14–7 Cell-mediated immunity. See text for description of events.

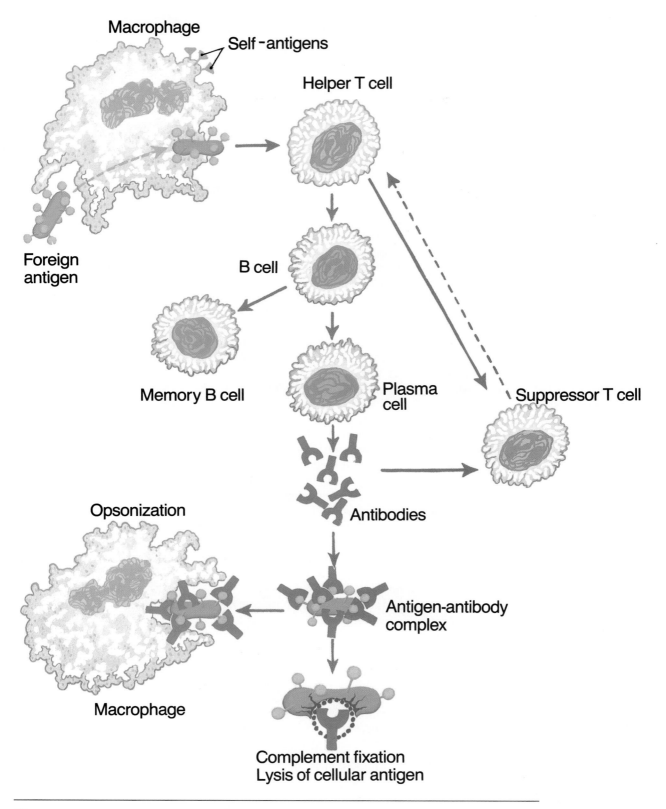

Figure 14–8 Humoral immunity. See text for description of events.

Box 14–2 AIDS

In 1981, young homosexual men in New York and California were diagnosed with Kaposi's sarcoma and *Pneumocystis carinii* pneumonia. At that time, Kaposi's sarcoma was known as a rare, slowly growing malignancy in elderly men. Pneumocystis pneumonia was almost unheard of; *P. carinii* is a protozoan that does not cause disease in healthy people. That in itself was a clue. These young men were not healthy; their immune systems were not functioning normally. As the number of patients increased rapidly, the disease was given a name (Acquired Immune Deficiency Syndrome—AIDS) and the pathogen was found. Human Immunodeficiency Virus (HIV) is a retrovirus that infects helper T cells, macrophages, and other human cells. Once infected, the human cells contain HIV genes for the rest of their lives. Without sufficient helper T cells, the immune system is seriously impaired. Foreign antigens are not recognized, B cells are not activated, and killer T cells are not stimulated to proliferate.

The person with AIDS is susceptible to opportunistic infections, that is, those infections caused by fungi and protozoa that would not affect average healthy adults. Some of these infections may be treated with medications and even temporarily cured, but the immune system cannot prevent the next infection, or the next. As of this writing, AIDS is considered an incurable disease, although with proper medical treatment, some people with AIDS may live for many years.

Where did this virus come from? We really do not know. Some researchers believe that a mutation in a previously harmless virus produced HIV, with devastating consequences for those infected with this "new" virus.

The incubation period of AIDS is highly variable, ranging from a few months to several years. An infected person may unknowingly spread HIV to others before any symptoms appear. It should be emphasized that AIDS, although communicable, is not a contagious disease. It is not spread by casual contact as is measles or the common cold. Transmission of AIDS occurs through sexual contact, contact with infected blood, or by placental transmission of the virus from mother to fetus.

In the United States, most of the cases of AIDS during the 1980s were in homosexual men and IV drug users who shared infected syringes contaminated with their blood. By 1993, however, it was clear that AIDS was becoming more of a heterosexually transmitted disease, with rapidly increasing case rates among women and teenagers. In much of the rest of the world, especially Africa and Asia, the transmission of AIDS has always been primarily by heterosexual contact, with equal numbers of women and men infected.

How can the spread of HIV be stopped? At present we still have no antiviral medications that will eradicate this virus, and the development of new antiviral drugs is an unfortunately slow process.

Development of an AIDS vaccine is unlikely before the turn of the century, although more than a dozen vaccines are undergoing clinical trials. A vaccine stimulates antibody production to a specific pathogen, but everyone who has died of AIDS had antibodies to HIV, and those antibodies were not protective. Why not? The most likely explanation is that HIV is a mutating virus; it constantly changes itself, making previously produced antibodies ineffective.

If we cannot cure AIDS and we cannot prevent it by vaccination, what recourse is left? Education. Everyone should know how AIDS is spread. The obvious reason is to be able to avoid the high-risk behaviors that make acquiring HIV more likely. Yet another reason, however, is that everyone should know that they need not fear casual contact with people with AIDS. Healthcare personnel have a special responsibility, not only to educate themselves, but to provide education about AIDS for their patients and the families of their patients.

Several important laboratory tests involve antibodies and may be very useful to confirm a diagnosis.

Complement Fixation Test—determines the presence of a particular antibody in the patient's blood.

Antibody Titer—determines the level or amount of a specific antibody in the patient's blood. If another titer is done 1 to several weeks later, an increase in the antibody level shows the infection to be current.

Fluorescent Antibody Test—uses antibodies tagged with fluorescent dyes, which are added to a clinical specimen such as blood, sputum, or a biopsy of tissue. If the suspected pathogen is present, the fluorescent antibodies will bond to it and the antigen-antibody complex will "glow" when examined with a fluorescent microscope.

Tests such as these are used in conjunction with patient history and symptoms to arrive at a diagnosis.

the person will have clinical measles. Most people who get measles recover, and upon recovery have antibodies and memory cells that are specific for the measles virus.

On a second exposure to this virus, the memory cells initiate rapid production of large amounts of antibodies, enough to prevent a second case of measles. This is the reason why we develop immunity to certain diseases, and this is also the basis for the protection given by **vaccines** (see Box 14–4: Vaccines).

As mentioned above, antibodies label pathogens or other foreign antigens for phagocytosis or complement fixation. More specifically, antibodies cause agglutination or neutralization of pathogens before their eventual destruction. **Agglutination** means "clumping," and this is what happens when antibodies bond to bacterial cells. The bacteria that are clumped together by attached antibodies are more easily phagocytized by macrophages.

The activity of viruses may be neutralized by antibodies. A virus must get inside a living cell in order to reproduce itself. However, a virus with antibodies attached to it is unable to enter a cell, cannot reproduce, and will soon be phagocytized. (Another aspect of antiviral defense is a chemical called **interferon,** which is discussed in Box 14–5.) Bacterial toxins may also be neutralized by attached an-

tibodies. The antibodies change the shape of the toxin, prevent it from exerting its harmful effects, and promote its phagocytosis by macrophages.

Allergies are also the result of antibody activity (see Box 14–6).

TYPES OF IMMUNITY

There are two major categories of immunity: genetic immunity, which is conferred by our DNA, and acquired immunity, which we must develop or acquire by natural or artificial means.

Genetic immunity does not involve antibodies or the immune system; it is the result of our genetic makeup. What it means is that some pathogens cause disease in certain host species but not in others. Dogs and cats, for example, have genetic immunity to the measles virus, which is a pathogen only for people. Plant viruses affect only plants, not people; we have genetic immunity to them. This is not due to antibodies against these plant viruses, but rather that our genetic makeup makes it impossible for such pathogens to reproduce in our cells and tissues. Since this is a genetic characteristic programmed in DNA, genetic immunity always lasts a lifetime.

Acquired immunity does involve antibodies. **Passive immunity** means that the antibodies are

Box 14–4 VACCINES

The purpose of vaccines is to prevent disease. A vaccine contains an antigen that the immune system will respond to, just as it would to the actual pathogen. The types of vaccine antigens are a killed or weakened (**attenuated**) pathogen, part of a pathogen such as a bacterial capsule, or an inactivated bacterial toxin called a **toxoid.**

Since the vaccine itself does not cause disease (with very rare exceptions), the fact that antibody production to it is slow is not detrimental to the person. The vaccine takes the place of the first exposure to the pathogen and stimulates production of antibodies and memory cells. On exposure to the pathogen itself, the memory cells initiate rapid production of large amounts of antibody, enough to prevent disease.

We now have vaccines for many diseases. The tetanus and diphtheria vaccines contain toxoids, the inactivated toxins of these bacteria. Vaccines for pneumococcal pneumonia and meningitis contain bacterial capsules. These vaccines cannot cause disease because the capsules are nontoxic and nonliving; there is nothing that can reproduce. Influenza and rabies vaccines contain killed viruses. Measles and the oral polio vaccines contain attenuated (weakened) viruses.

Although attenuated pathogens are usually strongly antigenic and stimulate a protective immune response, there is a very small chance that the pathogen may regain its virulence and cause the disease. For example, a few cases each year of polio are caused by the oral polio vaccine itself. The risk is small, however (1 in 500,000), and the vaccine is considered highly effective.

Box 14–5 INTERFERON

Interferon is produced by cells infected with viruses and by T cells. Viruses must be inside a living cell to reproduce, and although interferon cannot prevent the entry of viruses into cells, it does block their reproduction. When viral reproduction is blocked, the viruses cannot infect new cells and cause disease. Interferon is probably a factor in the self-limiting nature of many viral diseases.

There are several types of human interferon (alpha, beta, gamma), which were first produced in amounts sufficient for clinical research in the 1970s. At that time there was hope that interferon would be an effective anticancer therapy, but results against the most common forms of cancer were disappointing. Alpha interferon is effective, however, in the treatment of a rare type of leukemia called hairy-cell leukemia and has also been approved for use in cases of genital warts and Kaposi's sarcoma. Most recently, interferon has proved useful in the treatment of hepatitis B and C and multiple sclerosis, although it is not equally effective for all patients.

The research on the interferons is a good example of the way science works. Discovery, research, dead-ends, more research, other possibilities, and so on. Our knowledge is gained piece by piece and often requires many years of careful study. In the future, the interferons will probably become more useful theraputic tools, but there is much to be learned first.

Box 14–6 ALLERGIES

An **allergy** is a hypersensitivity to a particular foreign antigen, called an **allergen**. Allergens include plant pollens, foods, chemicals in cosmetics, antibiotics such as penicillin, dust, and mold spores. Such allergens are not themselves harmful. Most people, for example, can inhale pollen, eat peanuts, or take penicillin with no ill effects.

Hypersensitivity means that the immune system overresponds to the allergen, and produces tissue damage by doing so. Allergic responses are characterized by the production of IgE antibodies, which bond to mast cells. Mast cells are specialized cells that differentiate from basophils, and are numerous in the connective tissue of the skin and mucous membranes. One of several chemicals in mast cells is histamine, which is released by the bonding of IgE antibodies or when tissue damage occurs.

Histamine contributes to the process of inflammation by increasing the permeability of capillaries and venules. When tissue is damaged this promotes greater tissue fluid formation and brings more WBCs to the damaged area.

In an allergic reaction, the effects of histamine and other inflammatory chemicals create symptoms such as watery eyes and runny nose (hay fever) or the more serious wheezing and difficult breathing that characterize asthma. People with seasonal hay fever may take antihistamines to counteract these effects (see Chapter 15 for a description of asthma).

Anaphylactic shock is an extreme allergic response which may be elicited by exposure to penicillin or insect venoms. On the first exposure the person becomes highly sensitized to the foreign antigen. On the second exposure, histamine is released from mast cells throughout the body and causes a drastic decrease in blood volume. The resulting drop in blood pressure may be fatal in only a few minutes. People who know they are allergic to bee stings, for example, may obtain a self-contained syringe of epinephrine to carry with them. Epinephrine can delay the progression of anaphylactic shock long enough for the person to seek medical attention.

from another source, while **active immunity** means that the individual produces his or her own antibodies.

One type of naturally acquired passive immunity is the placental transmission of antibodies (IgG) from maternal blood to fetal circulation. The baby will then be born temporarily immune to the diseases the mother is immune to. Such passive immunity may be prolonged by breast-feeding, since breast milk also contains maternal antibodies (IgA).

Artificially acquired passive immunity is obtained by the injection of immune globulins (gamma glob-

ulins or preformed antibodies) after presumed exposure to a particular pathogen. Such immune globulins are available for German measles, hepatitis A and B, tetanus and botulism (anti-toxins), and rabies. These are *not* vaccines; they do not stimulate immune mechanisms, but rather provide immediate antibody protection. Passive immunity is always temporary, lasting a few weeks to a few months, because antibodies from another source eventually break down.

Active immunity is the production of one's own antibodies and may be stimulated by natural or ar-

Box 14–7   VACCINES THAT HAVE CHANGED OUR LIVES

In 1797, Edward Jenner (in England) used the cowpox virus called Vaccinia as the first vaccine for smallpox, a closely related virus. (He was unaware of this, since viruses had not yet been discovered, but he had noticed that milkmaids who got cowpox rarely got smallpox.) In 1980, the World Health Organization declared that smallpox had been eradicated throughout the world. A disease that had killed or disfigured millions of people throughout recorded history is now considered part of history.

In the 19th century in the northern United States, thousands of children died of diphtheria every winter. Today there are fewer than 10 cases of diphtheria each year in the entire country. In the early 1950s, 50,000 cases of paralytic polio were reported in the United States each year. Today, fewer than 10 cases per year are reported.

Smallpox, diphtheria, and polio are no longer the terrible diseases they once were, and this is because of the development and widespread use of vaccines. When people are protected by a vaccine, they are no longer possible reservoirs or sources of the pathogen for others, and the spread of disease may be greatly limited.

Other diseases that have been controlled by the use of vaccines are whooping cough, tetanus, mumps, influenza, measles, and German measles. A new vaccine for hepatitis B has already significantly decreased the number of cases of this disease among healthcare workers, and the CDC recommends the vaccine for all children. People who have been exposed to rabies, which is virtually always fatal, can be protected by a new and safe vaccine.

Without such vaccines our lives would be very different. Infant mortality or death in childhood would be much more frequent, and all of us would have to be much more aware of infectious diseases. In many parts of the world this is still true; many of the developing countries in Africa and Asia still cannot afford extensive vaccination programs for their children. Many of the diseases mentioned above, which we may rarely think of, are still a very significant part of the lives of millions of people.

tificial means. Naturally acquired active immunity means that a person has recovered from a disease and now has antibodies and memory cells specific for that pathogen. Artificially acquired active immunity is the result of a vaccine that has stimulated production of antibodies and memory cells (see Box 14–7: Vaccines That Have Changed Our Lives). No general statement can be made about the duration of active immunity. Recovering from plague, for example, confers lifelong immunity, but the plague vaccine does not. Duration of active immunity, therefore, varies with the particular disease or vaccine.

The types of immunity are summarized in Table 14–2.

SUMMARY

The preceding discussions of immunity will give you a small idea of the complexity of the body's defense system. However, there is still much more to be learned, especially about the effects of the nervous system and endocrine system on immunity. For example, it is known that people under great stress have immune systems that may not function as they did when stress was absent. One such study showed that students in the midst of final exam week had fewer killer T cells and lower levels of interferon than they had a few weeks previously.

Table 14–2　TYPES OF IMMUNITY

Type	Description
Genetic	• Does not involve antibodies; is programmed in DNA • Some pathogens affect certain host species but not others
Acquired	• Does involve antibodies
Passive 　NATURAL	• Antibodies from another source • Placental transmission of antibodies from mother to fetus • Transmission of antibodies in breast milk
ARTIFICIAL	• Injection of preformed antibodies (gamma globulins or immune globulins) after presumed exposure
Active 　NATURAL	• Production of one's own antibodies • Recovery from a disease, with production of antibodies and memory cells
ARTIFICIAL	• A vaccine stimulates production of antibodies and memory cells

At present, there is much research being done in this field. The goal is not to eliminate all disease, for that would not be possible. Rather, the aim is to enable people to live healthier lives by preventing certain diseases.

STUDY OUTLINE

Functions of the Lymphatic System
1. To return tissue fluid to the blood to maintain blood volume (see Fig. 14–1).
2. To protect the body against pathogens and other foreign material.

Parts of the Lymphatic System
1. lymph
2. lymph vessels
3. lymph nodes and nodules
4. spleen
5. thymus

Lymph—the tissue fluid that enters lymph capillaries
1. Similar to plasma, but more WBCs are present.
2. Must be returned to the blood to maintain blood volume and blood pressure.

Lymph Vessels
1. Dead-end lymph capillaries are found in most tissue spaces; collect tissue fluid and proteins (see Fig. 14–2).
2. The structure of larger lymph vessels is like that of veins; valves prevent the backflow of lymph.

3. Lymph is kept moving in lymph vessels by:
   • constriction of the lymph vessels
   • the skeletal muscle pump
   • the respiratory pump
4. Lymph from the lower body and upper left quadrant enters the thoracic duct and is returned to the blood in the left subclavian vein (see Fig. 14–3).
5. Lymph from the upper right quadrant enters the right lymphatic duct and is returned to the blood in the right subclavian vein.

Lymph Nodes—masses of lymphatic tissue; produce lymphocytes and monocytes
1. Found in groups along the pathways of lymph vessels.
2. As lymph flows through the nodes:
   • lymphocytes and monocytes enter the lymph
   • foreign materials are phagocytized by fixed macrophages
   • fixed plasma cells produce antibodies to foreign antigens (see Fig. 14–4)
3. The major paired groups of lymph nodes are the cervical, axillary, and inguinal groups. These are at the junctions of the head and extremities with the trunk; remove pathogens from the lymph from the extremities before the lymph is returned to the blood.

Lymph Nodules—small masses of lymphatic tissue; produce lymphocytes and monocytes
1. Found beneath the epithelium of all mucous membranes, that is, the tracts that have natural openings to the environment.
2. Destroy pathogens that penetrate the epithelium of the respiratory, digestive, urinary, or reproductive tracts.
3. Tonsils are the lymph nodules of the pharynx; Peyer's patches are those of the small intestine.

Spleen—located in the upper left abdominal quadrant behind the stomach
1. The fetal spleen produces RBCs.
2. Functions after birth:
   • production of lymphocytes and monocytes
   • fixed plasma cells produce antibodies
   • fixed macrophages (RE cells) phagocytize

pathogens and old RBCs; bilirubin is formed and sent to the liver for excretion in bile

Thymus—in the fetus and infant the thymus is large and inferior to the thyroid gland; with age the thymus shrinks (see Fig. 14–5)
1. Produces T lymphocytes (T cells).
2. Produces thymosin and other hormones that make T cells immunologically competent: able to recognize foreign antigens and provide immunity.

Immunity
1. The ability to destroy foreign antigens and prevent future cases of certain infectious diseases.
2. Foreign antigens include bacteria, viruses, fungi, protozoa, and malignant cells.

Lymphocytes
1. T lymphocytes (T cells)—in the embryo are produced in the thymus and RBM; they require the hormones of the thymus for maturation; migrate to the spleen, lymph nodes and nodules.
2. B lymphocytes (B cells)—in the embryo are produced in the RBM; migrate to the spleen, lymph nodes and nodules.

Antigens
1. Chemical markers that identify cells.
2. Human cells have "self" antigens—the HLA types.
3. Foreign antigens stimulate antibody production or other immune responses.

Antibodies—immune globulins or gamma globulins (see Table 14–1 and Fig. 14–6)
1. Proteins produced by plasma cells in response to foreign antigens.
2. Each antibody is specific for only one foreign antigen.
3. Bond to the foreign antigen to label it for phagocytosis (opsonization).

Mechanisms of Immunity
1. The antigen must first be recognized as foreign; this is accomplished by helper T cells that com-

pare the foreign antigen to "self" antigens present on macrophages.

2. Helper T cells then initiate one or both of the immune mechanisms: cell-mediated immunity and humoral immunity.

Cell-Mediated Immunity (see Fig. 14–7)

1. Does not involve antibodies; is effective against intracellular pathogens, malignant cells, and grafts of foreign tissue.
2. Helper T cells recognize the foreign antigen, become antigen specific, and begin to divide to form different groups of T cells.
3. Memory T cells will remember the specific foreign antigen.
4. Killer (cytotoxic) T cells chemically destroy foreign cells and produce lymphokines to attract macrophages.
5. Suppressor T cells stop the immune response once the foreign antigen has been destroyed.

Humoral Immunity (see Fig. 14–8)

1. Does involve antibody production; is effective against pathogens and foreign cells.
2. Helper T cells recognize the foreign antigen and present it to B cells, which begin to divide.
3. Memory B cells will remember the specific foreign antigen.
4. Other B cells become plasma cells that produce antigen-specific antibodies.

5. An antigen-antibody complex is formed, which attracts macrophages (opsonization).
6. Complement fixation is stimulated by antigen-antibody complexes. The complement proteins bind to the antigen-antibody complex and lyse cellular antigens or enhance the phagocytosis of noncellular antigens.
7. Suppressor T cells stop the immune response when the foreign antigen has been destroyed.

Antibody Responses and Functions

1. On the first exposure to a foreign antigen, antibodies are produced slowly and in small amounts, and the person may develop clinical disease.
2. On the second exposure, the memory cells initiate rapid production of large amounts of antibodies, and a second case of the disease may be prevented. This is the basis for the protection given by vaccines, which take the place of the first exposure.
3. Antibodies cause agglutination (clumping) of bacterial cells; clumped cells are easier for macrophages to phagocytize.
4. Antibodies neutralize viruses by bonding to them and preventing their entry into cells.
5. Antibodies neutralize bacterial toxins by bonding to them and changing their shape.

Types of Immunity (see Table 14–2)

REVIEW QUESTIONS

1. Explain the relationships among plasma, tissue fluid, and lymph, in terms of movement of water throughout the body. (p. 310)

2. Describe the system of lymph vessels. Explain how lymph is kept moving in these vessels. Into which veins is lymph emptied? (pp. 310, 313)

3. State the locations of the major groups of lymph nodes, and explain their functions. (pp. 310, 312)

4. State the locations of lymph nodules, and explain their functions. (p. 312)

5. Describe the location of the spleen and explain its functions. If the spleen is removed, what organs will compensate for its functions? (p. 315)

6. Explain the function of the thymus, and state when (age) this function is important. (p. 315)

7. Name the different kinds of foreign antigens that the immune system responds to. (p. 316)

8. State the functions of helper T cells, killer T cells, memory T cells, and suppressor T cells. (p. 317)

9. Plasma cells differentiate from which type of lymphocyte? State the function of plasma cells. What other type of cell comes from B lymphocytes? (pp. 317, 319)

10. Explain how a foreign antigen is recognized as foreign. Which mechanism of immunity involves antibody production? Explain what opsonization means. (pp. 317, 319)

11. What is the stimulus for complement fixation? How does this process destroy cellular antigens and non-cellular antigens? (p. 319)

12. Explain the antibody reactions of agglutination and neutralization. (p. 323)

13. Explain how a vaccine provides protective immunity in terms of first and second exposures to a pathogen. (pp. 319, 323)

14. Explain the difference between the following: (pp. 323, 325, 327)
    a. genetic immunity and acquired immunity
    b. passive acquired immunity and active acquired immunity
    c. natural and artificial passive acquired immunity
    d. natural and artificial active acquired immunity

Chapter 15
The Respiratory System

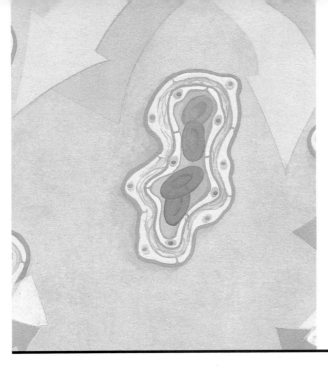

Chapter 15

Chapter Outline

Student Objectives

- State the general function of the respiratory system.
- Describe the structure and functions of the nasal cavities and pharynx.
- Describe the structure of the larynx and explain the speaking mechanism.
- Describe the structure and functions of the trachea and bronchial tree.
- State the locations of the pleural membranes, and explain the functions of serous fluid.
- Describe the structure of the alveoli and pulmonary capillaries, and explain the importance of surfactant.
- Name and describe the important air pressures involved in breathing.
- Describe normal inhalation and exhalation and forced exhalation.
- Explain the diffusion of gases in external respiration and internal respiration.
- Describe how oxygen and carbon dioxide are transported in the blood.
- Name the pulmonary volumes and define each.
- Explain the nervous and chemical mechanisms that regulate respiration.
- Explain how respiration affects the pH of body fluids.

The Respiratory System

New Terminology

Alveoli (al–**VEE**–oh–lye)
Bronchial tree (**BRONG**–kee–uhl TREE)
Epiglottis (Ep–i–**GLAH**–tis)
Glottis (**GLAH**–tis)
Intrapleural pressure (IN–trah–**PLOOR**–uhl **PRES**–shur)
Intrapulmonic pressure (IN–trah–pull–**MAHN**–ik **PRES**–shur)
Larynx (**LA**–rinks)
Partial pressure (**PAR**–shul **PRES**–shur)
Phrenic nerves (**FREN**–ik NURVZ)
Pulmonary surfactant (**PULL**–muh–ner–ee sir–**FAK**–tent)
Residual air (ree–**ZID**–yoo–al AYRE)
Respiratory acidosis (RES–pi–rah–**TOR**–ee ass–i–**DOH**–sis)
Respiratory alkalosis (RES–pi–rah–**TOR**–ee al–kah–**LOH**–sis)

Soft palate (SAWFT **PAL**–uht)
Tidal volume (**TIGH**–duhl **VAHL**–yoom)
Ventilation (VEN–ti–**LAY**–shun)
Vital capacity (**VY**–tuhl kuh–**PASS**–i–tee)

Related Clinical Terminology

Cyanosis (SIGH–uh–**NO**–sis)
Dyspnea (**DISP**–nee–ah)
Emphysema (EM–fi–**SEE**–mah)
Heimlich maneuver (**HIGHM**–lik ma–**NEW**–ver)
Hyaline membrane disease (**HIGH**–e–lin **MEM**–brain dis–EEZ)
Pneumonia (new–**MOH**–nee–ah)
Pneumothorax (NEW–moh–**THAW**–raks)
Pulmonary edema (**PULL**–muh–ner–ee uh–**DEE**–muh)G1.

Terms that appear in **bold type** in the chapter text are defined in the glossary, which begins on page 549.

Sometimes a person will describe a habit as being "as natural as breathing." Indeed, what could be more natural? We rarely think about breathing, and it isn't something we look forward to, as we would a good dinner. We just breathe, usually at the rate of 12 to 20 times per minute, and faster when necessary (such as during exercise). You may have heard of trained singers "learning how to breathe," but they are really learning how to make their breathing more efficient.

Most of the **respiratory system** is concerned with what we think of as breathing: moving air into and out of the lungs. The lungs are the site of the exchanges of oxygen and carbon dioxide between the air and the blood. Both of these exchanges are important. All our cells must obtain oxygen to carry out cell respiration to produce ATP. Just as crucial is the elimination of the CO_2 produced as a waste product of cell respiration, and, as you already know, the proper functioning of the circulatory system is essential for the transport of these gases in the blood.

DIVISIONS OF THE RESPIRATORY SYSTEM

The respiratory system may be divided into the upper respiratory tract and the lower respiratory tract. The **upper respiratory tract** consists of the parts outside the chest cavity: the air passages of the nose, nasal cavities, pharynx, larynx, and upper trachea. The **lower respiratory tract** consists of the parts found within the chest cavity: the lower trachea and the lungs themselves, which include the bronchial tubes and alveoli. Also part of the respiratory system are the pleural membranes and the respiratory muscles that form the chest cavity: the diaphragm and intercostal muscles.

Have you recognized some familiar organs and structures thus far? There will be more, for this chapter includes material from Chapters 1 through 9, 11, and 12. Even though we are discussing the body system by system, the respiratory system is an excellent example of the interdependent functioning of all the body systems.

NOSE AND NASAL CAVITIES

Air enters and leaves the respiratory system through the **nose,** which is made of bone and cartilage covered with skin. Just inside the nostrils are hairs, which help block the entry of dust.

The two **nasal cavities** are within the skull, separated by the **nasal septum,** which is a bony plate made of the ethmoid bone and vomer. The **nasal mucosa** (lining) is ciliated epithelium, with goblet cells that produce mucus. The surface area of the nasal mucosa is increased by the conchae, shelf-like bones on the lateral wall of each nasal cavity (Fig. 15–1). As air passes through the nasal cavities it is warmed and humidified, so that air that reaches the lungs is warm and moist. Bacteria and particles of air pollution are trapped on the mucus; the cilia continuously sweep the mucus toward the pharynx. Most of this mucus is eventually swallowed, and any bacteria present will be destroyed by the hydrochloric acid in the gastric juice.

In the upper nasal cavities are the **olfactory receptors,** which detect vaporized chemicals that have been inhaled. The olfactory nerves pass through the ethmoid bone to the brain.

You may also recall the **paranasal sinuses,** air cavities in the maxillae, frontal, sphenoid, and ethmoid bones (see Figs. 15–1 and 6–9). These sinuses are lined with ciliated epithelium, and the mucus produced drains into the nasal cavities. The functions of the paranasal sinuses are to lighten the skull and provide resonance for the voice.

PHARYNX

The **pharynx** is a muscular tube posterior to the nasal and oral cavities and anterior to the cervical vertebrae. For descriptive purposes, the pharynx may be divided into three parts: the nasopharynx, oropharynx, and laryngopharynx (see Fig. 15–1).

The uppermost portion is the **nasopharynx,** which is behind the nasal cavities. The **soft palate**

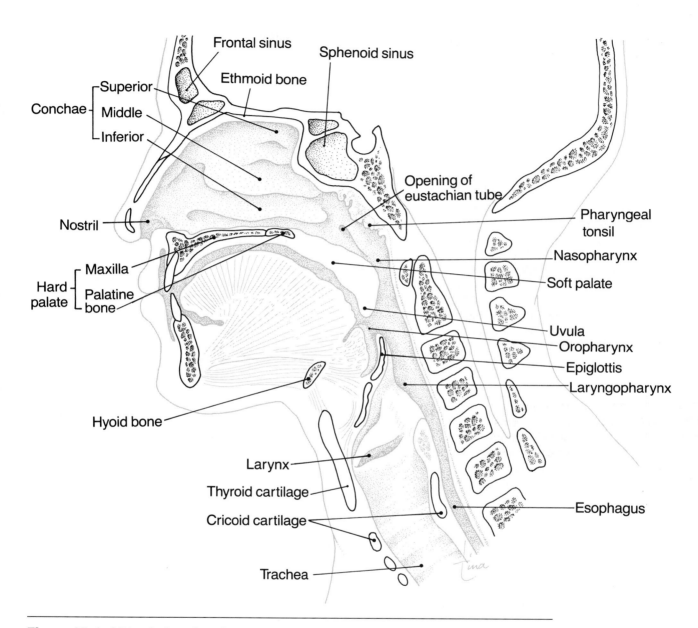

Figure 15–1 Midsagittal section of the head and neck showing the structures of the upper respiratory tract.

is elevated during swallowing to block the naso-pharynx and prevent food or saliva from going up rather than down. The uvula is the part of the soft palate you can see at the back of the throat. On the posterior wall of the nasopharynx is the adenoid or pharyngeal tonsil, a lymph nodule that contains macrophages. Opening into the nasopharynx are the two eustachian tubes, which extend to the middle ear cavities. The purpose of the eustachian tubes is to permit air to enter or leave the middle ears, allowing the eardrums to vibrate properly.

The nasopharynx is a passageway for air only, but the remainder of the pharynx serves as both an air and food passageway, although not for both at the same time. The **oropharynx** is behind the mouth; its mucosa is stratified squamous epithelium, continuous with that of the oral cavity. On its lateral walls are the palatine tonsils, also lymph nodules. Together with the adenoid and the lingual tonsils on the base of the tongue, they form a ring of lymphatic tissue around the pharynx to destroy pathogens that penetrate the mucosa.

The **laryngopharynx** is the most inferior portion of the pharynx. It opens anteriorly into the larynx and posteriorly into the esophagus. Contraction of the muscular wall of the oropharynx and laryngopharynx is part of the swallowing reflex.

LARYNX

The **larynx** is often called the voice box, a name that indicates one of its functions, which is speaking. The other function of the larynx is to be an air passageway between the pharynx and the trachea. Air passages must be kept open at all times, and so the larynx is made of nine pieces of cartilage connected by ligaments. Cartilage is a firm yet flexible tissue that prevents collapse of the larynx. In comparison, the esophagus is a collapsed tube except when food is passing through it.

The largest cartilage of the larynx is the **thyroid cartilage** (Fig. 15–2), which you can feel on the anterior surface of your neck. The **epiglottis** is the

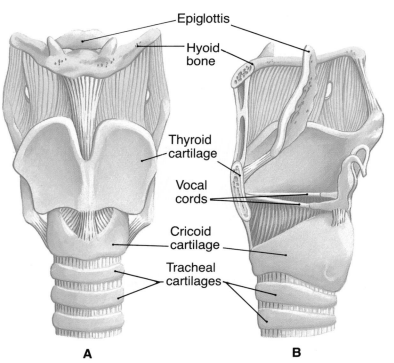

Epiglottis
Hyoid bone
Thyroid cartilage
Vocal cords
Cricoid cartilage
Tracheal cartilages

A   **B**

Figure 15–2 Larynx. **(A)**, Anterior view. **(B)**, Midsagittal section through the larynx, viewed from the left side.

uppermost cartilage. During swallowing, the larynx is elevated, and the epiglottis closes over the top to prevent the entry of food into the larynx.

The mucosa of the larynx is ciliated epithelium, except for the vocal cords (stratified squamous epithelium). The cilia of the mucosa sweep upward to remove mucus and trapped dust and microorganisms.

The **vocal cords** (or vocal folds) are on either side of the **glottis,** the opening between them. During breathing, the vocal cords are held at the sides of the glottis, so that air passes freely into and out of the trachea (Fig. 15–3). During speaking, the intrinsic muscles of the larynx pull the vocal cords across the glottis, and exhaled air vibrates the vocal cords to produce sounds which can be turned into speech. It is also physically possible to speak while inhaling, but this is not what we are used to. The cranial nerves that are motor nerves to the larynx for speaking are the vagus and accessory nerves.

TRACHEA AND BRONCHIAL TREE

The **trachea** is about 4 to 5 inches (10 to 13 cm) long and extends from the larynx to the primary bronchi. The wall of the trachea contains 16 to 20 C-shaped pieces of cartilage, which keep the trachea open. The gaps in these incomplete cartilage rings are posterior, to permit the expansion of the esophagus when food is swallowed. The mucosa of the trachea is ciliated epithelium with goblet cells. As in the larynx, the cilia sweep upward toward the pharynx.

The right and left **primary bronchi** (Fig. 15–4) are the branches of the trachea that enter the lungs. Within the lungs, each primary bronchus branches into secondary bronchi leading to the lobes of each lung (three right, two left). The further branching of the bronchial tubes is often called the **bronchial tree.** Imagine the trachea as the trunk of an upside-down tree with extensive branches that become smaller and smaller; these smaller branches are the **bronchioles.** No cartilage is present in the walls of the bronchioles; this becomes clinically important in asthma (see Box 15–1: Asthma). The smallest bronchioles terminate in clusters of alveoli, the air sacs of the lungs.

LUNGS AND PLEURAL MEMBRANES

The **lungs** are located on either side of the heart in the chest cavity and are encircled and protected

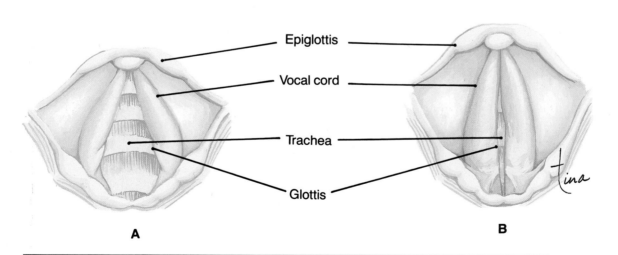

Figure 15–3 Vocal cords and glottis. **(A),** Position of the vocal cords during breathing. **(B),** Position of the vocal cords during speaking.

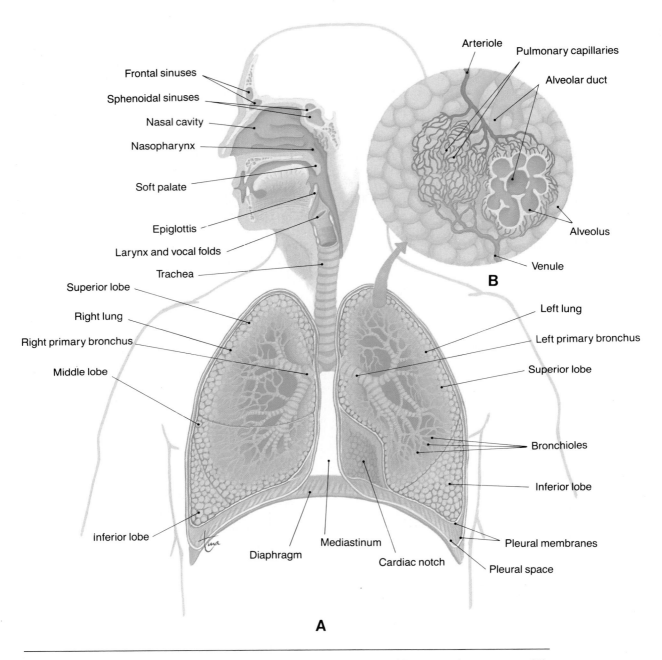

Figure 15–4 Respiratory system. **(A)**, Anterior view of the upper and lower respiratory tracts. **(B)**, Microscopic view of alveoli and pulmonary capillaries. (The colors represent the vessels, not the oxygen content of the blood within the vessel.)

Box 15–1   ASTHMA

Asthma is usually triggered by an allergic reaction that affects the smooth muscle and glands of the bronchioles. Allergens include foods and inhaled substances such as dust and pollen. Wheezing and dyspnea (difficult breathing) characterize an asthma attack, which may range from mild to fatal.

As part of the allergic response, the smooth muscle of the bronchioles constricts. Since there is no cartilage present in their walls, the bronchioles may close completely. The secretion of mucus increases, perhaps markedly, so the already constricted bronchioles may become clogged or completely obstructed with mucus.

Chronic asthma is a predisposing factor for emphysema. When obstructed bronchioles prevent ventilation of alveoli, the walls of the alveoli begin to deteriorate and break down, leaving large cavities that do not provide much surface area for gas exchange.

by the rib cage. The base of each lung rests on the diaphragm below; the apex (superior tip) is at the level of the clavicle. On the medial surface of each lung is an indentation called the **hilus,** where the primary bronchus and the pulmonary artery and veins enter the lung.

The pleural membranes are the serous membranes of the thoracic cavity. The **parietal pleura** lines the chest wall, and the **visceral pleura** is on the surface of the lungs. Between the pleural membranes is serous fluid, which prevents friction and keeps the two membranes together during breathing.

Alveoli

The functional units of the lungs are the **alveoli,** which are made of simple squamous epithelium. In the spaces between clusters of alveoli is elastic connective tissue, which is important for exhalation. There are millions of alveoli in each lung, and each alveolus is surrounded by a network of pulmonary capillaries (see Fig 15–4). Recall that capillaries are also made of simple squamous epithelium, so there are only two cells between the air in the

alveoli and the blood in the pulmonary capillaries, which permits efficient diffusion of gases (Fig. 15–5).

Each alveolus is lined with a thin layer of tissue fluid, which is essential for the diffusion of gases, because a gas must dissolve in a liquid in order to enter or leave a cell (the earthworm principle—an earthworm breathes through its moist skin, and will suffocate if its skin dries out). Although this tissue fluid is necessary, it creates a potential problem in that it would make the walls of an alveolus stick together internally. Imagine a plastic bag that is wet inside; its walls would stick together because of the surface tension of the water. This is just what would happen in alveoli, and inflation would be very difficult.

This problem is overcome by **pulmonary surfactant,** a lipoprotein secreted by alveolar cells. Surfactant mixes with the tissue fluid within the alveoli and decreases its surface tension, permitting inflation of the alveoli (see Box 15–2: Hyaline Membrane Disease). Normal inflation of the alveoli in turn permits the exchange of gases, but before we discuss this process, we will first see how air gets into and out of the lungs.

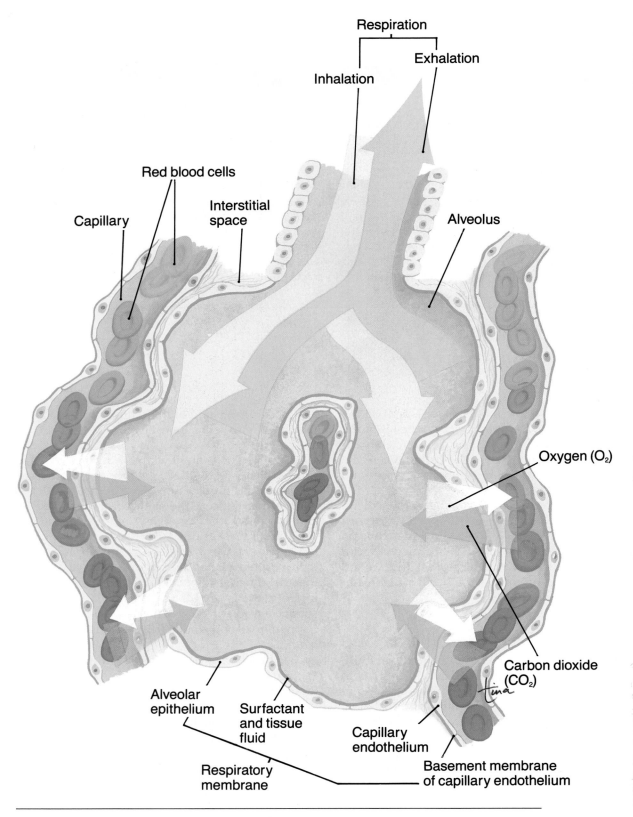

Figure 15–5 The respiratory membrane: the structures and substances through which gases must pass as they diffuse from air to blood (oxygen) or from blood to air (CO₂).

Box 15–2 HYALINE MEMBRANE DISEASE

Hyaline membrane disease is also called Respiratory Distress Syndrome (RDS) of the Newborn, and most often affects premature infants whose lungs have not yet produced sufficient quantities of pulmonary surfactant.

The first few breaths of a newborn inflate most of the previously collapsed lungs, and the presence of surfactant permits the alveoli to remain open. The following breaths become much easier, and normal breathing is established.

Without surfactant, the surface tension of the tissue fluid lining the alveoli causes the air sacs to collapse after each breath rather than remain inflated. Each breath, therefore, is difficult, and the newborn must expend a great deal of energy just to breathe.

Premature infants may require respiratory assistance until their lungs are mature enough to produce surfactant. Recent clinical trials of a synthetic surfactant have shown that some infants are helped significantly, and because they can breathe more normally, their dependence on respirators is minimized. Still undergoing evaluation are the effects of the long-term use of this surfactant in the most premature babies, who may require it for much longer periods of time.

MECHANISM OF BREATHING

Ventilation is the term for the movement of air to and from the alveoli. The two aspects of ventilation are inhalation and exhalation, which are brought about by the nervous system and the respiratory muscles. The respiratory centers are located in the medulla and pons. Their specific functions will be covered in a later section, but it is the medulla that generates impulses to the respiratory muscles.

These muscles are the diaphragm and the external and internal intercostal muscles (Fig. 15–6). The **diaphragm** is a dome-shaped muscle below the lungs; when it contracts, the diaphragm flattens and moves downward. The intercostal muscles are found between the ribs. The **external intercostal muscles** pull the ribs upward and outward, and the **internal intercostal muscles** pull the ribs downward and inward. Ventilation is the result of the respiratory muscles producing changes in the pressure within the alveoli and bronchial tree.

With respect to breathing, the important pressures are these three:

1. **Atmospheric Pressure**—the pressure of the air around us. At sea level, atmospheric pressure is 760 mmHg. At higher altitudes, of course, atmospheric pressure is lower.
2. **Intrapleural Pressure**—the pressure within the potential pleural space between the parietal pleura and visceral pleura. This is a potential rather than a real space. A thin layer of serous fluid causes the two pleural membranes to adhere to one another. Intrapleural pressure is always slightly below atmospheric pressure (about 756 mmHg). This is called a "negative" pressure because the elastic lungs are always tending to collapse and pull the visceral pleura away from the parietal pleura. The serous fluid, however, prevents separation of the pleural membranes (see Box 15–3: Pneumothorax).
3. **Intrapulmonic Pressure**—the pressure within the bronchial tree and alveoli. This pressure fluctuates below and above atmospheric pressure during each cycle of breathing.

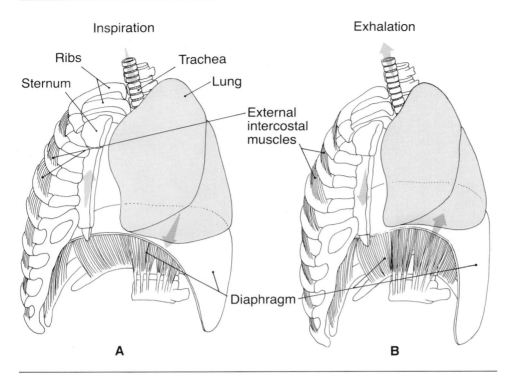

Inspiration

Exhalation

Ribs
Sternum
Trachea
Lung
External
intercostal
muscles
Diaphragm

A

B

Figure 15–6 Actions of the respiratory muscles. **(A)**, Inhalation: diaphragm contracts downward; external intercostal muscles pull rib cage upward and outward; lungs are expanded. **(B)**, Normal exhalation: diaphragm relaxes upward; rib cage falls down and in as external intercostal muscles relax; lungs are compressed.

Box 15–3   PNEUMOTHORAX

Pneumothorax is the presence of air in the pleural space, which causes collapse of the lung on that side. Recall that the pleural space is only a potential space because the serous fluid keeps the pleural membranes adhering to one another, and the intrapleural pressure is always slightly below atmospheric pressure. Should air at atmospheric pressure enter the pleural cavity, the suddenly higher pressure outside the lung will contribute to its collapse (the other factor is the normal elasticity of the lungs).

A spontaneous pneumothorax, without apparent trauma, may result from rupture of weakened alveoli on the lung surface. Pulmonary diseases such as emphysema may weaken alveoli.

Puncture wounds of the chest wall also allow air into the pleural space, with resulting collapse of a lung. In severe cases, large amounts of air push the heart, great vessels, trachea, and esophagus toward the opposite side (mediastinal shift), putting pressure on the other lung and making breathing difficult. This is called tension pneumothorax, and requires rapid medical intervention to remove the trapped air.

INHALATION

Inhalation, also called **inspiration,** is a precise sequence of events that may be described as follows:

Motor impulses from the medulla travel along the **phrenic nerves** to the diaphragm and along the **intercostal nerves** to the external intercostal muscles. The diaphragm contracts, moves downward, and expands the chest cavity from top to bottom. The external intercostal muscles pull the ribs up and out, which expands the chest cavity from side to side and front to back.

As the chest cavity is expanded, the parietal pleura expands with it. Intrapleural pressure becomes even more negative as a sort of suction is created between the pleural membranes. The adhesion created by the serous fluid, however, permits the visceral pleura to be expanded too, and this expands the lungs as well.

As the lungs expand, intrapulmonic pressure falls below atmospheric pressure, and air enters the nose and travels through the respiratory passages to the alveoli. Entry of air continues until intrapulmonic pressure is equal to atmospheric pressure; this is a normal inhalation. Of course, inhalation can be continued beyond normal, that is, a deep breath.

This requires a more forceful contraction of the respiratory muscles to further expand the lungs, permitting the entry of more air.

EXHALATION

Exhalation may also be called **expiration** and begins when motor impulses from the medulla decrease, and the diaphragm and external intercostal muscles relax. As the chest cavity becomes smaller, the lungs are compressed, and their elastic connective tissue, which was stretched during inhalation, recoils and also compresses the alveoli. As intrapulmonic pressure rises above atmospheric pressure, air is forced out of the lungs until the two pressures are again equal.

Notice that inhalation is an active process that requires muscle contraction, but normal exhalation is a passive process, depending to a great extent on the normal elasticity of healthy lungs. In other words, under normal circumstances we must expend energy to inhale but not to exhale (see Box 15–4: Emphysema).

We can, however, go beyond a normal exhalation and expel more air, as when talking, singing,

---

Box 15–4 EMPHYSEMA

Emphysema is a degenerative disease in which the alveoli lose their elasticity and cannot recoil. Perhaps the most common (and avoidable) cause is cigarette smoking; other causes are long-term exposure to severe air pollution or industrial dusts, or chronic asthma. Inhaled irritants damage the alveolar walls and cause deterioration of the elastic connective tissue surrounding the alveoli. As the alveoli break down, larger air cavities are created that are not efficient in gas exchange.

In progressive emphysema, damaged lung tissue is replaced by fibrous connective tissue (scar tissue), which further limits the diffusion of gases. Blood oxygen level decreases, and blood carbon dioxide level increases. Accumulating carbon dioxide decreases the pH of body fluids; this is a respiratory acidosis.

One of the most characteristic signs of emphysema is that the affected person must make an effort to exhale. The loss of lung elasticity makes normal exhalation an active process, rather than the passive process it usually is. The person must expend energy to exhale in order to make room in the lungs for inhaled air. This extra "work" required for exhalation may be exhausting for the person and contribute to the debilitating nature of emphysema.

Box 15–5 THE HEIMLICH MANEUVER

The **Heimlich maneuver** has received much well-deserved publicity in recent years, and indeed it is a life-saving technique.

If a person is choking on a foreign object (such as food) lodged in the pharynx or larynx, the air in the lungs may be utilized to remove the object. The physiology of this technique is illustrated at right.

The person performing the maneuver stands behind the choking victim and puts both arms around the victim's waist. One hand forms a fist that is placed between the victim's navel and rib cage (below the diaphragm), and the other hand covers the fist. It is important to place hands correctly, in order to avoid breaking the victim's ribs. With both hands, a quick, forceful upward thrust is made and repeated if necessary. This forces the diaphragm upward to compress the lungs and force air out. The forcefully expelled air is often sufficient to dislodge the foreign object.

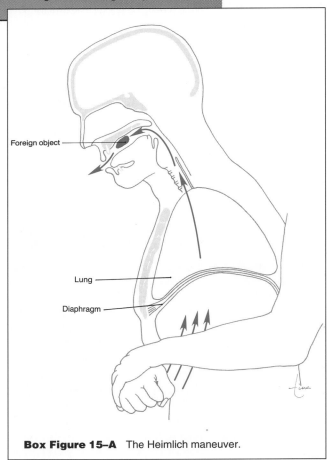

Foreign object

Lung

Diaphragm

Box Figure 15–A The Heimlich maneuver.

or blowing up a balloon. Such a forced exhalation is an active process that requires contraction of other muscles. Contraction of the internal intercostal muscles pulls the ribs down and in and squeezes even more air out of the lungs. Contraction of abdominal muscles, such as the rectus abdominus, compresses the abdominal organs and pushes the diaphragm upward, which also forces more air out of the lungs (see Box 15–5: The Heimlich Maneuver).

EXCHANGE OF GASES

There are two sites of exchange of oxygen and carbon dioxide: the lungs and the tissues of the body. The exchange of gases between the air in the alveoli and the blood in the pulmonary capillaries is called **external respiration.** This term may be a bit confusing at first, because we often think of "external" as being outside the body. In this case, however, "external" means the exchange that involves air from the external environment. **Internal respiration** is the exchange of gases between the blood in the systemic capillaries and the tissue fluid (cells) of the body.

The air we inhale (the earth's atmosphere) is approximately 21% oxygen and 0.04% carbon dioxide. Although most (78%) of the atmosphere is nitrogen, this gas is not physiologically available to us, and we simply exhale it. This exhaled air also contains about 16% oxygen and 4.5% carbon dioxide, so it is apparent that some oxygen is retained within the body, and the carbon dioxide produced by cells is exhaled.

DIFFUSION OF GASES— PARTIAL PRESSURES

Within the body, a gas will diffuse from an area of greater concentration to an area of lesser concentration. The concentration of each gas in a particular site (alveolar air, pulmonary blood, and so on) is expressed in a value called **partial pressure.** The partial pressure of a gas, measured in mmHg, is the pressure it exerts within a mixture of gases,

whether the mixture is actually in a gaseous state or is in a liquid such as blood. The partial pressures of oxygen and carbon dioxide in the atmosphere and in the sites of exchange in the body are listed in Table 15–1. The abbreviation for partial pressure is "P," which is used, for example, on hospital lab slips for blood gases and will be used here.

The partial pressures of oxygen and carbon dioxide at the sites of external respiration (lungs) and internal respiration (body) are shown in Fig. 15–7. Since partial pressure reflects concentration, a gas will diffuse from an area of higher partial pressure to an area of lower partial pressure.

The air in the alveoli has a high P_{O_2} and a low P_{CO_2}. The blood in the pulmonary capillaries, which has just come from the body, has a low P_{O_2} and a high P_{CO_2}. Therefore, in external respiration, oxygen diffuses from the air in the alveoli to the blood, and carbon dioxide diffuses from the blood to the air in the alveoli. The blood that returns to the heart now has a high P_{O_2} and a low P_{CO_2} and is pumped by the left ventricle into systemic circulation.

The arterial blood that reaches systemic capillar-

Table 15–1 PARTIAL PRESSURES

Site	P_{O_2} (mmHg)	P_{CO_2} (mmHg)
Atmosphere	160	0.15
Alveolar air	104	40
Pulmonary blood (venous)	40	45
Systemic blood (arterial)	100	40
Tissue fluid	40	50

Partial pressure is calculated as follows:
  % of the gas in the mixture × total pressure = Pgas

  Example: **O_2 in the atmosphere**
    21% × 760 mmHg = 160 mmHg (P_{O_2})
  Example: **CO_2 in the atmosphere**
    0.04% × 760 mmHg = 0.15 mmHg (P_{CO_2})

Notice that alveolar partial pressures are not exactly those of the atmosphere. Alveolar air contains significant amounts of water vapor and the CO_2 diffusing in from the blood. Oxygen also diffuses readily from the alveoli into the pulmonary capillaries. Therefore, alveolar P_{O_2} is lower than atmospheric P_{O_2}, and alveolar P_{CO_2} is significantly higher than atmospheric P_{CO_2}.

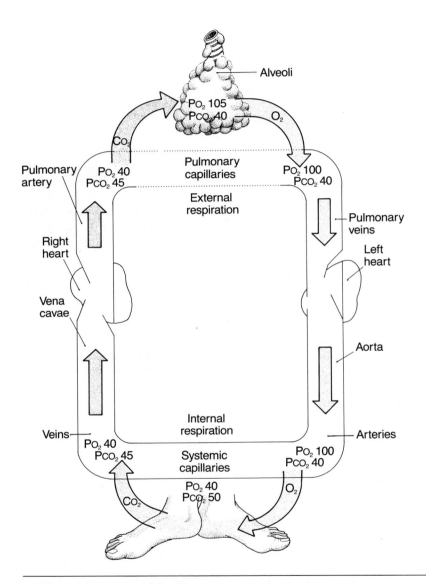

Alveoli

P_{O_2} 105
P_{CO_2} 40

O_2

CO_2

Pulmonary
artery

P_{O_2} 40
P_{CO_2} 45

Pulmonary
capillaries

P_{O_2} 100
P_{CO_2} 40

External
respiration

Pulmonary
veins

Right
heart

Left
heart

Vena
cavae

Aorta

Internal
respiration

Veins

Arteries

P_{O_2} 40
P_{CO_2} 45

Systemic
capillaries

P_{O_2} 100
P_{CO_2} 40

P_{O_2} 40
P_{CO_2} 50

CO_2

O_2

Figure 15–7 External respiration in the lungs and internal respiration in the body. The partial pressures of oxygen and carbon dioxide are shown at each site.